Soil Quality and Innovation in Agriculture: Dynamics, Indicators, and Sustainability

Soil Quality and Innovation in Agriculture: Dynamics, Indicators, and Sustainability

Guest Editor

Eleftherios Evangelou

MDPI

Basel • Beijing • Wuhan • Barcelona • Belgrade • Novi Sad • Cluj • Manchester

Guest Editor
Eleftherios Evangelou
Institute of Industrial and
Forage Crops
Hellenic Agricultural
Organization "Demeter"
Larisa
Greece

Editorial Office
MDPI AG
Grosspeteranlage 5
4052 Basel, Switzerland

This is a reprint of the Special Issue, published open access by the journal *Sustainability* (ISSN 2071-1050), freely accessible at: https://www.mdpi.com/journal/sustainability/special_issues/QN8JCCIU7J.

For citation purposes, cite each article independently as indicated on the article page online and as indicated below:

Lastname, A.A.; Lastname, B.B. Article Title. *Journal Name* **Year**, *Volume Number*, Page Range.

ISBN 978-3-7258-5399-1 (Hbk)
ISBN 978-3-7258-5400-4 (PDF)
https://doi.org/10.3390/books978-3-7258-5400-4

Cover image courtesy of Eleftherios Evangelou
Pinios River Basin, Greece.

Contents

About the Editor

Eleftherios Evangelou

Eleftherios Evangelou is an agronomist/soil scientist working as a researcher in the Institute of Industrial and Forage Crops of the Hellenic Agricultural Organization "DIMITRA". His focus is on improving soil quality through soil management. His main research interests include soil quality monitoring, soil organic carbon pools, re-use of agricultural and municipal wastes on soils, and use of innovative technologies in precision agriculture. He is trained as an agronomist (Aristotelian University of Thessaloniki, Greece) and holds a Ph.D. in Soil Science (Department of Environment, University of the Aegean). His Ph.D. research deals with the land use effect on soil quality. He is working on international/national soil-related research projects, and the results of his work have been published in peer-reviewed journals and at international conferences.

sustainability

MDPI

Editorial

Soil Quality and Innovation in Agriculture

Eleftherios Evangelou

Institute of Industrial and Forage Crops, Hellenic Agricultural Organization "Dimitra", 41335 Larisa, Greece; evangelou@elgo.gr

1. Soil Quality: A Fundamental Concept for Sustainability

Life on Earth depends on healthy soils. As the living skin of our planet, soil lies at the heart of Earth's terrestrial ecosystems, playing a crucial role in agricultural productivity, carbon sequestration, water filtration, and biodiversity. However, soils are fragile, and the impact of our actions is often overlooked. Increasing pressures from intensification, climate variability, pollution, and land-use change, challenge the sustainability of this vital resource. To ensure a healthy and green future for current and future generations, we must protect and take care of our soils.

The soil quality concept provides a comprehensive framework for understanding the interactions between soil's biological, chemical, and physical properties [1]. This holistic approach is essential for sustainable use and effective management of non-renewable soil resources [1,2]. Pioneering work by Larson and Pierce (1991) introduced this concept and proposed a "minimum data set" of indicators for its assessment [3]. Building on this, Doran and Parkin (1994) offered a widely accepted definition, characterizing soil quality as "the capacity of a specific kind of soil to function, within natural or managed ecosystem boundaries, to sustain plant and animal productivity, maintain or enhance water and air quality, and promote plant, animal, and human health." [4]. While the concept has faced ongoing debate and criticism [5], a rising number of studies emphasize its fundamental importance for environmental sustainability and human well-being [6,7].

This Special Issue, "Soil Quality and Innovation in Agriculture: Dynamics, Indicators, and Sustainability," addresses the vital concept of soil quality, examining the relationships between biological, chemical, and physical soil parameters in the context of sustainable land use and management. Soil quality degradation poses a significant challenge to the agricultural sector, and effective soil management is essential for improving soil health [8]. Assessing soil quality involves considering physical, chemical, and biological properties as indicators, which can help identify factors that inhibit soil function and monitor the effects of various management practices [9].

This Special Issue brings together 13 original studies from diverse regions and soil environments to explore the state, monitoring, and enhancement of soil quality, integrating emerging innovations and sustainability perspectives.

2. Key Themes from the Research

The collected works span experimental, methodological, and applied frameworks that deepen our understanding of soil quality indicators, carbon and nutrient dynamics, biological function, and sustainable land use. From Croatia, China, and Poland to the Mediterranean and Peru, the research captures regional variations while aligning with global efforts to restore and sustain soil health. Several key themes emerge from the research presented in this Special Issue:

- **Soil quality as affected by Soil Carbon and Nutrient Dynamics in Agricultural Systems:** Several papers emphasize the influence of land management on carbon fluxes and nutrient cycling. Bhandari et al. (2025) evaluated the biological soil quality indicator of soil respiration in maize, wheat, and barley, demonstrating crop-specific dynamics that highlight the influence of crop type, temperature, and moisture on carbon cycling. Xin et al. (2025) conducted a detailed spatial assessment of soil organic carbon, Nitrogen, and Phosphorus stocks across cropping systems in arid Northwest China, identifying paddy and orchard systems as the most effective for nutrient retention. Similarly, Kondratowicz-Maciejewska et al. (2025) highlighted how organic amendments affected enzymatic activity and labile carbon pools in sandy soils, while Lemanowicz et al. (2025) illustrated how tillage regimes shape both physical soil structure and microbial function.
- **Effect of Organic Amendments and Soil Restoration Management on Soil Quality:** In East-Central Poland, Malinowska and Kania (2025) demonstrated the dual benefits and risks of applying waste-derived organic matter—enhancing nitrogen and microbial activity but increasing heavy metal content under some treatments. In the U.S., Young and Sherman (2024) compared solid dairy manure sources and their greenhouse gas profiles relative to corn yield, showing the need to align organic input strategies with soil aeration status to optimize climate outcomes. In Turkey, Bilen et al. (2025) brought an industrial lens to soil quality, examining how cement dust emissions interact with tillage systems and microbial indicators.
- **Indicators of Soil Health and Long-Term Practices:** Long-term data underscore the cumulative effects of soil management. Hewelke et al. (2024) documented how no-till systems over decades increased soil organic carbon, aggregate stability, and water retention in Polish soils. Samaniego et al. (2025) aimed to develop and spatially evaluate a Soil Quality Index (SQI) tailored to a mountainous region in Peru, using a combination of Principal Component Analysis (PCA) and Expert Opinion (EO) with linear and non-linear scoring functions to assess soil health. Evangelou and Giourga (2024) applied multivariate methods to Mediterranean agroecosystems, pinpointing total nitrogen and C/N ratio as reliable indicators of land use impacts among 23 soil physical, chemical, and biological soil quality indicators. Such work provides tools for simplified yet meaningful monitoring, especially in data-scarce regions.
- **Innovations for Soil Enhancement and Pollution Mitigation:** Several papers explored innovative interventions and remediation techniques. Liu et al. (2024) analyzed how compounding sandy soil with Pisha sandstone altered electrochemical properties and structural stability, a promising technique for erosion-prone areas. Espada et al. (2024) performed a life cycle assessment comparing phytoremediation strategies for copper-contaminated soils, finding that alfalfa with biomass cogeneration yielded the best environmental performance. This systemic view adds a much-needed climate and energy lens to soil restoration practices.
- **Urbanization and Land Transformation. Impacts on Soil Quality**: Finally, Wang et al. (2024) examined rural urbanization's impact on soil organic matter in Northeast China, using spatial analysis to show both gains and losses associated with urban encroachment and mechanization. These findings highlight the importance of contextualizing soil quality within socio-economic transitions, especially in rapidly changing peri-urban landscapes.

3. Conclusions

Collectively, the contributions to this Special Issue illustrate the multidimensional nature of soil quality and the array of emerging innovations, from UAV-based precision agriculture to organic recycling, long-term conservation practices, and life cycle tools. The papers reaffirm that soil is not merely a substrate, but a living, dynamic system shaped by biological, chemical, physical, and human forces.

The research presented here identifies appropriate indicators and implements innovative management practices to maintain and improve soil health. From carbon cycling to tillage systems, waste management to pollution impacts, the papers in this issue provide a comprehensive overview of the challenges and opportunities in soil science today.

As pressures on soils intensify globally, the works presented here offer critical knowledge for guiding evidence-based policy, adaptive management, and technological advancement. The diversity of methods and settings showcased in this collection reflects the complexity of soil systems—and our collective responsibility to sustain them. We thank the contributors and reviewers for their valuable work in making this Special Issue a success.

We hope this Special Issue offers valuable reference for researchers, practitioners, and decision-makers working toward resilient, productive, and climate-smart soil systems that enhance soil quality and environmental health.

Conflicts of Interest: The authors declare no conflicts of interest.

List of Contributions:

1. Bhandari, D.; Bilandžija, N.; Krička, T.; Zdunić, Z.; Ghimire, S.; Piskáčková, T.R.; Bilandžija, D. Soil Respiration in Maize, Wheat, and Barley Across a Growing Season: Findings from Croatia's Continental Region. *Sustainability* **2025**, *17*, 4207. https://doi.org/10.3390/su17094207.
2. Xin, H.; Lv, C.; Li, N.; Peng, L.; Chang, M.; Li, Y.; Geng, Q.; Chen, S.; Lai, N. Analysis of Soil Nutrient Content and Carbon Pool Dynamics Under Different Cropping Systems. *Sustainability* **2025**, *17*, 3881. https://doi.org/10.3390/su17093881.
3. Kondratowicz-Maciejewska, K.; Lemanowicz, J.; Jaskulska, I. Effects of Long-Term Multi-Treatment Experiments on Organic Matter and Enzymatic Activity in Sandy Soil. *Sustainability* **2025**, *17*, 3252. https://doi.org/10.3390/su17073252.
4. Lemanowicz, J.; Balontayová, E.; Dębska, B.; Bartkowiak, A.; Wasilewski, P. Tillage System as a Practice Affecting the Quality of Soils and Its Sustainable Management. *Sustainability* **2025**, *17*, 2867. https://doi.org/10.3390/su17072867.
5. Malinowska, E.; Kania, P. The Effect of Waste Organic Matter on the Soil Chemical Composition After Three Years of *Miscanthus* × *giganteus* Cultivation in East-Central Poland. *Sustainability* **2025**, *17*, 2532. https://doi.org/10.3390/su17062532.
6. Young, E.; Sherman, J. Effects of Solid Dairy Manure Application on Greenhouse Gas Emissions and Corn Yield in the Upper Midwest, USA. *Sustainability* **2024**, 16, 11171. https://doi.org/10.3390/su162411171.
7. Bilen, S.; Bilen, M.; Ozgul, M.; Ozlu, E.; Simsek, U. Response of Soil Chemical and Biological Properties to Cement Dust Emissions: Insights for Sustainable Soil Management. *Sustainability* **2025**, *17*, 1409. https://doi.org/10.3390/su17041409.
8. Hewelke, E.; Mielnik, L.; Weber, J.; Perzanowska, A.; Jamroz, E.; Gozdowski, D.; Szacki, P. Chemical and Physical Aspects of Soil Health Resulting from Long-Term No-Till Management. *Sustainability* **2024**, *16*, 9682. https://doi.org/10.3390/su16229682.
9. Samaniego, T.; Sales, B.; Solórzano, R. Assessment of Soil Quality in Peruvian Andean Smallholdings: A Comparative Study of PCA and Expert Opinion Approaches. *Sustainability* **2025**, *17*, 7610. https://doi.org/10.3390/su17177610.
10. Evangelou, E.; Giourga, C. Identification of Soil Quality Factors and Indicators in Mediterranean Agro-Ecosystems. *Sustainability* **2024**, *16*, 10717. https://doi.org/10.3390/su162310717.

11. Liu, Z.; Zhou, L.; Zhang, Y.; Han, J.; Sun, Y.; Zhang, R.; Li, X.; Hu, F. Internal Force Mechanism of Pisha Sandstone as a Soil Amendment to Improve Sandy Soil Structural Stability in Mu Us Sandy Land. *Sustainability* **2024**, *16*, 4415. https://doi.org/10.3390/su16114415.
12. Espada, J.J.; Rodríguez, R.; Delgado, A.; Vicente, G.; Bautista, L.F. Assessing Environmental Sustainability of Phytoremediation to Remove Copper from Contaminated Soils. *Sustainability* **2024**, *16*, 2441. https://doi.org/10.3390/su16062441.
13. Wang, X.; Fei, L.; An, Y.; Liu, X.; Zhang, M. The Influence of Rural Urbanization on the Change in Soil Organic Matter of Farmland in Northeast China. *Sustainability* **2024**, *16*, 4683. https://doi.org/10.3390/su16114683.

References

1. Karlen, D.L.; Ditzler, C.A.; Andrews, S.S. Soil quality: Why and how? *Geoderma* **2003**, *114*, 145–156. [CrossRef]
2. Karlen, D.L.; Mausbach, M.J.; Doran, J.W.; Cline, R.G.; Harris, R.F.; Schumann, G.F. Soil quality: A concept, definition and framework for evaluation. *Soil Sci. Soc. Am. J.* **1997**, *61*, 4–10. [CrossRef]
3. Larson, W.E.; Pierce, F.J. Conservation and Enhancement of Soil Quality. In *Evaluation for Sustainable Land Management in the Developing World*; Dumanski, J., Pushparajah, E., Latham, M., Myers, R., Eds.; International Board for Soil Research and Management, Bangkok, Thailand Technical Papers; Proceeding International Workshop: Chiang Rai, Thailand, 1991; Volume 2 pp. 15–21.
4. Doran, J.W.; Parkin, T.B. Defining and Assessing Soil Quality. In *Defining Soil Quality for Sustainable Environment*; Doran, J.W., Ed. SSSA Special Publication: Madison, WI, USA, 1994; pp. 3–23.
5. Sojka, R.E.; Upchurch, D.R. Reservations regarding the soil quality concept. *Soil Sci. Soc. Am. J.* **1999**, *63*, 1039–1054. [CrossRef]
6. Doran, J.W. Soil health and global sustainability: Translating science into practice. *Agric. Ecosyst. Environ.* **2002**, *88*, 119–127 [CrossRef]
7. Lehmann, J.; Bossio, D.A.; Kögel-Knabner, I.; Rillig, M.C. The concept and future prospects of soil health. *Nat. Rev. Earth Environ* **2020**, *1*, 544–553. [CrossRef] [PubMed]
8. Lal, R. Restoring Soil Quality to Mitigate Soil Degradation. *Sustainability* **2015**, *7*, 5875–5895. [CrossRef]
9. Andrews, S.; Karlen, D.L.; Cambardella, C.A. The soil management-assessment framework: A quantitative soil quality evaluation method. *Soil Sci. Soc. Am. J.* **2004**, *68*, 1945–1962. [CrossRef]

sustainability

MDPI

Article

Assessment of Soil Quality in Peruvian Andean Smallholdings: A Comparative Study of PCA and Expert Opinion Approaches

Tomás Samaniego [1], Beatriz Sales [1] and Richard Solórzano [2,3,*]

1 Estación Experimental Agraria Donoso, Dirección de Servicios Estratégicos Agrarios, Instituto Nacional de Innovación Agraria (INIA), Lima 15200, Peru; td.samaniego@gmail.com (T.S.)
2 Centro Experimental La Molina, Dirección de Servicios Estratégicos Agrarios, Instituto Nacional de Innovación Agraria (INIA), Lima 15024, Peru
3 Facultad de Ciencias Ambientales, Universidad Científica del Sur (UCSUR), Lima 15024, Peru
* Correspondence: investigacion_labsaf@inia.gob.pe

Abstract: Soil degradation poses a significant threat to the sustainability of agroecosystems, particularly in mountainous regions where environmental conditions are highly variable and management practices are often suboptimal. In this context, soil quality assessment emerges as a key tool for guiding sustainable land use and informing decision-making processes. This study aimed to develop and spatially evaluate a Soil Quality Index (SQI) tailored to the northeast sector of Jangas district, Ancash, Peru. A total of 24 soil indicators were initially considered and reduced using Spearman's correlations to avoid multicollinearity. Depending on the weighting strategy applied, the final SQI configurations incorporated between 14 and 15 indicators. Two weighting strategies—Principal Component Analysis (PCA) and Expert Opinion (EO)—were combined with linear and non-linear (sigmoidal) scoring functions, resulting in four distinct SQI configurations. The spatial performance of each index was tested using Geographically Weighted Regression Kriging (GWRK), incorporating covariates like NDMI, elevation, slope, and aspect. The SQI constructed using PCA combined with non-linear scoring achieved the highest performance, effectively minimizing skewness and while achieving the highest predictive accuracy under GWRK. By contrast, although the EO-based index with linear scoring demonstrated similar statistical robustness, it failed to achieve comparable effectiveness in terms of spatial predictive accuracy. The SQIs generated offer a practical framework for local institutions to identify and prioritize areas requiring intervention. Through the interpretation of complex soil data into accessible, spatially explicit maps, these indices facilitate the targeted application of inputs—such as organic amendments in low-SQI zones—and support the implementation of improved management practices, including crop rotation and soil conservation, without necessitating advanced technical expertise.

Keywords: principal component analysis; agroecosystems; spatial modeling

1. Introduction

Soil quality is defined as "the capacity of a soil to function within ecosystem boundaries, to sustain biological productivity, maintain environmental quality, and promote plant, animal, and human health" [1]. SQIs function as a decision-making instrument by enabling the assessment of the sustainability and impact of soil management practices [2–4]. They also facilitate the systematic organization of commonly evaluated soil parameters, shifting the focus of soil quality from solely productivity to a more comprehensive approach centered on sustainable soil management [5]. The application of an SQI as a tool for assessing management practices has been explored across various spatial scales, including crop-level, experimental field, and regional contexts [6–9]. This approach is valuable because isolated measurements of soil properties may offer limited practical relevance to farmers when interpreted without contextual understanding. A comprehensive evaluation of soil conditions can empower farmers to devise effective management strategies for optimizing crop production [10].

Soil quality indices and indicators should be carefully selected to ensure their congruence with soil functions that cannot be directly measured [11]. The selection of indicators must be aligned with the specific soil functions of interest and the defined management objectives of the system to ensure a comprehensive evaluation of soil health and functional performance [12]. Developing a reliable SQI requires first identifying the most relevant indicators and then assigning weights that reflect their relative importance in representing particular soil functions [13]. In most cases, the formulation of an integrated SQI follows three main stages: determining the set of indicators, transforming these indicators into scores, and combining the scores to produce a single index value [8].

Soil properties frequently exhibit interdependencies; however, under diverse agricultural management practices, their influence on soil quality and productivity may vary significantly, complicating the interpretation of their effects across different management regimes [14]. An effective soil quality indicator should exhibit strong correlations with essential soil functions. In addition, it should be scientifically robust, methodologically precise, readily measurable, and responsive to variations in management practices [15]. In addition, their selection needs to reflect site-specific conditions, including geographic location, climatic regime, and management goals [16].

Given the collinearity of many soil attributes and the high demands of time and resources for exhaustive sampling, it becomes essential to define a Minimum Dataset (MDS) that preserves the most relevant information while eliminating redundancy [17,18]. Traditionally, the development of such datasets has relied heavily on expert judgment and statistical approaches, like PCA, to identify the most representative indicators [19]. A soil function–based approach incorporating EO offers a practical and reliable method for regional-scale assessments, provided that the derived SQI is validated against parameters aligned with specific management objectives [20]. On the contrary, a significance constraint of the PCA approach is its tendence to exclusively consider those Principal Components (PCs) that account for at least 5% of the total variance and have eigenvalues greater than 1.0. Consequently, this criterion may lead to the exclusion of important parameters that are critical for addressing specific management objectives [9].

In the context of SQI development, once the MDS is established, the selected indicators must be standardized to a common scale before integration into the index. Among the most widely used transformation methods are linear and non-linear (sigmoidal) scoring functions. The linear approach assumes a constant rate of change between an indicator and soil quality, offering simplicity and ease of interpretation, although it may fail to reflect threshold effects and non-linear ecological responses [21]. On the other hand, the sigmoidal approach models gradual changes at both low and high indicator values, with greater sensitivity near optimal ranges, thereby capturing ecological thresholds and diminishing returns more effectively [10]. While the latter can provide a more ecologically representative assessment, it demands additional parameterization and may be less intuitive for non-specialist users.

The choice of scoring function is therefore critical, as it not only determines how raw indicators are normalized—helping to mitigate issues such as non-normality and skewness [22]—but also influences the performance of subsequent spatial analyses. This becomes particularly important when SQI computation is integrated with spatial interpolation techniques, such as GWRK, which leverage the spatial structure of residuals to refine predictions [23,24] and generate high-resolution soil quality maps for more targeted land management decisions.

Traditional agricultural systems in the Peruvian Andes are highly vulnerable, largely due to the intrinsic fragility of their soils—marked by low fertility, high susceptibility to erosion, and limited water-holding capacity—conditions that are further aggravated by steep topography and intensive land use [25,26]. Climate change intensifies these challenges by increasing the frequency and severity of extreme events—such as droughts and frosts—that disrupt hydrological dynamics and exacerbate soil degradation processes [27]. Recent research indicates a progressive decline in soil organic matter and increased compaction across Andean landscapes, posing significant risks to the food security of Andean communities [28,29]. Smallholder farmers, who typically cultivate plots smaller than two hectares, often rely on empirical indicators—such as soil color and texture—to guide their management decisions, thereby limiting the integration of scientific knowledge into sustainable land use strategies.

Developing an SQI enables a comprehensive assessment of soil conditions, allowing for the identification of both limitations and potentials of the soil resource. This, in turn, facilitates more precise and effective agronomic decision making in regions with diverse topographic and edaphic conditions, such as Jangas, located in the Ancash region of Peru. A locally adapted SQI is critical for optimizing land use, conserving natural resources, and promoting long-term agroecosystem sustainability. In this study, four SQIs were developed based on two methodological frameworks (PCA and EO), each combined with two scoring functions: linear and non-linear (sigmoidal). These approaches were evaluated in terms of their statistical and spatial performance. Additionally, spatial modeling and mapping of soil quality were conducted using GWRK to explore how soil quality varies across the northeastern part of the Jangas district. This integrated methodology aims to identify the most reliable SQI formulation for supporting sustainable agricultural planning and decision making at the local scale.

2. Materials and Methods

2.1. Study Area

The study was conducted in the Jangas district, located in the Ancash region of Peru, within a tropical montane dry forest ecosystem. The area is primarily used for agricultural production, with key crops including avocado, maize, and beans. Geographically, the study area is located between latitudes 9°27′ and 9°23′ S and longitudes 77°36′ and 77°32′ W, covering an estimated surface area of approximately 2000 hectares (Figure 1). The study site is situated at elevations between 2865 and 2880 m. Climatic records from the Yungay weather station (9°8′30.79″ S; 77°44′59.91″ W; 2466 m) report an average annual temperature of 15.3 °C, with substantial variation between daily maximum and minimum values. The mean annual precipitation is approximately 616 mm, with a marked dry season from May to August, and two rainfall peaks, the main one occurring between January and March.

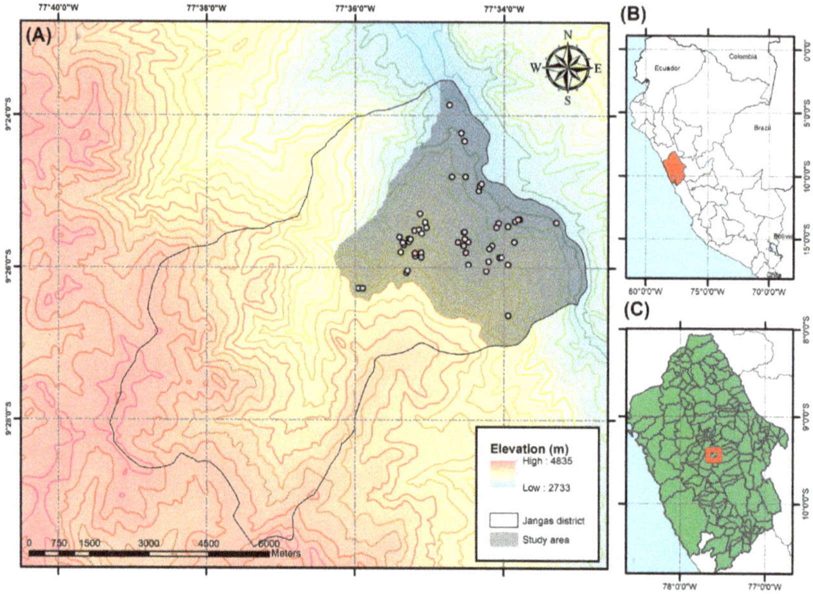

Figure 1. Location of the study area in the northwest of Jangas district, Ancash, Peru. (**A**) Elevation map with sampling points; (**B**) location of Ancash region in Peru; and (**C**) study area within the Ancash region.

2.2. Soil Sampling and Analysis

Georeferenced soil samples were taken from 55 different sampling points. Within each sampling point, five sub-samples were extracted at a depth of 20 cm. Areas with atypical moisture conditions, plot edges, and zones with heavy traffic were excluded from sampling. The collected sub-samples were thoroughly homogenized to produce a single composite sample. Around 1 kg loose soil was collected, bagged, and stored under cool, dark conditions before transport to the laboratory. After air-drying, samples were sieved to 2 mm for subsequent analyses at INIA's Soil, Water, and Foliar Laboratories network. The evaluated variables formed part of a comprehensive soil characterization based on standardized reference methods. Soil texture (sand, silt, and clay percentages) was determined using the Bouyoucos hydrometer method [30]. The pH was measured following EPA guidelines [31], while electrical conductivity (EC) was determined using the saturation extract method [32]. Soil organic carbon (SOC) was analyzed using NOM-021-RECNAT-2000 [30]. Available phosphorus for both neutral and acidic soils was measured using the Olsen method [30], and available potassium was measured according to Bazán [33]. Exchangeable cations (H^+, Al^{+3}, Ca^{+2}, Mg^{+2}, K^+, and Na^+) and micronutrient contents (Fe, Cu, Zn, and Mn) were also quantified [30]. Additional variables included particulate organic matter, determined using the wet sieving technique and divided into two size categories: fine particulate organic matter (fPOM, 0.053–0.25 mm) and coarse particulate organic matter (cPOM, 0.25–2.0 mm) [34]; labile soil organic carbon (POXC), measured using the $KMnO_4$ oxidation method [35]; aggregate stability in water, assessed using the successive sieving method [36]; as well as leaf biomass and litter production from the cover crops.

2.3. Soil Quality Evaluation

Evaluation of an SQI consists of three main stages: (1) identifying representative indicators from the complete set of measured soil indicators to establish the MDS; (2) transforming the MDS indicators into scores; and (3) integrating these scores into a single index [4,37].

2.3.1. Selecting the MDS

The selection of indicators was performed using two methods viz., PCA and EO methods.

Principal Component Analysis Approach

PCA is a statistical method used to reduce the dimensionality of a dataset by minimizing the number of variables while retaining most of the original variability. PCs that have higher eigenvalues are considered to be the most representative, accounting for the greatest proportion of variance within the data [4,21].

In this study, PCA was applied to the 24 measured soil indicators. The number of PCs was determined using the eigenvalue criterion, retaining those with values ≥ 1 for MDS identification. Varimax rotation was applied to the retained PCs to enhance their correlation with soil indicators through variance redistribution [8]. The selection of indicators was guided by the weighted loadings for each component, retaining those with absolute values exceeding 0.60 [4,38]. A multivariate correlation analysis was performed to assess redundancy and the degree of association among variables. For variable pairs showing strong correlations ($r \geq 0.60$), the variable with the highest absolute factor loading was retained as the representative indicator. Conversely, when highly weighted variables were not significantly correlated—indicating distinct functional roles—all such variables were included in the MDS to preserve their contributions [39].

Expert Opinion Aproach

In this approach, the selection of primary soil properties was based on the criteria described by Lenka et al. [39], as well as on their recognized influence on soil fertility [40,41]. Indicators representing four key soil functions—(1) soil structure and water retention, (2) nutrient supply capacity, and (3) fundamental soil characteristics that may constrain productive use—were identified through expert judgment, relevant literature, and site-specific edaphic conditions (Table 1).

Table 1. Soil functions, associated indicators, and their respective weights.

Function	Weight	Function Indicators	Weight	Scoring Function
Soil structural stability and water storage	0.35	SOC	0.20	More is better
		LAgre	0.10	More is better
		Sand	0.05	More is better
Nutrient supply function	0.30	POXC	0.0375	More is better
		fPOM	0.0375	More is better
		Ava_P	0.0375	More is better
		Ava_K	0.0375	More is better
		Fe	0.0375	More is better
		Cu	0.0375	More is better
		Zn	0.0375	More is better
		Mn	0.0375	More is better
Soil basic properties, potential to limit production	0.35	pH	0.10	Optimum is better
		EC	0.10	Less is better
		CEC	0.10	More is better
		Carb	0.05	Less is better

SOC: Soil organic carbon; LAgre: Large soil aggregates (2.00–0.25 mm); POXC: Permanganate oxidizable carbon, fPOM: Fine particulate organic matter; Ava_P: Available P; Ava_K: Available K; EC: Electric conductivity; CEC: Cation Exchange Capacity; Carb: Carbonates percentage.

2.3.2. Transformation of the MDS Indicators and Weight Assignment

Due to differences in measurement units among the selected soil parameters, values were standardized to a 0–1 scale using appropriate linear or non-linear transformation techniques. The scoring was based on each indicator's relationship with soil fertility. Indicators that positively influence soil fertility were classified under a "more is better" function, while those with negative effects were assessed using a "less is better" approach. Indicators exhibiting both beneficial and detrimental effects depending on their levels were evaluated using an "optimum range" scoring method [5,42].

In the linear scoring approach, two types of functions were applied: "more is better" (Equation (1)) and "less is better" (Equation (2)):

$$S_L = (x - l)/(h - l) \tag{1}$$

$$S_L = 1 - ((x - l)/(h - l)) \tag{2}$$

where S_L is the normalized linear score (0–1), x is the measured variable, l is the minimum, and h the maximum value [8].

In the non-linear scoring approach, a sigmoidal function was applied (Equation (3)) [5,10]:

$$S_{NL} = a/(1 + (x/x_o)^b) \tag{3}$$

where S_{NL} is the normalized nonlinear score (0–1), a is the maximum score (set to 1 in this study), x is the measured soil variable, x_0 is its mean value, and b is the slope (−2.5 for "more is better" and +2.5 for "less is better") [43,44].

The indicator scores were integrated into indices using a weighted additive approach (Equation (4)) [4]:

$$SQI_W = \sum_{n=1}^{n} WiSi \tag{4}$$

where Si represents the score assigned to each indicator (whether obtained through linear or non-linear transformation), n denotes the total number of variables included in the index, and Wi corresponds to the weight assigned to each indicator.

In the PCA-based approach, the weighting of the indicators forming the MDS was determined according to the variance distribution obtained from the PCA results. Each PC accounted for a specific proportion of the total variance in the dataset. The weighting of each indicator in a PC was computed as the proportion of the variance it explained within that component. It was explained by the cumulative variance interpreted by all retained PCs with eigenvalues greater than 1, thereby reflecting the relative contribution of each indicator to the overall dataset variability [39].

Following the expert-based procedure, weighting values were attributed to the key soil functions (Table 1) to indicate their contribution to total soil performance. These weights were then proportionally distributed to the individual indicators within each function, as determined by expert judgment and supported by evidence from the scientific literature [9,39].

2.4. Spatial Mapping Using GWRK

To generate spatial predictions of the four developed IQS, the GWRK method was applied. GWRK is a hybrid geostatistical technique that integrates the strengths of both geographically weighted regression and ordinary kriging. This method consists of a deterministic component—modeled using geographically weighted regression, which captures spatially varying relationships between the target variable and a set of covariates—and a stochastic component, represented by the spatial interpolation of the regression residuals via kriging [23]. GWRK was implemented using the set of covariates listed in Table 2. This approach allowed for the generation of maps for each IQS, capturing both spatial trends and local variations in soil quality across the study area.

Table 2. Environmental covariates used in the GWRK.

Covariate	Source
NDMI	Image collection "COPERNICUS/S2_SR"
Elevation	Shuttle Radar Topographic Mission (SRTM)
Slope	Shuttle Radar Topographic Mission (SRTM)
Aspect	Shuttle Radar Topographic Mission (SRTM)

The flowchart shown in Figure 2 summarizes the methodology used for the development of the four SQIs—EOLinear, EONLinear, PCALinear, and PCANLinear—and the subsequent spatial mapping. It outlines the sequence from soil sampling and laboratory analysis to indicator selection and weighting, using either the EO or PCA approach, each combined with linear or non-linear scoring methods. The resulting indices were then integrated into a spatial modeling process to generate high-resolution soil quality maps.

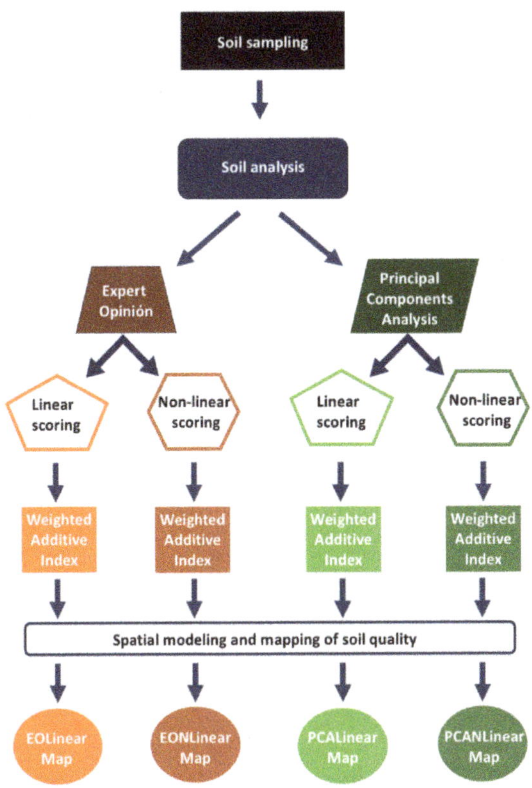

Figure 2. Flowchart summarizing the methodological framework for the development of four SQIs: EOLinear, EONLinear, PCALinear, and PCANLinear.

2.5. Statistical Analysis

The associations between the 24 soil parameters were examined using Spearman's rank correlation coefficients [45]. A PCA was carried out to reduce the dimensionality of the data and to identify the MDS. Differences among the developed soil quality indices were evaluated using the non-parametric Friedman test, followed by Nemenyi's post hoc test for pairwise comparisons [46]. All statistical analyses, including the implementation of GWRK, were performed using R software version 4.4.1.for Windows (R Core Team, Vienna, Austria, 2023) [47].

3. Results

3.1. Statistical Summary of Soil Properties

The parameters of interest included principal particles of soil (sand, silt, and clay), aggregate stability (in 2 sizes), and organic carbon due to their relevance in maintaining soil structure and enhancing water infiltration (see Table 3). The sand fraction in the soils under study varied between 36.70% and 79.06%, with an average content of 57.49% and an approximately symmetrical distribution (skewness = 0.12). The silt content ranged from 18.78% to 60.29%, with a mean of 36.84%, also displaying a near-symmetrical distribution (skewness = 0.16). By contrast, the clay content was markedly lower, averaging 5.66%, but exhibited substantial variability (CV = 94.09%) and a positively skewed distribution (1.08), suggesting the occurrence of occasional high values. In terms of soil structural composition, the proportion of large aggregates ranged from 8.50% to 83.88%, with a mean value of 42.89%. The distribution was symmetrical but platykurtic (kurtosis = −1.17), indicating a flatter-than-normal distribution. Small aggregates averaged 28.30%, with a slightly right-skewed distribution (skewness = 0.27). The soil organic carbon content was relatively low overall, with a mean of 1.27%. However, it was highly variable (66.01%) and exhibited moderate positive skewness (skewness = 0.50), reflecting a predominance of low-carbon samples with some locations showing elevated organic matter levels.

Table 3. Statistical characteristics of parameters that maintain soil structure and water storage.

Variable	Min	Max	Mean	Median	Std. Deviation	CV	Kurtosis	Skewness
Sand (%)	36.70	79.06	57.49	56.98	7.82	13.59	1.32	0.12
Silt (%)	18.78	60.29	36.84	37.63	8.64	23.46	0.49	0.16
Clay (%)	1.46	22.46	5.66	2.94	5.33	94.09	1.83	1.08
LAgre (%)	8.50	83.88	42.89	44.51	20.84	48.60	−1.17	−0.04
SAgre (%)	5.24	59.61	28.30	28.46	13.22	46.73	−0.51	0.27
SOC (%)	0.12	3.31	1.27	1.04	0.84	66.01	−0.27	0.50

LAgre: Large aggregates (2.00−0.25 mm); Sagre: Small aggregates (0.25−0.053 mm), SOC: Soil organic carbon.

Spearman correlation analysis revealed significant relationships among soil physical and chemical properties (see Figure 3). A robust negative correlation was observed between sand and clay contents (−0.68, ***), indicating a compensatory relationship typical of soil textural components. Similarly, sand was moderately and negatively correlated with silt (−0.41, *). The proportion of large aggregates (LAgre) was strongly and negatively correlated with small aggregates (SAgre) (−0.79, ***), suggesting a clear structural differentiation. Additionally, LAgre showed a moderate positive correlation with SOC (0.45, **), highlighting the role of organic matter in promoting soil aggregation. By contrast, SOC did not exhibit significant correlations with texture components. Overall, the results suggested that while soil texture fractions are interrelated, the structural stability of the soil, particularly the formation of large aggregates, is more closely associated with organic carbon content than with texture.

Descriptive statistics for labile carbon fractions, available nutrients, and exchangeable cations revealed notable variability across soil samples (see Table 4). The POXC ranged from 203.64 to 1011.81 mg kg^{-1}, showing moderate variability (28.47%) and a near-normal distribution. By contrast, both particulate organic matter fractions—cPOM and fPOM—exhibited high variability (64.91% and 76.63%, respectively) and highly leptokurtic distributions (kurtosis = 11.32 and 32.38, respectively), indicating strong data concentration near the mean and the presence of outliers. fPOM also showed marked positive skewness (3.39), reflecting a tail toward higher values. The available phosphorus and potassium levels showed considerable variation (CV > 60%). Ava_K, in particular, displayed a highly skewed and leptokurtic distribution, suggesting the influence of extreme values in some samples. Nonetheless, the majority of Ava_P and Ava_K values fell within ranges that are generally considered adequate for plant nutrition. Exchangeable cations, including Ca^{2+} and Mg^{2+}, had moderate coefficients of variation (34.49% and 47.88%, respectively), whereas Na^{+} and K^{+} were more variable and right-skewed (CV > 70%). Micronutrient concentrations—Fe, Cu, Zn, and Mn—were especially variable, with CVs exceeding 100%; however, most values for these micronutrients also fell within levels considered adequate for soils, indicating generally favorable nutrient status despite the observed variability. All micronutrients exhibited highly skewed and leptokurtic distributions, indicating the presence of extreme concentrations in a subset of the samples. These findings suggested a heterogeneous distribution of micronutrients and labile organic matter in the soils, possibly influenced by localized management practices or parent material variation.

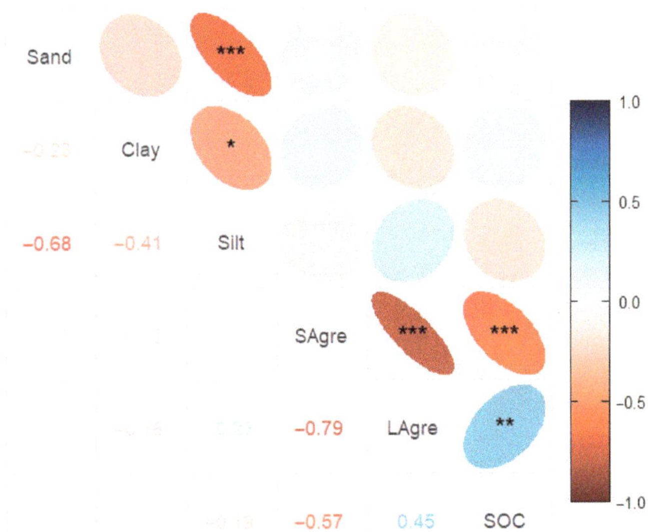

Figure 3. Spearman correlation matrix of parameters that maintain soil structure and water storage. LAgre: Large aggregates (2.00–0.25 mm); SAgre: Small aggregates (0.25–0.053 mm), SOC: Soil organic carbon; (*); p-value < 0.05; (**): p-value < 0.01; (***): p-value < 0.001.

The Spearman correlation matrix revealed multiple significant associations among labile carbon fractions, macronutrients, exchangeable bases, and micronutrients (see Figure 4). A strong and highly significant positive correlation was observed between Ex_K and Ava_K (Spearman's ρ = 0.94, ***), indicating a high consistency between these two potassium pools. The organic carbon fractions (POXC, fPOM, and cPOM) were weakly correlated with each other and showed only modest associations with nutrient variables. Micronutrient elements exhibited strong intercorrelations. For example, Cu was highly correlated with Zn (0.81, ***) and Mn (0.54, **), indicating co-accumulation or shared geochemical behavior.

Zn was also positively associated with Mn (0.51, **). Among exchangeable bases, Ca was negatively correlated with Mn (−0.60, ***).

Table 4. Statistical characteristics of parameters related to nutrient availability.

Variable	Min	Max	Mean	Median	Std. Deviation	CV	Kurtosis	Skewness
POXC mg kg^{-1}	203.64	1011.81	716.31	768.55	203.97	28.47	0.09	−0.51
cPOM (g kg^{-1})	0.62	11.32	2.74	2.24	1.78	64.91	11.32	1.94
fPOM (g kg^{-1})	1.68	28.15	4.75	4.10	3.64	76.63	32.38	3.39
Ava_P (mg kg^{-1})	2.37	44.02	16.68	13.45	10.59	63.46	−0.36	0.50
Ava_K (mg kg^{-1})	35.60	766.00	179.20	151.20	133.08	74.26	8.00	1.71
Ex_Ca (cmol$_{(+)}$ kg^{-1})	3.68	16.51	8.52	8.48	2.94	34.49	−0.29	0.21
Ex_Mg (cmol$_{(+)}$ kg^{-1})	0.13	3.03	1.17	1.10	0.56	47.88	3.10	0.80
Ex_Na (cmol$_{(+)}$ kg^{-1})	0.06	1.56	0.42	0.39	0.31	75.06	1.72	0.72
Ex_K (cmol$_{(+)}$ kg^{-1})	0.03	1.01	0.25	0.18	0.21	83.34	5.01	1.42
Fe (mg kg^{-1})	10.24	339.52	48.20	25.28	56.64	117.51	12.79	2.06
Cu (mg kg^{-1})	0.08	4.24	0.82	0.40	1.07	130.22	4.08	1.45
Zn (mg kg^{-1})	0.10	8.62	1.74	0.64	2.17	124.71	2.11	1.11
Mn (mg kg^{-1})	1.84	173.08	24.68	15.36	28.86	116.93	12.60	2.02

POXC: Permanganate oxidizable carbon, cPOM: Coarse particulate organic matter; fPOM: Fine particulate organic matter; Ava_P: Available P; Ava_K: Available K; Ex_Ca: Exchangeable Ca; Ex_Mg: Exchangeable Mg; Ex_Na: Exchangeable Na; Ex_K: Exchangeable K.

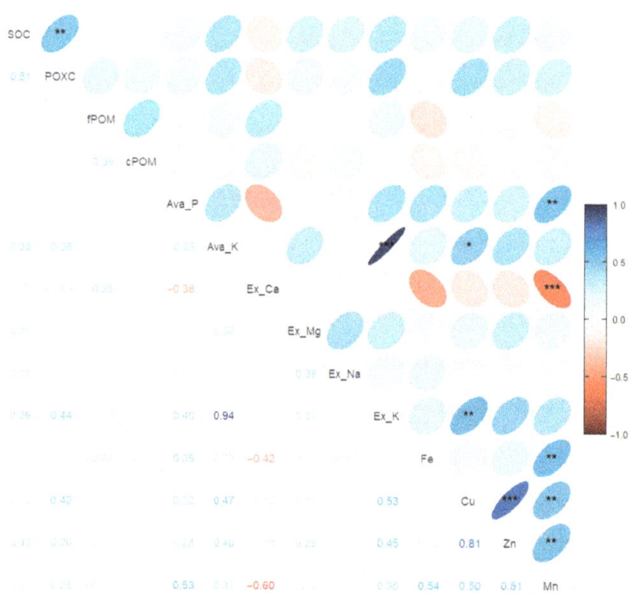

Figure 4. Spearman correlation matrix of parameters related to nutrient availability. POXC: Permanganate oxidizable carbon, cPOM: Coarse particulate organic matter; fPOM: Fine particulate organic matter; Ava_P: Available P; Ava_K: Available K; Ex_Ca: Exchangeable Ca; Ex_Mg: Exchangeable Mg; Ex_Na: Exchangeable Na; Ex_K: Exchangeable K; (*); p-value < 0.05; (**): p-value < 0.01; (***): p-value < 0.001.

The results showed that the majority of these soils possessed optimal pH levels, were devoid of salinity issues, and did not exhibit sodium-related problems, making them well suited for cultivating a wide range of crops. Soil pH values ranged from 4.57 to 7.90, with an average of 6.77 and a low coefficient of variation (9.16%), indicating relatively consistent soil low acidity levels. The distribution of pH was slightly platykurtic and negatively skewed, suggesting a tendency toward lower pH values in some samples (Table 5). Electrical conductivity values ranged from 0.29 to 2.72 dS m^{-1}, with a mean of 0.95 dS m^{-1} and a high CV (64.58%), reflecting heterogeneity in soil salinity. Similarly, CEC varied from 4.77 to 19.60 $cmol_{(+)}$ kg^{-1}, with moderate variability (30.73%) and a near-normal distribution. Sodium exchange percentage and carbonate content (%) also displayed high variability. SEP values were slightly right-skewed and platykurtic, while the carbonate content showed a leptokurtic and positively skewed distribution, suggesting the presence of localized carbonate accumulation. These results indicated diverse soil chemical conditions across the sampled sites, with implications for nutrient availability and soil management.

Table 5. Statistical characteristics of basic parameters, potential to limit production.

Variable	Min	Max	Mean	Median	Std. Deviation	CV	Kurtosis	Skewness
pH	4.57	7.90	6.77	6.76	0.62	9.16	1.85	−0.51
EC (dS m^{-1})	0.29	2.72	0.95	0.74	0.62	64.58	0.84	0.81
CEC ($cmol_{(+)}$ kg^{-1})	4.77	19.60	10.35	9.98	3.18	30.73	0.23	0.38
ESP (%)	0.53	11.64	4.19	3.18	3.05	72.81	−0.48	0.47
Carb. (%)	0.36	10.83	3.23	2.33	2.09	64.74	1.75	0.83

EC: Electric conductivity; CEC: Cation exchange capacity; ESP: Exchangeable sodium percentage; Carb: Carbonates percentage.

Soil pH showed a moderately strong positive correlation with CEC (0.63) and carbonate content (0.56), indicating that higher pH levels are associated with increased nutrient retention capacity and carbonate accumulation. A strong positive correlation was also observed between carbonate content and CEC (0.72), suggesting a close linkage between soil alkalinity components and exchangeable base capacity. By contrast, EC and ESP showed no significant correlations with any of the other measured soil properties (Figure 5). These results highlighted the important interactions among pH, carbonates, and CEC in determining soil chemical fertility.

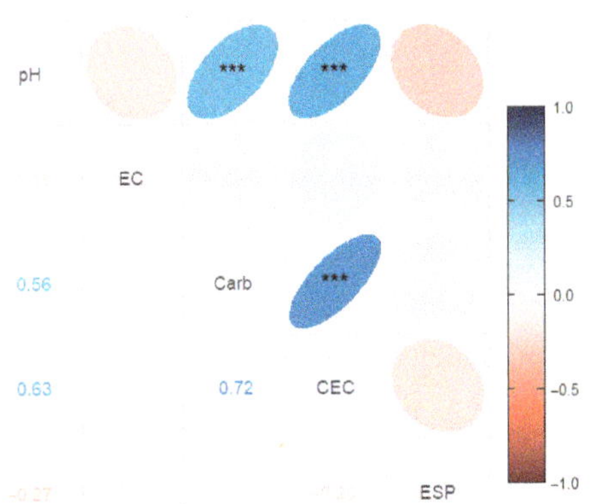

Figure 5. Spearman correlation matrix of basic parameters, potential to limit production. EC: Electric conductivity; CEC: Cation exchange capacity; ESP: Exchangeable sodium percentage; Carb: Carbonates percentage; (***): *p*-value < 0.001.

3.2. Principal Component Analysis

Based on the results of the Spearman correlation analysis among the initial set of variables (see Figures 3–5), the number of indicators was reduced from 24 to 16 by applying an absolute correlation threshold of 0.60 in order to minimize multicollinearity. PCA extracted six components with eigenvalues greater than one, cumulatively explaining 74% of the total variance in the dataset (Table 6). PC1 accounted for the highest proportion of variance (15.4%), followed by PC2 (15.2%), PC3 (12.0%), PC4 (11.5%), PC5 (11.3%), and PC6 (8.5%). Varimax rotation of the factor loading matrix revealed distinct associations among variables within each component. PC1 was strongly influenced by clay content, Fe, and ESP. PC2 was associated with high loadings of Zn, LAgre, SOC, and POXC. PC3 was primarily defined by carbonate content and exchangeable magnesium (Mg), while PC4 reflected high loadings for EC and available phosphorus. PC5 was dominated by fractions of organic matter, specifically cPOM and fPOM. Finally, PC6 was solely characterized by sand content, which had a factor loading greater than 0.60.

Table 6. PCA output including eigenvalues, percentage of variance explained, and factor loadings of component matrix variables.

Principal Components	PC1	PC2	PC3	PC4	PC5	PC6
Eigenvalue	2.464	2.437	1.918	1.846	1.808	1.357
% variance	0.154	0.152	0.120	0.115	0.113	0.085
% cumulative variance	0.154	0.306	0.426	0.542	0.655	0.739
Weightage factor	0.21	0.42	0.58	0.73	0.89	1.00
Factor Loadings from the Rotated Component Matrix						
Sand	−0.08	−0.06	−0.09	0.09	0.05	**−0.91**
Clay	**0.82**	0.05	−0.09	0.03	−0.23	0.17
LAgre	−0.16	**0.77**	0.16	−0.15	0.42	0.12
SOC	0.36	**0.72**	0.16	0.14	0.01	−0.25
POXC	0.12	**0.60**	0.13	0.26	0.24	0.06
fPOM	−0.24	0.05	0.17	0.16	**0.77**	0.11
cPOM	0.23	0.12	−0.15	−0.01	**0.85**	−0.17
Ava_P	0.13	0.15	−0.32	**0.69**	−0.15	−0.02
Ava_K	0.22	0.42	0.08	0.45	0.11	0.48
Ex_Mg	0.06	0.21	**0.78**	0.17	−0.12	0.09
Fe	**0.78**	0.19	−0.22	0.09	0.15	0.11
Zn	−0.05	**0.79**	−0.07	0.04	−0.18	0.19
pH	−0.51	0.16	0.36	−0.54	−0.07	0.27
EC	−0.04	0.10	0.23	**0.78**	0.25	−0.03
Carb	−0.16	0.03	**0.79**	−0.28	0.18	0.05
ESP	**0.74**	−0.04	0.45	0.06	0.04	−0.10

Indicators shown in bold were those selected in the corresponding principal component.

3.3. Expert Opinion

The selection of properties for the minimum soil dataset was based on the available dataset, a consensus among the authors, and supporting literature. For indicators related to soil structure and water retention, sand content was chosen because it serves as a fundamental indicator of soil texture. The percentage of large stable aggregates was included due to its relevance in representing soil structural integrity. Organic carbon was also selected for its essential role in stabilizing aggregates and enhancing water-holding capacity. Regarding nutrient supply, the most reactive fractions of organic matter—POXC and fPOM—were prioritized, as they reflect the mineralization potential of soil organic matter. Additionally, direct indicators of macro- and micronutrient availability were included, such as P and K, along with Fe, Cu, Zn, and Mn. Among general soil chemical parameters, pH and EC were included for their influence on nutrient dynamics and plant growth. CEC was selected due to its central role in soil fertility and ion exchange processes, while carbonate content was considered for its importance as a limiting factor in nutrient availability and soil chemical behavior.

3.4. Indicator Scores

The MDS indicators were scored using both linear and non-linear functions, as described in Equations (1)–(3). The indicators were categorized into three functional classes based on their relationship with soil quality. The first group, categorized under a "more is better" function, included sand, clay, LAgre, SOC, POXC, fPOM, cPOM, Ava_P, Ava_K, Ex_Mg, Fe, Zn, Cu, Mn, and CEC, as higher values of these variables are generally associated with improved soil function and productivity. The second group followed a "less is better" function and included EC and Carb, given their potential negative impacts on soil quality when present in excess. The third group included pH, which was assessed using an "optimum" function, recognizing that both excessively low and high values can adversely affect soil health and crop performance.

3.5. Results of Weighted Soil Quality Index

In the PCA-based approach, weights for the 14 selected indicators were determined according to the proportion of variance each PC contributed, as presented in Table 5. As an example, the 21% variance explained by PC1 was distributed among the indicators with significant loadings—Clay, Fe, and ESP—by proportionally allocating this total variance based on the absolute values of their respective loading coefficients. The same procedure was applied to each of the six retained components. The integration of the index was based on the following formula:

$$\begin{aligned} IQS_{PCA} = {} & 0.074S_{Clay} + 0.070S_{Fe} + 0.066S_{ESP} + 0.056S_{Lagre} + 0.053S_{SOC} + 0.058S_{Zn} + \\ & 0.044S_{POXC} + 0.079S_{Ex\_Mg} + 0.083S_{Carb} + 0.085S_{EC} + 0.077S_{Ava\_P} + 0.071S_{fPOM} + \\ & 0.079S_{cPOM} + 0.12S_{Sand} \end{aligned} \quad (5)$$

Under the EO approach, 15 parameters were identified based on their importance to fundamental soil physical, chemical, and biological processes. The relative importance of each variable was quantified, and their respective weights are detailed in Table 2. The integration of the index was based on the following formula:

$$\begin{aligned} IQS_{EO} = {} & 0.20S_{SOC} + 0.10S_{LAgre} + 0.05S_{Sand} + 0.0375S_{POXC} + 0.0375S_{fPOM} + 0.0375S_{Ava\_P} + \\ & 0.0375S_{Ava\_K} + 0.0375S_{Fe} + 0.0375S_{Cu} + 0.0375S_{Zn} + 0.0375S_{Mn} + 0.10S_{pH} + 0.10S_{EC} \\ & + 0.10S_{CEC} + 0.05S_{Carb} \end{aligned} \quad (6)$$

3.6. Soil Quality Index

The SQI values derived from the four assessment methods—PCALinear, PCANLinear, EOLinear, and EONLinear—showed considerable variability across the 55 soil samples (Table 7). Mean values ranged from 0.43 (PCALinear and EOLinear) to 0.47 (PCANLinear), with median values following a similar trend. The highest SQI was recorded under the PCANLinear approach (0.80), while the lowest value was observed using the EOLinear method (0.23). The coefficient of variation was lowest for PCALinear (16.68%), suggesting greater consistency, and highest for EONLinear (22.19%), indicating more heterogeneity among samples. Distributional statistics showed that all methods exhibited positive skewness, particularly PCALinear (1.666), implying a concentration of lower index values with fewer higher value outliers. Similarly, the kurtosis values suggested a leptokurtic distribution in PCALinear (5.142), contrasting with the slightly platykurtic pattern in EONLinear (−0.356).

Table 7. Soil quality index by four methods in 55 soil samples.

	PCALinear	PCANLinear	EOLinear	EONLinear
Min	0.31	0.31	0.23	0.25
Max	0.73	0.80	0.67	0.72
Mean	0.43	0.47	0.43	0.46
Median	0.42	0.46	0.42	0.44
Std. deviation	0.071	0.088	0.087	0.101
CV	16.683	18.965	20.280	22.194
Kurtosis	5.142	2.374	0.362	−0.356
Skewness	1.666	1.171	0.971	0.881
Friedman Test	Friedman χ^2 (df = 3)		*p*-value	
	33.48		<0.001 ***	
	Comparison		Nemenyi *p*-value	
Nemenyi Test	PCALinear vs. PCANLinear		<0.001 (***)	
	PCALinear vs. EOLinear		0.998 (ns)	
	PCALinear vs. EONLinear		<0.001 (***)	
	PCANLinear vs. EOLinear		<0.001 (***)	
	PCANLinear vs. EONLinear		0.921 (ns)	
	EOLinear vs. EONLinear		0.002 (**)	

(**): *p*-value < 0.01; (***): *p*-value < 0.001; (ns): non significative.

The Friedman test detected significant differences among the four methods ($\chi^2 = 33.48$, $p < 0.001$). Post hoc comparisons using the Nemenyi test indicated that PCANLinear differed significantly from PCALinear and EOLinear. Significant differences were also found between EOLinear and EONLinear, while no significant difference was observed between PCALinear and EOLinear (Figure 6).

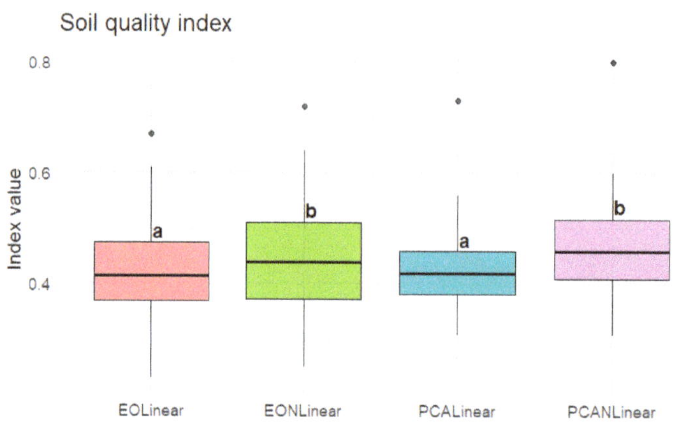

Figure 6. Boxplots showing the distribution of SQI scores across the 55 sampling locations. Different lowercase letters denote statistically significant differences among methods based on the Nemenyi post hoc test ($p < 0.05$).

3.7. Spatial Mapping of Soil Quality Indices

The spatial distribution of SQI values derived from the four methodological approaches is illustrated in Figure 7. While the general point patterns were broadly consistent across methods, localized discrepancies were observed at specific sampling points. Notably, all four indices displayed a clear trend: lower SQI values tended to cluster in the southwestern part of the study area, while higher values were primarily found in the northern zone. This pattern suggested that, despite methodological variations, there was strong agreement among the indices in identifying areas of relatively high and low soil quality.

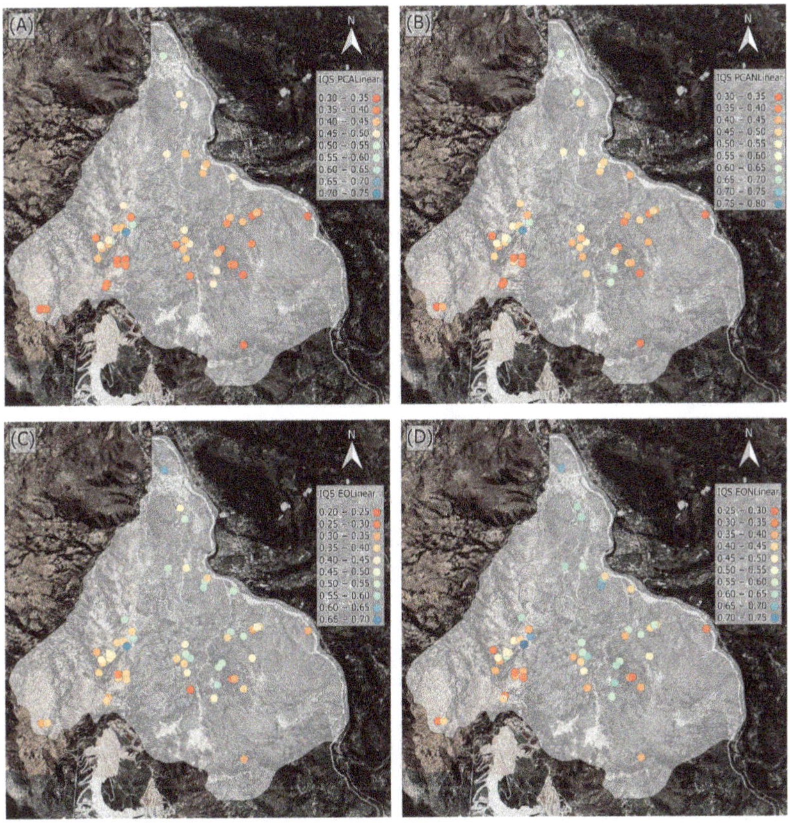

Figure 7. Spatial distribution of the SQI across 54 points using four different assessment methods: (**A**) PCALinear, (**B**) PCANLinear, (**C**) EOLinear, and (**D**) EONLinear.

The GWRK performance of the four SQI calculation methods was evaluated using five statistical metrics (Table 8). The PCANLinear method demonstrated the highest predictive accuracy, with the greatest R^2 value (0.736), the lowest MAE (0.0233), and the lowest CV_RMSE (9.77), indicating a better fit and more reliable estimation across the dataset. Although PCALinear showed a slightly lower R^2 (0.588), it exhibited the lowest RMSE (0.0451) and a competitive AIC value (−132.32), suggesting a good balance between model fit and parsimony. The EOLinear method presented intermediate performance, with an R^2 of 0.689 and moderate error metrics. Conversely, the EONLinear method showed the weakest performance among the four, with the highest RMSE (0.0558), MAE (0.0435), and CV_RMSE (12.13), indicating lower precision and greater variability in its estimations. These results supported the use of PCANLinear as the most robust approach for SQI prediction in this study.

The spatial interpolation maps of adjusted SQI values generated using the four methodological approaches are shown in Figure 8. These maps reflect the extrapolation of SQI values across the study area based on the model performances summarized previously. As observed, the spatial distribution patterns were broadly consistent across all four methods, reinforcing the trends previously noted in Figure 7 with the raw point data. The PCALinear and PCANLinear maps (Figure 7A,B) exhibited particularly high spatial agreement, highlighting very similar patterns of SQI variation across the landscape. Both indicated higher soil quality values concentrated in the northern and northeastern portions of the study area, with a gradual decrease in quality toward the southwestern and central zones. Similarly, the EOLinear and EONLinear maps (Figure 7C,D) also presented comparable spatial patterns, albeit with slightly smoother transitions between classes and marginally broader areas showing intermediate quality values. These similarities within method pairs (PCA-based and EO-based) likely reflected the influence of the underlying indicator selection and weighting criteria, as well as their model-based performance metrics. This spatial coherence supported the reliability of the SQI indices for guiding site-specific land management and soil improvement strategies in the region.

Table 8. Performance metrics of GWRK models.

	R^2	RMSE	MAE	AIC	CV_RMSE
PCALinear	0.588	0.0451	0.0242	−132.32	10.49
PCANLinear	0.736	0.0459	0.0233	−136.51	9.77
EOLinear	0.689	0.0478	0.0346	−134.58	11.12
EONLinear	0.688	0.0558	0.0435	−127.33	12.13

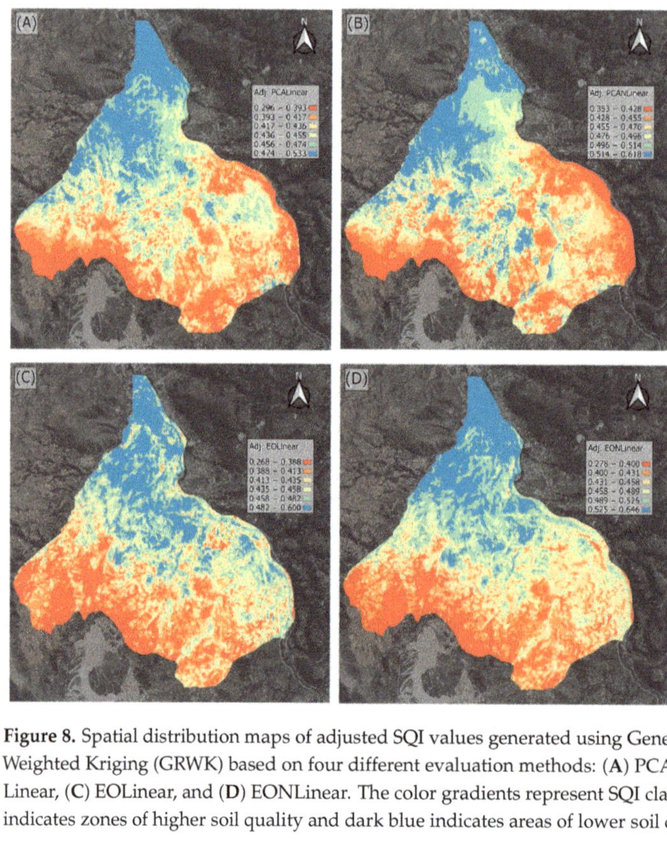

Figure 8. Spatial distribution maps of adjusted SQI values generated using Generalized Regression Weighted Kriging (GRWK) based on four different evaluation methods: (**A**) PCALinear, (**B**) PCANLinear, (**C**) EOLinear, and (**D**) EONLinear. The color gradients represent SQI classes, where yellow indicates zones of higher soil quality and dark blue indicates areas of lower soil quality.

4. Discussion

4.1. Soil Properties and Their Role in Soil Quality

The results obtained for the soil structural parameters indicated that the percentage of stable aggregates, in both large and small fractions, was more strongly related to SOC content than to the proportion of sand, silt, or clay [48–50]. This trend may be attributed to the low clay content observed in the study area (median = 2.94%), as clay is one of the primary physical agents involved in soil aggregation through mechanisms such as zeta potential, which influence the dispersion or aggregation of soil particles [51]. SOC plays a pivotal role in the formation of macroaggregates, as evidenced by studies by Zhou et al. [52], conducted under different soil management systems, and Wang et al. [53], conducted under various fertilization regimes. The percentage of strong organo-mineral bonds is higher in large aggregates, highlighting the binding strength of SOC, which can potentially reduce the bioavailability of the nutrients it contains [54]. Despite the relatively low levels of SOC observed in the studied soil, the positive effect on aggregate stability suggested that high concentrations of SOC may not be necessary to promote aggregation. Instead, the effectiveness of SOC appears to depend on the clay/SOC ratio, as reported by Soinne et al. [55], who found that a low clay/SOC ratio significantly enhanced aggregate formation. Conversely, small-sized stable aggregates demonstrated an absence of statistically significant relationships with any of the evaluated variables, with the exception of an inverse correlation with SOC. This phenomenon may be attributable to the influence of other factors, such as the specific mineralogy of the clay fraction, which plays a more prominent role in microaggregate formation [56].

The results obtained for POXC were consistent with the findings reported by various authors [35,57]. As one of the most active fractions of SOC, POXC showed a direct and significant relationship with SOC, which was also supported by other studies [58,59]. However, recent research has questioned whether POXC accurately represents the most labile fractions of SOC [60]. Regarding the different fractions of POM, the results for both fPOM (0.053–0.25 mm) and cPOM (0.25–2.0 mm) were consistent with those reported in other studies [61–63]. Nevertheless, in this study, neither fraction showed significant correlations with SOC or POXC. This contrasted with the results of other research, which reported significant associations [35,64,65]. One potential explanation is that those studies were conducted under controlled experimental conditions with defined treatments, which tend to reduce data variability. By contrast, our data were derived from a non-experimental context, characterized by high variability, elevated coefficients of variation, and the presence of extreme values (see Table 4).

Samaniego et al. [66] reported litter decomposition curves for the region's primary crops, finding that nutrient release rates were generally rapid. This may explain why most of the assessed nutrients fell within adequate concentration ranges. On the other hand, the concentrations and significant correlations observed among the micronutrients may be associated with similar patterns previously documented by other authors [67–70]. All of these elements exist in cationic form in the soil solution, which leads to comparable interactions with the soil's colloidal phase [40]. This assertion was further supported by the significant correlations observed between the micronutrients and exchangeable soil cations (see Table 4). In addition, given their low concentrations and high susceptibility to sorption and desorption processes [71], as well as potential interactions with the soil microbiome [72], these elements demonstrated high coefficients of variation and skewed distributions with long tails toward higher values.

Soil pH is a significant parameter that influences numerous chemical and biological processes. In this study, the pH showed significant correlations with CEC, calculated as the sum of exchangeable cations, thereby corroborating previous findings [73–75]. This relationship can be attributed to the fact that higher pH levels increase the number of negatively charged sites on the clay–humic complex, thereby enhancing the soil's capacity to retain cations and increasing its CEC [40]. Additionally, soils with higher carbonate contents were significantly correlated with elevated pH values, likely due to the neutralization of H^+ ions when carbonates react with the soil solution [76,77]. However, the low EC observed, along with its weak correlation with carbonate content, suggested that the presence of carbonates in these soils may originate from the parent material rather than from secondary precipitation via irrigation water—an accumulation mechanism reported in other environments [78].

Most of the evaluated indicators are commonly included in agricultural soil assessments. Some parameters, such as sand, silt, and clay contents, exhibit high temporal stability and are not easily influenced by management interventions [79]. By contrast, other indicators—such as soil SOC, EC, pH, CEC, and aggregate stability—are more responsive to medium-term management practices [80]. Finally, certain variables, such as the balance of exchangeable cations and the availability of micronutrients, can respond more rapidly to targeted soil fertilization strategies [81].

4.2. Comparative Analysis of SQI Construction

The development of SQIs using both the PCA- and EO-based approaches revealed distinct patterns in the selection and weighting of indicators, reflecting their underlying conceptual and methodological differences. In the PCA-based SQI, indicator weights were derived from the proportion of variance each contributed within their respective PCs. As a result, variables such as Sand (0.12), Ex_Mg (0.079), Carb (0.083), and EC (0.085) received relatively high weights. However, it has been observed that many of these variables are not typically representative of key soil functions, a phenomenon commonly observed in other studies. For instance, Damiaba et al. [82] reported high weights for Mg and moisture content, while Vasu et al. [9] found the sodium adsorption ratio and clay to dominate the PCA-derived SQI. Nonetheless, certain studies have shown that PCA places greater emphasis on functionally relevant indicators, such as SOC, macronutrients, and aggregate stability [5,83]. This suggests that the utility of PCA can vary depending on the dataset's structure and soil conditions.

By contrast, the EO-based SQI prioritized indicators based on their relevance to essential soil functions, ensuring balanced representation across physical (e.g., LAgre and Sand), chemical (e.g., pH, EC, CEC, and Ava_K), and biological (e.g., SOC, POXC, and fPOM) domains. The highest weight was assigned to SOC (0.20), emphasizing its central role in supporting soil structure, nutrient retention, and microbial activity. Additionally, pH, EC, and CEC each contributed 0.10 to the index, thereby aligning with their well-established importance in regulating chemical balance and buffering capacity in agricultural soils. This function-oriented selection approach is strongly supported in the literature [10,39,84], and it allows for greater ecological interpretability and management relevance.

Agronomically significant indicators included in the EO approach—such as pH and Ava_K—were not retained in the PCA-based index. Their exclusion suggested either limited statistical variance across samples or collinearity with other variables, which would cause PCA to discard them during dimensional reduction. However, the omission of functionally essential variables represents a limitation of purely statistical selection methods. As shown in previous studies [4,85,86], PCA's objectivity can sometimes come at the cost of ecological interpretability, particularly in systems where management goals dictate the need for certain indicator types, regardless of statistical loading.

The observed differences in SQI values between linear (PCALinear, EOLinear) and non-linear (PCANLinear, EONLinear) normalization approaches stemmed from the mathematical nature of their respective scoring functions. Linear normalization, using simple range rescaling (Equations (1) and (2)), tends to assign scores evenly across the data distribution, which can overemphasize the contribution of extreme values and underestimate those in proximity to the mean [87,88]. This leads to more skewed and peaked distributions, as evidenced in PCALinear.

By contrast, non-linear scoring through sigmoidal functions (Equation (3)) has been demonstrated to moderate the influence of extreme values and amplify mid-range differences. This result yields distributions that are more balanced and reflective of functional performance. This is illustrated by the flatter, less skewed distribution of EONLinear. Previous studies emphasized the superiority of non-linear transformations for environmental indices, noting that soil functions do not always improve proportionally in conjunction with increases in indicator values [5,10,89]. Consequently, scoring methods that incorporate thresholds or diminishing returns—such as sigmoid curves—more accurately reflect ecological realities.

This consideration is particularly important in edaphic datasets, which often exhibit high variability, skewness, and occasional values near or below detection limits [90,91]. These data characteristics can hinder statistical analyses and spatial interpolations, especially when using methods such GWRK that rely on residual modeling [92]. By attenuating the disproportionate effect of outliers and enhancing sensitivity in the mid-range, non-linear scoring facilitates the development of more robust soil quality indices and spatial representations.

The significant differences found in the Friedman and Nemenyi tests further supported that the adoption of scoring method—not just variable selection—can meaningfully alter soil quality assessments. In summary, while linear methods offer simplicity, non-linear approaches provide greater ecological realism and statistical robustness in SQI construction. They are particularly effective at capturing nuanced differences in mid-range soil quality and preventing the distortion caused by outliers.

4.3. Soil Quality Maps

A variable may exhibit statistical skewness or non-normality yet still display spatially coherent patterns that enhance model performance. Conversely, a variable with a statistically ideal distribution may lack spatial structure, rendering it less predictable in spatially explicit models. This distinction is especially relevant when employing GWRK, which relies on both spatial autocorrelation and covariate relationships to generate localized predictions [93,94]. For example, if the values are very uneven or have extreme outliers, the model may become less reliable in some areas. However, when there is a strong spatial pattern in the data, this can help the model stay consistent and accurate, even if the overall distribution is not ideal [95].

Despite the similar statistical robustness between PCANLinear and EONLinear, only PCANLinear achieved superior predictive performance when GWRK was implemented. The PCANLinear index (see Figure 8B) manifested the most extensive range of values and a more nuanced spatial structure, especially in the central-eastern part of the study area, which aligns with its superior performance. This map delineated a finer gradient of soil quality, particularly in transitional zones. By contrast, PCALinear (Figure 8A) displayed a more compressed value distribution, with a predominance of mid-range classes. This suggested that linear scoring may underrepresent variability at the extremes, leading to a loss of spatial detail in areas with marginal or exceptional soil conditions. The EOLinear (Figure 8B) and EONLinear (Figure 8D) maps shared some similarities with their PCA counterparts. However, they also exhibited increased spatial fragmentation and patchiness. In particular, EONLinear, despite its use of non-linear scoring, showed reduced spatial coherence and more noise in the southern region, which was consistent with its higher prediction error. As illustrated in Figure 7, the spatial distributions of the indicators derived from PCANLinear and EONLinear followed analogous trends However, minor discrepancies at specific points resulted in substantial variation in the performance of the GWRK model. Overall, the maps illustrated that non-linear scoring combined with PCA yielded the most consistent and spatially informative results, reinforcing the idea that both statistical distribution and spatial behavior must be considered when designing soil quality indices.

The objective of constructing SQIs is to create tools specifically designed for the management goals and environmental conditions of a study area. The application of an identical index—with similar indicators and weights—in diverse contexts can prove arduous and is frequently discouraged [96]. Nevertheless, this study provides important theoretical insights by demonstrating that the combination of PCA-based variable selection and non-linear normalization generates indices that are not only statistically robust but also ecologically interpretable. This integration of methodological rigor with ecological relevance strengthens the scientific basis for SQI development by showing how indices can more effectively represent non-linear responses and threshold dynamics in soil functions. From a practical standpoint, the framework developed also provides a transferable reference for agroecosystems with similar challenges, especially in regions marked by heterogeneity and scarce data resources. Because indicator selection is data-driven while remaining sensitive to local conditions, the approach can be readily adjusted to different regions without relying on a rigid or uniform scheme. Likewise, the use of non-linear scoring functions mitigates the influence of outliers and enhances the interpretability of mid-range variability, making the indices highly applicable to soils in tropical, arid, or degraded landscapes where such data features are common [97,98]. In this sense, the research contributes not only to advancing soil quality theory but also to providing a flexible and transferable tool for guiding sustainable land management in diverse agroecological contexts.

5. Conclusions

Among the four SQIs evaluated, the PCANLinear configuration exhibited superior performance, effectively reducing data skewness and kurtosis while achieving the highest predictive accuracy under GWRK. This finding underscores the benefits of objective weighting approaches such as PCA when combined with non-linear normalization. Nonlinear scoring functions were particularly effective in mitigating the influence of outliers and enhancing the representation of mid-range variability—an essential aspect analyzing highly dispersed edaphic datasets. SQIs should be integrated into public soil conservation policies, such as the development of targeted programs in the Ancash region that allocate subsidies for regenerative practices informed by edaphic quality maps and aligned with existing initiatives. This approach would promote sustainable land management aligned with local needs and climate resilience. Specifically, in areas with low SQI values (<0.4), farmers should be encouraged to apply organic amendments and mulching systems to improve soil conditions. Conversely, zones with high SQI values (>0.6) offer favorable conditions for the intensification of high-input crops, such as potatoes, alfalfa, and maize.

Author Contributions: Author Contributions: T.S.: methodology, conceptualization, investigation, validation, visualization, and writing—original draft. B.S.: methodology, conceptualization, and writing—review. R.S.: supervision, funding acquisition, and writing—review and editing. All authors have read and agreed to the published version of the manuscript.

Funding: This research was funded by the INIA project "Mejoramiento de los servicios de investigación y transferencia tecnológica en el manejo y recuperación de suelos agrícolas degradados y aguas para riego en la pequeña y mediana agricultura en los departamentos de Lima, Áncash, San Martín, Cajamarca, Lambayeque, Junín, Ayacucho, Arequipa, Puno y Ucayali" (CUI 2487112).

Institutional Review Board Statement: Not applicable.

Informed Consent Statement: Not applicable.

Data Availability Statement: The data presented in this study are available upon request from the corresponding author.

Acknowledgments: The authors gratefully acknowledge the Municipality of Jangas for providing essential collaboration and logistical support during the development of this research.

Conflicts of Interest: The authors have no conflicts of interest to declare.

Abbreviations

The following abbreviations are used in this manuscript:

SQI	Soil quality index
MDS	Minimum dataset
PCA	Principal component analysis
EO	Expert opinion
GWRK	Geographically weighted regression kriging
PC	Principal component
LAgre	Large aggregates
SAgre	Small aggregates
SOC	Soil organic carbon
POCX	Permanganate oxidizable carbon,
cPOM	Coarse particulate organic matter
fPOM	Fine particulate organic matter
Ava_P	Available P
Ava_K	Available K
Ex_Ca	Exchangeable Ca
Ex_Mg	Exchangeable Mg
Ex_Na	Exchangeable Na
Ex_K	Exchangeable K
EC	Electric conductivity
CEC	Cation exchange capacity
ESP	Exchangeable sodium percentage
Carb	Carbonates percentage

References

1. Doran, J.W.; Parkin, T.B. Defining and Assessing Soil Quality. In *Defining Soil Quality for a Sustainable Environment*; Doran, J.W., Coleman, D.C., Bezdicek, D.F., Stewart, B.A., Eds.; Soil Science Society of America Book Series: Madison, WI, USA, 1994.
2. Nakajima, T.; Lal, R.; Jiang, S. Soil Quality Index of a Crosby Silt Loam in Central Ohio. *Soil. Tillage Res.* **2015**, *146*, 323–328. [CrossRef]
3. Vasu, D.; Tiwari, G.; Sahoo, S.; Dash, B.; Jangir, A.; Sharma, R.P.; Naitam, R.; Tiwary, P.; Karthikeyan, K.; Chandran, P. A Minimum Data Set of Soil Morphological Properties for Quantifying Soil Quality in Coastal Agroecosystems. *Catena* **2021**, *198*, 105042. [CrossRef]
4. Andrews, S.S.; Karlen, D.L.; Mitchell, J.P. A Comparison of Soil Quality Indexing Methods for Vegetable Production Systems in Northern California. *Agric. Ecosyst. Environ.* **2002**, *90*, 25–45. [CrossRef]
5. Pouladi, N.; Jafarzadeh, A.A.; Shahbazi, F.; Ghorbani, M.A.; Greve, M.H. Assessing the Soil Quality Index as Affected by Two Land Use Scenarios in Miandoab Region. *SN Appl. Sci.* **2020**, *2*, 1875. [CrossRef]
6. Stott, D.E.; Cambardella, C.A.; Tomer, M.D.; Karlen, D.L.; Wolf, R. A Soil Quality Assessment within the Iowa River South Fork Watershed. *Soil. Sci. Soc. Am. J.* **2011**, *75*, 2271–2282. [CrossRef]
7. Gelaw, A.M.; Singh, B.R.; Lal, R. Soil Quality Indices for Evaluating Smallholder Agricultural Land Uses in Northern Ethiopia. *Sustainability* **2015**, *7*, 2322–2337. [CrossRef]

8. Askari, M.S.; Holden, N.M. Indices for Quantitative Evaluation of Soil Quality under Grassland Management. *Geoderma* **2014**, *230–231*, 131–142. [CrossRef]
9. Vasu, D.; Singh, S.K.; Ray, S.K.; Duraisami, V.P.; Tiwary, P.; Chandran, P.; Nimkar, A.M.; Anantwar, S.G. Soil Quality Index (SQI) as a Tool to Evaluate Crop Productivity in Semi-Arid Deccan Plateau, India. *Geoderma* **2016**, *282*, 70–79. [CrossRef]
10. Chaudhry, H.; Vasava, H.B.; Chen, S.; Saurette, D.; Beri, A.; Gillespie, A.; Biswas, A. Evaluating the Soil Quality Index Using Three Methods to Assess Soil Fertility. *Sensors* **2024**, *24*, 864. [CrossRef]
11. Evangelou, E.; Giourga, C. Identification of Soil Quality Factors and Indicators in Mediterranean Agro-Ecosystems. *Sustainability* **2024**, *16*, 10717. [CrossRef]
12. Luján Soto, R.; Cuéllar Padilla, M.; de Vente, J. Participatory Selection of Soil Quality Indicators for Monitoring the Impacts of Regenerative Agriculture on Ecosystem Services. *Ecosyst. Serv.* **2020**, *45*, 101157. [CrossRef]
13. Amorim, H.C.S.; Ashworth, A.J.; Wienhold, B.J.; Savin, M.C.; Allen, F.L.; Saxton, A.M.; Owens, P.R.; Curi, N. Soil Quality Indices Based on Long-Term Conservation Cropping Systems Management. *Agrosyst. Geosci. Environ.* **2020**, *3*, e20036. [CrossRef]
14. Raiesi, F. A Minimum Data Set and Soil Quality Index to Quantify the Effect of Land Use Conversion on Soil Quality and Degradation in Native Rangelands of Upland Arid and Semiarid Regions. *Ecol. Indic.* **2017**, *75*, 307–320. [CrossRef]
15. Garcia, Y.; Ramírez, W.; Sánchez, S. Soil Quality Indicators: A New Way to Evaluate This Resource. *Pastos Y Forrajes* **2012**, *35*, 125–138.
16. Shukla, M.K.; Lal, R.; Ebinger, M. Determining Soil Quality Indicators by Factor Analysis. *Soil. Tillage Res.* **2006**, *87*, 194–204. [CrossRef]
17. Doran, J.W.; Jones, A.J. *Methods for Assessing Soil Quality*; Doran, J.W., Jones, A.J., Eds.; Soil Science Society of America Book Series NO. 49; CABI: Madison, WI, USA, 1996.
18. Li, G.; Chen, J.; Sun, Z.; Tan, M. Establishing a Minimum Dataset for Soil Quality Assessment Based on Soil Properties and Land-Use Changes. *Acta Ecol. Sin.* **2007**, *27*, 2715–2724. [CrossRef]
19. El Behairy, R.A.; El Arwash, H.M.; El Baroudy, A.A.; Ibrahim, M.M.; Mohamed, E.S.; Kucher, D.E.; Shokr, M.S. How Can Soil Quality Be Accurately and Quickly Studied? A Review. *Agronomy* **2024**, *14*, 1682. [CrossRef]
20. Fernandes, J.C.; Gamero, C.A.; Rodrigues, J.G.L.; Mirás-Avalos, J.M. Determination of the Quality Index of a Paleudult under Sunflower Culture and Different Management Systems. *Soil. Tillage Res.* **2011**, *112*, 167–174. [CrossRef]
21. Mukherjee, A.; Lal, R. Comparison of Soil Quality Index Using Three Methods. *PLoS ONE* **2014**, *9*, e105981. [CrossRef]
22. Wu, L.C.; Zhang, Z.Z.; Xu, D.K. Variable Selection in Joint Location and Scale Models of the Skew-Normal Distribution. *J. Stat Comput. Simul.* **2013**, *83*, 1266–1278. [CrossRef]
23. Kumar, S.; Lal, R.; Liu, D. A Geographically Weighted Regression Kriging Approach for Mapping Soil Organic Carbon Stock *Geoderma* **2012**, *189–190*, 627–634. [CrossRef]
24. Hengl, T.; Heuvelink, G.B.M.; Stein, A. A Generic Framework for Spatial Prediction of Soil Variables Based on Regression-Kriging *Geoderma* **2004**, *120*, 75–93. [CrossRef]
25. Bedoya Garland, E.; Aramburú, C.E.; Burneo, Z. Una Agricultura Insostenible y La Crisis Del Barbecho: El Caso de Los Agricultores Del VRAE. *Anthropologica* **2017**, *35*, 211–240. [CrossRef]
26. Taboada-Hermoza, R.; Martínez, A.G. "No One Is Safe": Agricultural Burnings, Wildfires and Risk Perception in Two Agropastoral Communities in the Puna of Cusco, Peru. *Fire* **2025**, *8*, 60. [CrossRef]
27. Lal, R. Climate Change and Soil Degradation Mitigation by Sustainable Management of Soils and Other Natural Resources. *Agric. Res.* **2012**, *1*, 199–212. [CrossRef]
28. Coaguila, L.; Mataix-Solera, J.; Nina, S.; García-Carmona, M.; Salazar, E.T. Soil Degradation Evidence Following a Wildfire in Arequipa's Andean Region, Peru. *Span. J. Soil. Sci.* **2025**, *15*, 13983. [CrossRef]
29. Alavi-Murillo, G.; Diels, J.; Gilles, J.; Willems, P. Soil Organic Carbon in Andean High-Mountain Ecosystems: Importance, Challenges, and Opportunities for Carbon Sequestration. *Reg. Environ. Change* **2022**, *22*, 128. [CrossRef]
30. Natural Resources and Environment Secretary Mexican Official Standard. NOM-021-RECNAT-2000: Especificaciones De Fertilidad, Salinidad Y Clasificación De Suelos. Available online: https://www.fao.org/faolex/results/details/es/c/LEX-FAOC050674/ (accessed on 6 May 2025).
31. United States Environmental Protection Agency (USEPA). Soil and Wate PH (9045D). Available online: https://www.epa.gov/sites/default/files/2015-12/documents/9045d.pdf (accessed on 6 May 2025).
32. *ISO 11265:1994*; Soil Quality—Determination of the Specific Electrical Conductivity. International Organization for Standardization: Geneva, Switzerland, 1994. Available online: https://www.iso.org/standard/19243.html (accessed on 6 May 2025).
33. Bazán, R. *Manual De Procedimientos De Los Análisis De Suelos Y Agua Con Fines De Riego*; Instituto Nacional de Innovación Agraria -INIA: Lima, Peru, 2017.
34. Cambardella, C.A.; Elliott, E.T. Particulate Soil Organic-Matter Changes across a Grassland Cultivation Sequence. *Soil. Sci. Soc. Am. J.* **1992**, *56*, 777–783. [CrossRef]

5. Culman, S.W.; Snapp, S.S.; Freeman, M.A.; Schipanski, M.E.; Beniston, J.; Lal, R.; Drinkwater, L.E.; Franzluebbers, A.J.; Glover, J.D.; Grandy, A.S.; et al. Permanganate Oxidizable Carbon Reflects a Processed Soil Fraction That Is Sensitive to Management. *Soil. Sci. Soc. Am. J.* **2012**, *76*, 494. [CrossRef]

6. NRCS-USDA. *Guía Para La Evaluación De La Calidad Y Salud Del Suelo*; Natural Resources Conservation Service-USDA: Washington, DC, USA, 1999.

7. Martín-Sanz, J.P.; de Santiago-Martín, A.; Valverde-Asenjo, I.; Quintana-Nieto, J.R.; González-Huecas, C.; López-Lafuente, A.L. Comparison of Soil Quality Indexes Calculated by Network and Principal Component Analysis for Carbonated Soils under Different Uses. *Ecol. Indic.* **2022**, *143*, 109374. [CrossRef]

8. Yao, R.; Yang, J.; Gao, P.; Zhang, J.; Jin, W. Determining Minimum Data Set for Soil Quality Assessment of Typical Salt-Affected Farmland in the Coastal Reclamation Area. *Soil. Tillage Res.* **2013**, *128*, 137–148. [CrossRef]

9. Lenka, N.K.; Meena, B.P.; Lal, R.; Khandagle, A.; Lenka, S.; Shirale, A.O. Comparing Four Indexing Approaches to Define Soil Quality in an Intensively Cropped Region of Northern India. *Front. Environ. Sci.* **2022**, *10*, 865473. [CrossRef]

0. Porta Casanellas, J.; López-Acevedo Reguerín, M.; Roquero de Laburu, C. *Edafología: Para La Agricultura Y El Medio Ambiente*, 3rd ed.; Ediciones Mundi-Prensa: Madrid, Spain, 2003; ISBN 8484761487.

1. Mandal, N.; Dey, A.; Rakshit, R. *Soil Management For Sustainable Agriculture*, 1st ed.; Apple Academic Press: New York, NY, USA, 2022.

2. Karlen, D.L.; Mausbach, M.J.; Doran, J.W.; Cline, R.G.; Harris, R.F.; Schuman, G.E. Soil Quality: A Concept, Definition, and Framework for Evaluation. *Soil. Sci. Soc. Am. J.* **1997**, *61*, 4–10. [CrossRef]

3. Zhang, C.; Xue, S.; Liu, G.-B.; Song, Z.-L. A Comparison of Soil Qualities of Different Revegetation Types in the Loess Plateau, China. *Plant Soil.* **2011**, *347*, 163–178. [CrossRef]

4. Sinha, S.; Masto, R.E.; Ram, L.C.; Selvi, V.A.; Srivastava, N.K.; Tripathi, R.C.; George, J. Rhizosphere Soil Microbial Index of Tree Species in a Coal Mining Ecosystem. *Soil. Biol. Biochem.* **2009**, *41*, 1824–1832. [CrossRef]

5. Kozak, M.; Krzanowski, W.; Tartanus, M. Use of the Correlation Coefficient in Agricultural Sciences: Problems, Pitfalls and How to Deal with Them. *Anais da Academia Brasileira de Ciências* **2012**, *84*, 1147–1156. [CrossRef]

6. Jia, B.-B.; Liu, J.Y.; Zhang, M.L. Pairwise Statistical Comparisons of Multiple Algorithms. *Front. Comput. Sci.* **2025**, *19*, 1912372. [CrossRef]

7. R: The R Project for Statistical Computing. Available online: https://www.r-project.org/ (accessed on 5 August 2025).

8. Udom, B.E.; Nuga, B.O.; Adesodun, J.K. Water-Stable Aggregates and Aggregate-Associated Organic Carbon and Nitrogen after Three Annual Applications of Poultry Manure and Spent Mushroom Wastes. *Appl. Soil. Ecol.* **2016**, *101*, 5–10. [CrossRef]

9. Sui, Y.-Y.; Jiao, X.-G.; Liu, X.-B.; Zhang, X.-Y.; Ding, G.-W. Water-Stable Aggregates and Their Organic Carbon Distribution after Five Years of Chemical Fertilizer and Manure Treatments on Eroded Farmland of Chinese Mollisols. *Can. J. Soil. Sci.* **2012**, *92*, 551–557. [CrossRef]

0. Tang, X.; Qiu, J.; Xu, Y.; Li, J.; Chen, J.; Li, B.; Lu, Y. Responses of Soil Aggregate Stability to Organic C and Total N as Controlled by Land-Use Type in a Region of South China Affected by Sheet Erosion. *Catena* **2022**, *218*, 106543. [CrossRef]

1. He, Y.; Yang, M.; Huang, R.; Wang, Y.; Ali, W. Soil Organic Matter and Clay Zeta Potential Influence Aggregation of a Clayey Red Soil (Ultisol) under Long-Term Fertilization. *Sci. Rep.* **2021**, *11*, 20498. [CrossRef] [PubMed]

2. Zhou, M.; Liu, C.; Wang, J.; Meng, Q.; Yuan, Y.; Ma, X.; Liu, X.; Zhu, Y.; Ding, G.; Zhang, J.; et al. Soil Aggregates Stability and Storage of Soil Organic Carbon Respond to Cropping Systems on Black Soils of Northeast China. *Sci. Rep.* **2020**, *10*, 265. [CrossRef] [PubMed]

3. Wang, J.; Sun, X.; Zhang, Y.; Fang, Y.; Zhan, Y.; Guo, R.; Qian, R.; Cai, T.; Liu, T.; Jia, Z.; et al. Soil Aggregates and Aggregate-Associated Carbon and Nitrogen in Farmland in Relation to Long-Term Fertilization on the Loess Plateau, China. *Agronomy* **2023**, *13*, 1312. [CrossRef]

4. Feng, M.; Xiang, J.; Ji, X.; Jiang, J. Larger Soil Water-Stable Aggregate May Exert a Negative Effect on Nutrient Availability: Results from Red Soil (Ultisol), in South China. *Forests* **2023**, *14*, 975. [CrossRef]

5. Soinne, H.; Keskinen, R.; Tähtikarhu, M.; Kuva, J.; Hyväluoma, J. Effects of Organic Carbon and Clay Contents on Structure-Related Properties of Arable Soils with High Clay Content. *Eur. J. Soil. Sci.* **2023**, *74*, e13424. [CrossRef]

6. Krinari, G.A.; Shinkarev, A.A.; Giniyatullin, K.G. Mineralogy of the Clay Fraction of Water-Stable Aggregates from Dark Gray Forest Soil. *Eurasian Soil. Sci.* **2006**, *39*, 71–83. [CrossRef]

7. Hurisso, T.T.; Culman, S.W.; Horwath, W.R.; Wade, J.; Cass, D.; Beniston, J.W.; Bowles, T.M.; Grandy, A.S.; Franzluebbers, A.J.; Schipanski, M.E.; et al. Comparison of Permanganate-Oxidizable Carbon and Mineralizable Carbon for Assessment of Organic Matter Stabilization and Mineralization. *Soil. Sci. Soc. Am. J.* **2016**, *80*, 1352. [CrossRef]

8. Chambers, L.G.; Mirabito, A.J.; Brew, S.; Nitsch, C.K.; Bhadha, J.H.; Hurst, N.R.; Berkowitz, J.F. Evaluating Permanganate Oxidizable Carbon (POXC)'s Potential for Differentiating Carbon Pools in Wetland Soils. *Ecol. Indic.* **2024**, *167*, 112624. [CrossRef]

59. Bongiorno, G.; Bünemann, E.K.; Oguejiofor, C.U.; Meier, J.; Gort, G.; Comans, R.; Mäder, P.; Brussaard, L.; de Goede, R. Sensitivity of Labile Carbon Fractions to Tillage and Organic Matter Management and Their Potential as Comprehensive Soil Quality Indicators across Pedoclimatic Conditions in Europe. *Ecol. Indic.* **2019**, *99*, 38–50. [CrossRef]
60. Woodings, F.S.; Margenot, A.J. Revisiting the Permanganate Oxidizable Carbon (POXC) Assay Assumptions: POXC Is Lignin Sensitive. *Agric. Environ. Lett.* **2023**, *8*, e20108. [CrossRef]
61. Mikha, M.M.; Marake, M.V. Soil Organic Matter Fractions and Carbon Distribution under Different Management in Lesotho Southern Africa. *Soil. Sci. Soc. Am. J.* **2023**, *87*, 140–155. [CrossRef]
62. Dong, X.; Zhao, K.; Wang, J.; Gui, H.; Xiao, Y.; Chen, Z.; Miao, Y.; Han, S. Effects of Forest Types on Soil Carbon Content in Aggregate Faction under Climate Transition Zone. *Front. Environ. Sci.* **2023**, *10*, 1052175. [CrossRef]
63. Liu, D.; Li, S.; Zhu, W.; Wang, Y.; Zhang, S.; Fang, Y. Storage and Stability of Soil Organic Carbon in Two Temperate Forests in Northeastern China. *Land* **2023**, *12*, 1019. [CrossRef]
64. Semenov, V.M.; Lebedeva, T.N.; Pautova, N.B. Particulate Organic Matter in Noncultivated and Arable Soils. *Eurasian Soil. Sci.* **2019**, *52*, 396–404. [CrossRef]
65. Awale, R.; Emeson, M.A.; Machado, S. Soil Organic Carbon Pools as Early Indicators for Soil Organic Matter Stock Changes under Different Tillage Practices in Inland Pacific Northwest. *Front. Ecol. Evol.* **2017**, *5*, 96. [CrossRef]
66. Samaniego, T.; Ramirez, J.; Solórzano, R. Litter Decomposition Rates of Four Species of Agroecological Importance in the Peruvian Coast and Andean Highland. *Nitrogen* **2024**, *5*, 772–789. [CrossRef]
67. Dinkecha, K. Effects of Liming on Physicochemical Properties and Nutrient Availability of Acidic Soils in Welmera Woreda, Central Highlands of Ethiopia. *Biochem. Mol. Biol.* **2017**, *2*, 102. [CrossRef]
68. Mohiuddin, M.; Irshad, M.; Sher, S.; Hayat, F.; Ashraf, A.; Masood, S.; Bibi, S.; Ali, J.; Waseem, M. Relationship of Selected Soil Properties with the Micronutrients in Salt-Affected Soils. *Land* **2022**, *11*, 845. [CrossRef]
69. Obasi, S.N.; Jokthan, G.E.; Obasi, C.C.; Madueke, C.O. Micronutrient Dynamics in Relation to Soil Properties in Arable Soils of Rigachikun-Kaduna, Northern Guinea Savannah, Nigeria. *J. CleanWAS* **2022**, *6*, 14–22. [CrossRef]
70. Peng, Z.; Liu, Y.; An, Y.; Wang, Z.; Li, X.; Gao, H.; Wei, G.; Jiao, S. Contrasting Patterns and Drivers of Soil Micronutrient Availability in Paddy and Maize Fields of Eastern China. *Geoderma* **2023**, *431*, 116342. [CrossRef]
71. Covelo, E.F.; Vega, F.A.; Andrade, M.L. Competitive Sorption and Desorption of Heavy Metals by Individual Soil Components. *J. Hazard. Mater.* **2007**, *140*, 308–315. [CrossRef] [PubMed]
72. Dai, Z.; Guo, X.; Lin, J.; Wang, X.; He, D.; Zeng, R.; Meng, J.; Luo, J.; Delgado-Baquerizo, M.; Moreno-Jiménez, E.; et al. Metallic Micronutrients Are Associated with the Structure and Function of the Soil Microbiome. *Nat. Commun.* **2023**, *14*, 8456. [CrossRef] [PubMed]
73. Curtin, D.; Smillie, G.W. Origin of the PH-Dependent Cation Exchange Capacities of Irish Soil Clays. *Geoderma* **1979**, *22*, 213–224. [CrossRef]
74. Juhos, K.; Madarász, B.; Kotroczó, Z.; Béni, Á.; Makádi, M.; Fekete, I. Carbon Sequestration of Forest Soils Is Reflected by Changes in Physicochemical Soil Indicators–A Comprehensive Discussion of a Long-Term Experiment on a Detritus Manipulation. *Geoderma* **2021**, *385*, 114918. [CrossRef]
75. de Albuquerque, C.G.; Gavelaki, F.; Matera, H.B.; Motta, A.C.V.; Prior, S.A.; Ercole, T.M.; Araújo, E.M. Relationship between PH and Base Saturation Associated with Soil Cation Exchange Capacity in Soils of Mato Grosso Do Sul, Brazil. *Bragantia* **2024**, *83*, e20230291. [CrossRef]
76. Zamanian, K.; Pustovoytov, K.; Kuzyakov, Y. Pedogenic Carbonates: Forms and Formation Processes. *Earth Sci. Rev.* **2016**, *157*, 1–17. [CrossRef]
77. Wang, C.; Li, W.; Yang, Z.; Chen, Y.; Shao, W.; Ji, J. An Invisible Soil Acidification: Critical Role of Soil Carbonate and Its Impact on Heavy Metal Bioavailability. *Sci. Rep.* **2015**, *5*, 12735. [CrossRef]
78. Ortiz, A.C.; Jin, L.; Ogrinc, N.; Kaye, J.; Krajnc, B.; Ma, L. Dryland Irrigation Increases Accumulation Rates of Pedogenic Carbonate and Releases Soil Abiotic CO2. *Sci. Rep.* **2022**, *12*, 464. [CrossRef]
79. Yolcubal, I.; Brusseau, M.L.; Artiola, J.F.; Wierenga, P.J.; Wilson, L.G. Environmental Properties and Processes. In *Environmental Monitoring and Characterization*; Artiola, J.F., Pepper, I.L., Brusseau, M.L., Eds.; Academic Press: New York, NY, USA, 2004; pp. 207–239. ISBN 9780080491271.
80. Fagodiya, R.K.; Sharma, G.; Verma, K.; Rai, A.K.; Prajapat, K.; Singh, R.; Chandra, P.; Sheoran, P.; Yadav, R.K.; Biswas, A.K. Computation of Soil Quality Index after Fifteen Years of Long-Term Tillage and Residue Management Experiment (LT&RE) under Rice Wheat System. *Agric. Syst.* **2024**, *219*, 104039. [CrossRef]
81. Blanco, H.; Lal, R. Soil Fertility Management. In *Soil Conservation and Management*; Blanco, H., Lal, R., Eds.; Springer Nature: Cham, Switzerland, 2023; pp. 363–390. ISBN 978-3-031-30341-8.
82. Damiba, W.A.F.; Gathenya, J.M.; Raude, J.M.; Home, P.G. Soil Quality Index (SQI) for Evaluating the Sustainability Status of Kakia-Esamburmbur Catchment under Three Different Land Use Types in Narok County, Kenya. *Heliyon* **2024**, *10*, e25611. [CrossRef]

3. Li, Y.; Zhao, H.; Liu, J.; Chen, C.; Guo, Y. A Framework for Selecting and Assessing Soil Quality Indicators for Sustainable Soil Management in Waste Dumps. *Sci. Rep.* **2024**, *14*, 8491. [CrossRef]
4. Karlen, D.L.; Ditzler, C.A.; Andrews, S.S. Soil Quality: Why and How? *Geoderma* **2003**, *114*, 145–156. [CrossRef]
5. Qi, Y.; Darilek, J.L.; Huang, B.; Zhao, Y.; Sun, W.; Gu, Z. Evaluating Soil Quality Indices in an Agricultural Region of Jiangsu Province, China. *Geoderma* **2009**, *149*, 325–334. [CrossRef]
6. Vasu, D.; Tiwary, P.; Chandran, P. A Novel and Comprehensive Soil Quality Index Integrating Soil Morphological, Physical, Chemical, and Biological Properties. *Soil. Tillage Res.* **2024**, *244*, 106246. [CrossRef]
7. Kim, Y.-S.; Kim, M.K.; Fu, N.; Liu, J.; Wang, J.; Srebric, J. Investigating the Impact of Data Normalization Methods on Predicting Electricity Consumption in a Building Using Different Artificial Neural Network Models. *Sustain. Cities Soc.* **2025**, *118*, 105570. [CrossRef]
8. Muhammad Ali, P.J. Investigating the Impact of Min-Max Data Normalization on the Regression Performance of K-Nearest Neighbor with Different Similarity Measurements. *ARO Sci. J. Koya Univ.* **2022**, *10*, 85–91. [CrossRef]
9. Mahajan, G.R.; Das, B.; Morajkar, S.; Desai, A.; Murgaokar, D.; Patel, K.P.; Kulkarni, R.M. Comparison of Soil Quality Indexing Methods for Salt-Affected Soils of Indian Coastal Region. *Environ. Earth Sci.* **2021**, *80*, 725. [CrossRef]
0. Yuan, Y.; Yang, K.; Cheng, L.; Bai, Y.; Wang, Y.; Hou, Y.; Ding, A. Effect of Normalization Methods on Accuracy of Estimating Low- and High-Molecular Weight PAHs Distribution in the Soils of a Coking Plant. *Int. J. Environ. Res. Public Health* **2022**, *19*, 15470. [CrossRef]
1. Reimann, C. *Statistical Data Analysis Explained: Applied Environmental Statistics with R*; John Wiley & Sons: Hoboken, NJ, USA, 2008; ISBN 9780470985816.
2. Harris, P.; Charlton, M.; Fotheringham, A.S. Moving Window Kriging with Geographically Weighted Variograms. *Stoch. Environ. Res. Risk Assess.* **2010**, *24*, 1193–1209. [CrossRef]
3. Comber, A.; Brunsdon, C.; Charlton, M.; Dong, G.; Harris, R.; Lu, B.; Lü, Y.; Murakami, D.; Nakaya, T.; Wang, Y.; et al. A Route Map for Successful Applications of Geographically Weighted Regression. *Geogr. Anal.* **2023**, *55*, 155–178. [CrossRef]
4. Bahmani, S.; Naganna, S.R.; Ghorbani, M.A.; Shahabi, M.; Asadi, E.; Shahid, S. Geographically Weighted Regression Hybridized with Kriging Model for Delineation of Drought-Prone Areas. *Environ. Model. Assess.* **2021**, *26*, 803–821. [CrossRef]
5. Lloyd, C.D. *Spatial Data Analysis: An Introduction for GIS Users*, 2nd ed.; CRC Press: Boca Raton, FL, USA, 2010; ISBN 9780199554324.
6. Bünemann, E.K.; Bongiorno, G.; Bai, Z.; Creamer, R.E.; De Deyn, G.; de Goede, R.; Fleskens, L.; Geissen, V.; Kuyper, T.W.; Mäder, P.; et al. Soil Quality–A Critical Review. *Soil. Biol. Biochem.* **2018**, *120*, 105–125. [CrossRef]
7. Ellur, R.; Ankappa, A.M.; Dharumarajan, S.; Puttavenkategowda, T.; Nanjundegowda, T.M.; Sannegowda, P.S.; Pratap Mishra, A.; Đurin, B.; Dogančić, D. Soil Quality Assessment and Its Spatial Variability in an Intensively Cultivated Area in India. *Land* **2024**, *13*, 970. [CrossRef]
8. de Melo, D.M.A.; Pérez-Marin, A.M.; de Araújo, A.E.; de Oliveira, M.R.G.; Macedo, R.S.; do Nascimento, S.M.S.G. Soil Quality Indicators in Peasant Agroecosystems in Paraíba State, Brazil. *Trop. Conserv. Sci.* **2024**, *17*. [CrossRef]

sustainability

MDPI

Article

Soil Respiration in Maize, Wheat, and Barley Across a Growing Season: Findings from Croatia's Continental Region

Dija Bhandari [1,*], Nikola Bilandžija [2], Tajana Krička [2], Zvonimir Zdunić [3], Soni Ghimire [4], Theresa Reinhardt Piskáčková [5] and Darija Bilandžija [6,*]

1 Sustainability in Agriculture, Food Production and Food Technology in the Danube Region (DAFM), Faculty of Agriculture, University of Zagreb, Svetošimunska Cesta 25, 10 000 Zagreb, Croatia
2 Division of Agricultural Engineering and Technology, Faculty of Agriculture, University of Zagreb, Svetošimunska Cesta 25, 10 000 Zagreb, Croatia; nbilandzija@agr.hr (N.B.); tkricka@agr.hr (T.K.)
3 Agricultural Institute Osijek, Južno Predgrađe 17, 31 000 Osijek, Croatia; zvonimir.zdunic@poljinos.hr
4 Department of Soil and Environmental Sciences, University of Wisconsin-Madison, 1525 Observatory Drive, Madison, WI 53706, USA; sghimire3@wisc.edu
5 Department of Agroecology and Crop Production, Faculty of Agrobiology, Food and Natural Resources, Czech University of Life Sciences, Kamycka 129, 165000 Prague, Czech Republic; tpiskackova@afu.czu.cz
6 Division of Agroecology, Faculty of Agriculture, University of Zagreb, Svetošimunska Cesta 25, 10 000 Zagreb, Croatia
* Correspondence: bhandaridija27@gmail.com (D.B.); dbilandzija@agr.hr (D.B.); Tel.: +977-9843938018 (D.B.); +385-1-2394089 (D.B.)

Abstract: Soil respiration (Rs) in croplands is of primary importance in understanding the carbon (C) cycle mechanism and C balance of agroecosystems. This study examines the seasonal Rs dynamics in three predominant cereal crops, maize, wheat, and barley, in continental Croatia during the growing season 2021/2022. This study was conducted at the Agricultural Institute Osijek, featuring a continental climate and silty clay soil. Rs was measured monthly throughout the growing season by following an in situ closed static chamber method and using Infrared Gas Analyzers (IRGAs) with three replicates for each crop and a fallow control. This study found that crop type plays a prominent role in Rs dynamics, while temperature and moisture can have modifying effects. Significant ($p < 0.05$) temporal variation in Rs between months was found in wheat, barley, and maize. Mean seasonal Rs values for wheat, barley, and maize were, respectively, 14.73, 19.64, and 12.72 kg CO_2-C ha^{-1} day^{-1}. Cropped fields demonstrated two to three times higher Rs than no vegetation/fallow and indicated the significance of autotrophic respiration in cropped fields. There exists a seasonal dynamics of Rs governed by the complex interaction of biotic and abiotic factors that influences Rs. This necessitates a multifaceted examination for effective understanding of seasonal Rs dynamics and its integration to modeling studies.

Keywords: carbon cycle; cereal crops; soil respiration; temporal variation

1. Introduction

Soil is an important reservoir of carbon (C), and it contains twice the amount of C present in the atmosphere and thrice as much C as exists in vegetation [1,2]. Soil and atmosphere continuously exchange C and are vital components of the terrestrial C cycle. Soils release C in the form of carbon dioxide (CO_2) into the atmosphere during the process, called as soil respiration (Rs). Under natural conditions, the C efflux is balanced by similar opposite fluxes. However, over the past decades, Rs has been increasing [3–5] due to changes in climate [6], land cover/disturbance, and biogeochemical cycle change [5,7]. This increase disrupts the natural C cycle, causing atmospheric CO_2 to rise, which has increased

1.2 times from 1990 to 2022, and these trends are consistent with global temperature rise [8]. The increase in CO_2 along with greenhouse gases in the atmosphere have been identified as the principal reasons for climate change and global warming. To mitigate these effects, initiatives such as the worldwide '4p1000' initiative and the FAO's Global Assessment of Soil Organic Carbon Sequestration Potential (GSOCeq) program have the goal of enhancing soil C sequestration. Although adding to soil C storage is essential, to obtain net C accumulation in soils, it is equally important to monitor the release of C back into the atmosphere and utilize opportunities to prevent C losses from the soil. It is therefore important to understand Rs as a measure to mitigate climate change issues and challenges in agriculture.

Rs originates from the respiration of plant roots, the respiration activity of soil macroorganisms, and microbial mineralization and decomposition processes. Rs serves as an indicator of biological activities in the soil profile [9,10]. Rs is one of the most widely used biological indicators [11]; it estimates biological activity, nutrient mineralization [10], and biomass activity, and it provides early detection of the management effect on organic matter in the soil. As one of the key ecosystem processes, Rs is closely linked with ecosystem productivity, soil fertility, and decomposition processes, and it plays a key role in global and regional C cycling [12]. Measurements of Rs provide information on ecosystem metabolism [13]. Rs provides a crucial parameter in the determination of greenhouse gas emissions to the atmosphere and in modeling the C cycles.

Rs is highly variable at different spatial and temporal scales as it is affected by several factors. During the peak growing season, root activity and microbial processes are higher, which subsequently enhance Rs rates. However, Rs fluctuations in the temporal scale are primarily influenced by climatic factors, soil properties, crop growth stages, and land management practices. The estimation of Rs on global and regional scales is difficult, with many uncertainties. Therefore, understanding Rs requires an in-depth understanding of this dynamic across various ecosystems and climates. Rs values in agroecosystems vary substantially across the globe, as they are influenced by numerous types of cropland management. Agricultural ecosystems are mostly affected by human activities, and as such, sustainable management practices can be used to sequester carbon. Among the terrestrial ecosystems, the farmland ecosystem is the most active and can be adjusted by humans in the shortest period [14]. Understanding Rs in agricultural ecosystems thus becomes crucial for understanding the C cycle, the status of soil health, and the impact of various agricultural practices on emissions of CO_2. A comprehensive evaluation of the seasonal patterns and seasonal cumulative amount of soil respiration in various croplands has remained lacking [15] till recent times. Studies that compare Rs values and their seasonal dynamics in crops are not adequate in terms of executing quantification and modeling studies, therefore limiting the application of these studies to C sequestration and climate change mitigation. Identifying key factors that control Rs is desirable for informed soil management decisions and for promoting and scaling up soil health [15]. Understanding C emission in Rs studies supports the modeling of Rs and provides a scientific basis for implementing taxes on C [14]. In our study, we investigate the seasonal dynamics of Rs in three different arable crops in continental Croatia.

Agricultural production in Croatia is dominated by cereal crops, and maize, wheat, and barley rank as the three leading crops in terms of area coverage [16]. Despite their significance, little is known from research regarding Rs dynamics on Croatian arable land. Therefore, this study will help improve the understanding of the theory and practice of C dynamics in agroecosystems in continental Croatia. With the EU's target of lowering greenhouse gas (GHG) emissions, this research on Rs in Croatia is valuable to GHG estimation and in developing suitable policy.

2. Materials and Methods

2.1. Description of Site

The experiment was carried out during the year 2021/2022 at the Agriculture Institute Osijek (latitude: 45°31′56.47″ N, longitude: 18°44′16.07″ E, 90 m a.s.l), near Osijek City. The Osijek area is characterized by a continental climate. According to a study, the average annual air temperature and precipitation for the region were 11.7 °C and 707 mm, respectively, while the evapotranspiration amounted to 590 mm per year for the period 1991–2018 [17]. A soil water deficit was observed during July–September, with 72 mm, and a water surplus occurred during December–March [17]. For our analysis, during the experimental period (2021/2022), the agroclimatic factor and indicator calculations were based on climate elements data from the Osijek–Čepin main meteorological station (latitude = 45°30′9″ N, longitude = 18°33′41″ E; 89 m a.s.l) of the Croatian Meteorological and Hydrological Service Network.

2.2. Experimental Design

The experiment included 3 different cereal crops (wheat, barley, and maize) and a control plot with three repetitions. The experiment was a part of an experimental field at the Agriculture Institute Osijek which consisted of 3 different cereal crops with 4 different cereal varieties of each crop and a control plot (black fallow) in three repetitions (Figure 1). The size of each plot was 120 m^2 (8 m × 15 m) for wheat and barley and 150 m^2 for corn (10 m × 15 m). In this research, one variety of each cereal crop and a control plot were studied:

1. T0—control-bare soil.
2. T1—winter wheat (*Triticum aestivum* L.) Srpanjka cultivar—an old cultivar, a very early growing cultivar, with very low habitus (64 cm), and a plant density of 9,110,000 plants ha^{-1}.
3. T2—corn (*Zea mays*) OS SK515 cultivar—for production of grain, cob, and/or silage, pronounced grain vigor, FAO group 520, stems with higher growth, large and numerous leaves, deep and branched roots, and a plant density of 65,000 grain ha^{-1}.
4. T3—barley (*Hordeum vulgare* L.) Rex cultivar—a medium—late-growing two-rowed cultivar with an average yield of 10 t/ha, low habitus (87–92 cm), and a plant density of 6,440,000 plants ha^{-1}.

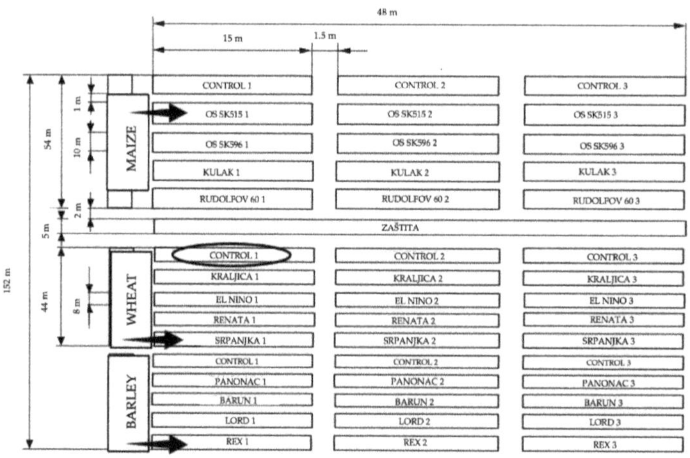

Figure 1. Scheme of the experimental field. (Plots marked with arrows represent the studied experimental plot of each crop, and the plot marked with an oval represents the control plot studied).

2.3. Soil Properties

Soil sampling (0–30 cm) was carried out before the beginning of the experiment in 2021, and the soil's physical and chemical properties were determined. The soil had a silty clay texture with 2.3% sand, 56% silt, and 41.7% clay. The soil bulk density and water holding capacity were found to be 1.39 g cm^{-3} and 37.7%, respectively. Soil pH_{KCL} was found to be 7.24. Soil had 0.11% total nitrogen, 1.25% total carbon, 0.06% total sulfur, 17.87 mg of P_2O_5, 15.50 mg of K_2O per 100 g of soil, 2.3% humus content, and 0.9% $CaCO_3$.

2.4. Maintenance of Experiment

The experimental period (2021/2022) covered one growing season for all the crops: October–July for barley and winter wheat and April–October for maize. Tilling of the field was carried out according to the principles of reduced tillage. Mineral fertilizer (N:P:K—7:20:30) at the rate of 400 kg/ha and urea 100 kg/ha was applied and spread with a mineral fertilizer spreader. Additional tillage was carried with a 4 m rotary harrow, which prepared the soil for sowing. The 1st sowing of the fields with wheat (Srpanjka cultivar) and barley (Rex cultivar) was carried out using a multirow mechanical seeder for small grains. The fertilization in the wheat and barley fields was supplemented through two top dressings with mineral fertilizer at rates of 100 kg/ha KAN and 150 kg/ha KAN in the second and third months, respectively. Herbicide was applied in the third month to suppress weed growth followed by fungicide application in the fourth month. At the same time, the 2nd sowing with maize (OS SK 515 cultivar) was performed following herbicide application. Harvesting was performed with a mechanical harvester.

2.5. Measurement of Soil CO_2 Efflux and Agroclimatic Elements

The field measurements of soil CO_2 efflux and agroclimatic elements (soil temperature and soil moisture) were conducted once every month during the growing season for all crops. These measurements were conducted thrice to obtain 3 repetitions. The measurement was not possible during the winter period (months—December and January) due to unfavorable weather conditions (snow cover). The number of measurements taken for winter wheat was 21 (7 months × 3 repetitions), for barley 18 (6 months × 3 repetitions), for corn 21 (7 months × 3 repetitions), and for control (12 × 3 = 36), resulting in a total of 96 observations. Soil CO_2 efflux was measured by the in situ closed static chamber method with a portable infrared carbon dioxide detector (GasAlerMicro5 IR, Staffordshire, UK 2011). Soil temperature and moisture conditions were measured continuously along with the Rs at a depth of 10 cm with a TDS sensor device, IMKO HD2 (Ettlingen, Germany 2011), placed close to the static chamber.

The calculation of CO_2 efflux was calculated as

$$F_{CO2} = [M \times P \times V \times (c_2 - c_1)]/[R \times T \times A \times (t_2 - t_1)]$$

where

F_{CO2}: soil CO_2 efflux (kg ha^{-1} day^{-1});
M: molar mass of the CO_2 (kg mol^{-1});
P: air pressure (Pa);
V: chamber volume (m^3);
$c_2 - c_1$: CO_2 concentration increase rate in the chamber for the incubation period (μmol mol^{-1});
R: gas constant (J mol^{-1} K^{-1});
T: air temperature (K);
A: chamber surface (m^2);

$t_2 - t_1$: incubation period (day).

2.6. Statistical Analysis

The effects of different covers, different months, and their interaction with Rs were evaluated using analysis of variance (ANOVA) at $\rho \leq 0.05$ with SAS 9.1 statistical software (SAS Inst. Inc., 2002–2004, Cary, NC, USA). Mean Rs values of individual cover types were compared in different months to examine the seasonal respiration, while the same across treatments (cover types) was also compared using ANOVA. A Least Significant Difference (LSD) test was used to determine the differences between means. Pearson correlation analysis was performed to evaluate the relationship between environmental variables (soil moisture and temperature) and Rs. Rs data were log-transformed to meet the normal distribution assumptions before carrying out the analysis.

3. Results

3.1. Seasonal Variation of Temperature, Moisture, and Carbon Fluxes

The average air temperature for the growing season 2021/2022 was 12.4 °C. The mean air temperature during the growing season and the observed soil temperature during the months (n = 1) when Rs was measured (n = 3) is shown in Figure 2a. A general seasonal fluctuation in mean air temperature was observed. Measurements of CO_2 flux over the experimental period had a range of soil temperature (8.9–45.9 °C) at a depth of 10 cm. The lowest mean air temperature was measured in January and the highest in July. The highest mean soil temperature (41.5 °C) was recorded in May and the lowest mean soil temperature (10.4 °C) was observed in November, when measurements were not made in December and January due to freezing soil conditions. At each observation date when soil temperature was measured, the mean observed soil temperature was higher than the monthly air temperature. This seems unusual at first since soil is known as a better insulator of temperature, but the soil temperature was only measured during one day in each month and within daylight hours. It is important to note that the soil temperature follows the same trend of seasonal variation but that the soil temperature at the time of data collection with other variables was higher than the average air temperature for the month to contextualize the data that will be discussed. In February, March, April, and May, differences in observed soil temperature were much higher (>15 °C) than the mean air temperature during those months. Due to the single observation per month, the observed soil temperature does not represent the actual temperature changes in the vegetative season.

The average amount of precipitation for the period of 2021/2022 was 638.7 mm, which was 9.6% less compared to the 1991–2018 period. Water deficit occurred during June, July, and August, whereas there was a water surplus in October, November, December, January, and February. Soil moisture was in the range of 5.0–34.9% measured at a depth of 10 cm. During December and January, measurements were not made. Soil moisture was more constant even though precipitation amounts greatly varied between the months, Figure 2b. The soil moisture did not exceed field capacity in any measurements; the soil moisture (%) never had flooding conditions.

Analysis of variance (ANOVA) was carried out to examine the monthly variation in Rs for each cover type (Table 1) This result indicated that the observations within each crop had at least one month which was significantly different than the values of other months, thereby supporting that respiration does have variation within the season.

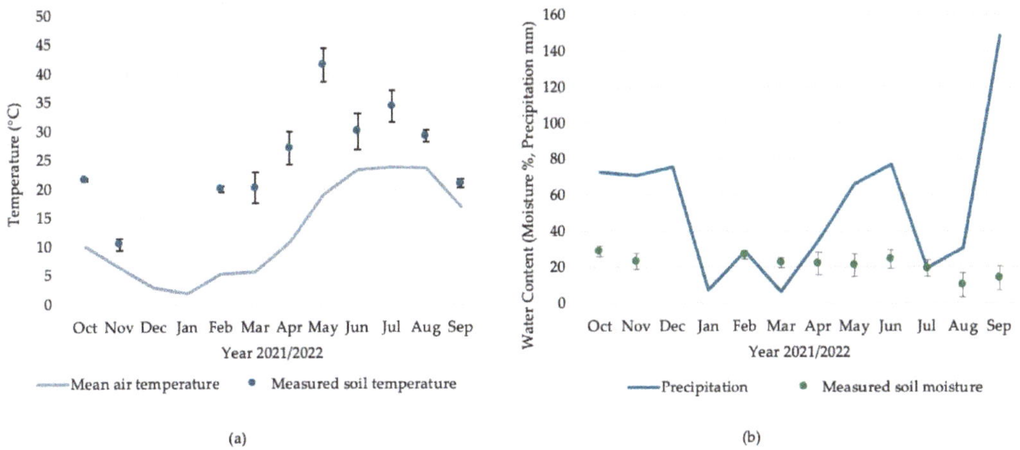

Figure 2. Average mean air temperature and corresponding soil temperature observed during the months (**a**) and average monthly precipitation (mm) and measured soil moisture (%) during the corresponding months (**b**). Error bars denote standard deviation.

Table 1. Analysis of variance for monthly Rs in wheat, barley, maize, and no vegetation among the observations throughout the season.

Source	DF	Sum of Squares	Mean Square	F value	Pr > F	R^2	C_v
			CO_2-C Wheat				
Model	6	722.03	120.34	18.42	<0.0001	0.88	17.35
Error	14	91.47	6.53				
Corrected total	20	813.50					
			CO_2-C Barley				
Model	5	2458.87	491.77	70.62	<0.0001	0.97	13.43
Error	12	83.56	6.96				
Corrected total	17	2542.44					
			CO_2-C Maize				
Model	6	1623.52	270.58	48.25	<0.0001	0.10	18.61
Error	14	78.51	5.60				
Corrected total	20	1702.03					
			CO_2-C No vegetation				
Model	11	437.46	39.77	8.63	<0.0001	0.79	31.77
Error	24	110.59	4.61				
Corrected total	35	548.06					

Seasonal variation is attributed to a number of biotic and abiotic factors, which include soil temperature, moisture conditions, and crop phenology. Under each crop and no vegetation, the *p*-values indicate a significant difference between the values observed throughout the season (Table 1). Since these values were found to be significant, LSD was used to separate the means (Table 2).

Table 2. Least Significant Difference for monthly Rs in wheat, barley, maize, and no vegetation among the observations throughout the season.

Cover Type	November	February	March	April	May	June	July	August	September	October
Barley	8.27 d	10.02 d	10.83 d	20.14 c	41.02 a	27.57 b				
Wheat	8.66 d	10.02 cd	13.15 c	23.50 a	18.23 b	21.49 ab	8.06 d			
Maize				7.08 c	9.46 bc	33.66 a	13.61 b	9.94 bc	7.90 c	7.38 c
No vegetation	8.27 bc	8.91 ab	6.97 bc	8.02 bc	5.25 bc	13.96 a	4.14 c	4.68 bc	4.89 bc	5.54 bc

Different letters denote a statistically significant difference between treatments; differences between months under the same crop type are according to the Fisher's LSD test at $p < 0.05$.

Significant temporal variation in Rs between months was found across the treatments in wheat, maize, barley, and no vegetation (Figure 3). Rs varied within a range of 8.27–41.02 kg CO_2-C ha^{-1} day^{-1} in barley, 8.06–23.50 kg CO_2-C ha^{-1} day^{-1} in wheat, and 7.08–33.66 kg CO_2-C ha^{-1} day^{-1} in maize. In barley, the Rs rates increased continually from November, reached a peak in May, and declined in June with the ending season. The pattern of monthly Rs in wheat showed an increasing trend from November and reached a peak from April to June before reducing drastically in July during senescence. Similarly, in maize, the Rs increased from the sowing season in April, reached the highest in June, and continuously declined until October. From the figure, despite marginal temperature variation between cover types, there is a significant variation in Rs, which indicates that besides temperature, several other factors govern Rs dynamics.

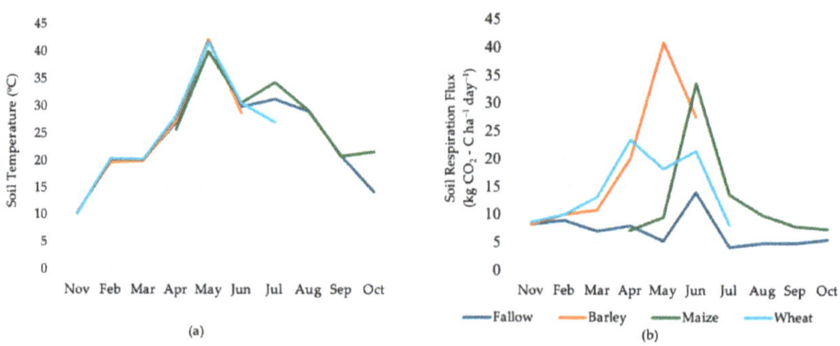

Figure 3. Monthly variation in soil temperature (**a**) and measured Rs (**b**). Values on the y-axis represent mean observed values taken in 3 replications of each crop once a day in each month.

3.2. Variation in Soil Respiration Among Cover Types

The temporal variations in Rs fluxes and two main environmental factors (soil temperature and soil moisture) affecting Rs flux with different cover types were compared (Table 3).

Average Rs fluxes of the vegetation season varied significantly among different crops (Figure 4). The mean flux of Rs for the growing season was found to be 19.64 kg CO_2-C ha^{-1} day^{-1} for barley, 12.72 kg CO_2-C ha^{-1} day^{-1} for maize, 14.73 kg CO_2-C ha^{-1} day^{-1} for wheat, and 6.75 kg CO_2-C ha^{-1} day^{-1} in the bare soil. A significant difference in Rs was present between barley and maize and between soil with crops and with no vegetation.

Table 3. Analysis of variance for Rs flux between different cover types.

Source	DF	Sum of Squares	Mean Square	F Value	Pr > F	R^2	C_v
			CO_2-C				
Model	3	2204.32	734.77	12.06	<0.0001	0.28	63.87
Error	92	5606.11	60.94				
Corrected total	95	7810.43					
			Soil Temperature				
Model	3	207.22	69.07	0.91	0.4389	0.03	33.16
Error	92	6975.19	75.82				
Corrected total	95	7182.42					
			Soil Moisture				
Model	3	408.06	136.02	3.32	0.0232	0.10	29.55
Error	92	3766.09	40.94				
Corrected total	95	4174.15					

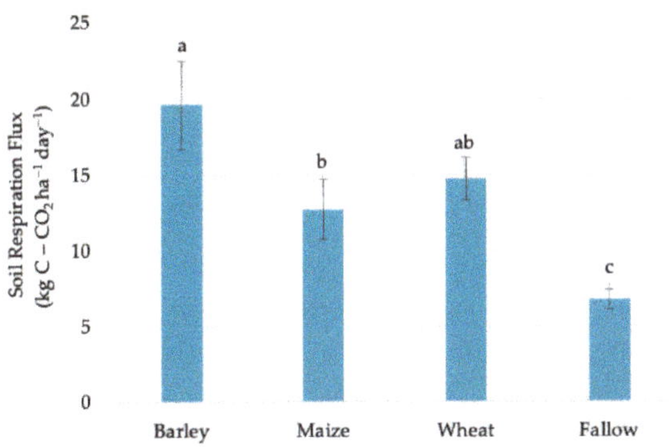

Figure 4. Average Rs under barley, maize, wheat, and no vegetation (fallow) observed during the vegetation season: Rs measured as kg CO_2-C $ha^{-1}day^{-1}$; different letters in bars indicate significant differences between cover types, error bars denote standard errors.

3.3. Correlation Between Temperature, Moisture, and Soil Respiration

A positive correlation (R = 0.26) was observed between Rs and soil temperature, while a negative correlation (R = −0.18) was observed between Rs and soil moisture. Linear regression analysis was performed, and the coefficient of determination was (R^2 = 0.07) with soil temperature and (R^2 = 0.13) with soil moisture, which indicated a poor relationship with Rs of the individual hydrothermal parameters in our study. It was observed that Rs in different covers responded differently with temperature (Figure A1). Likewise, the responses of soil respiration in different covers had differential responses with soil moisture (Figure A2). The interaction effects of soil temperature and soil moisture were determined using a linear model, and the coefficient of determination increased to (R^2 = 0.16), which indicated that the interaction of soil temperature and soil moisture had a greater role in influencing Rs than their individual effects.

4. Discussion

4.1. Temporal Interpretation of Temperature, Moisture, and Carbon Fluxes Control

There was significant monthly variation in Rs in all the cover types examined (Table 1), consistent with previous findings for winter wheat [9,18–20], summer maize [21–23], barley [20,24], cotton [21], and soybean [25]. In our study, temporal variation in Rs followed a pattern characteristic to many terrestrial ecosystems: low during initial growing phases, increased with crop growth, peaking at the time of maximum development, and declining after maturity. In a study by [26] in continental Croatia, the CO_2 efflux was higher during the second half of spring and the first half of summer, while lower CO_2 efflux was observed during autumn–winter, similar to our study.

Soil temperature and soil moisture are significant abiotic drivers of Rs that regulate the temporal variation in CO_2 efflux from soils. However, the impacts of these factors are not always direct due to interactions with biotic components such as plant growth and microbial activity. Non-cropped fields or fallow zones experience temporal Rs variation primarily due to heterotrophic respiration. However, in a cropped field, crop growth becomes a significant contributor to seasonal changes in Rs. In a study conducted in subtropical east China, a model based on soil temperature, moisture, and Leaf Area Index (LAI) accounted for 57.9–69.1% of the seasonal Rs variation in a winter wheat–soybean rotation cropland [27], indicating the leading role of crop growth factors. In a recent study conducted on summer maize in Songnen Plain in China, models based on soil hydrothermal factors and Net Primary Productivity (NPP) were able to explain 52.6 to 57.7% of the Rs variation [23]. Among them, changes in soil surface temperature and NPP (R^2 = 0.16) played dominant roles in influencing Rs [23].

Winter wheat Rs was in the range of 8.06–23.50 kg CO_2-C ha^{-1} day^{-1} with a maximum Rs in April and June. The values are comparable to those in continental Croatia, where the Rs in the same wheat variety was in the range of 8.35–20.23 kg CO_2-C ha^{-1} day^{-1}, with a maximum in June [20]. In another study conducted in the Tibetan Plateau, winter wheat had a minimum Rs value of 0.4 g m^{-2} d^{-1} and a maximum of 15.0 g m^{-2} d^{-1} on the 1st of July [18]. In the same study, the maximum Rs value was observed in the flowering and grain-filling stages of wheat in June/July. Though temperature was an important controlling factor of Rs, the maximum Rs did not necessarily coincide with the month of highest temperature but overlapped instead with times of maximum growth in wheat [18]. Measurements conducted by [9] included crop biomass and by [18] included live root biomass (LRB) and LAI to determine the relationship between crop phenological stages and Rs and concluded that crop activity was a major factor in controlling Rs. A similar pattern of seasonal variability was observed by [28]. Rs in maize was in the range of 7.08–33.66 kg CO_2-C ha^{-1} day^{-1} with a maximum Rs value reported in June. The minimum Rs was during the first month after seeding in April and in October before harvest.

A similar result was observed under summer maize cultivation in the subtropics, where the maximum Rs value of maize was observed at day 69 (around May/June), which corresponded with the silking stage of maize. The Rs value in maize ranged between 2.4 and 15.2 µmol m^{-2} s^{-1} during the growing season. The greatest Rs occurred in July and August, when maize was in the early reproductive stages [22]. Similarly, ref. [29] found a significant linear relationship between Rs rate and root biomass (R^2 = 0.73) in maize, indicating the dominant role of root biomass in influencing Rs. A similar conclusion as in winter wheat regarding the major influence of crop phenology on Rs can be made for maize.

Rs in barley was in the range of 8.27–41.06 kg CO_2-C ha^{-1} day^{-1} and was at the maximum in May around the ripening stage. The Rs value increased from sowing and was the highest during the ripening stage, and it declined toward harvest. The minimum Rs values occurred during the earlier growing period. The Rs corresponded with the month of highest temperature and the period of high growth. However, unlike wheat and maize, it did not necessarily correspond to the period of maximum growth but was increasingly in sync with temperature trends. In a previous study in continental Croatia, the Rs was in the range of 6.07–18.02 kg CO_2-C ha^{-1} day^{-1}, with a maximum Rs in June and suggesting lower Rs values [20], which indicates the effect of potential site-specific characteristics such as soil and microclimatic conditions. The Rs increases soon after sowing and reaches a maximum at about the time of the reproductive growth stage in barley [1]. In a study by [24] in Mediterranean Spain, during a 2-year period, the maximum Rs in barley showed a difference between years and did not necessarily correspond to the period of maximum LAI. During the year 2012, the maximum Rs was during tillering, whereas in 2013, the maximum value was during stem elongation, and both were one month earlier than the maximum LAI. It was concluded that the roles of temperature and moisture had limiting effects on the influence of crop growth compared to Rs [24].

The current work emphasizes the multivariate interactions of climate and vegetation development in regulating the temporal variation in Rs. Even if the Rs was generally in concurrence with crop phenology, temperature and moisture conditions played significant roles, sometimes overriding the immediate impact of vegetation growth. The Rs changes with different seasons, depending on climatological conditions, crop type, soil type, and many other factors without a defined pattern [30]. Substrate availability is the primary limiting factor for Rs, as it is the food for microbes [31]. Variations in organic carbon play a critical role in driving Rs changes. Plant growth strongly influences rhizosphere respiration, indicating that differences in growth variables influence Rs [1,9]. Substrate availability plays a critical role in affecting Rs, explained by amounts of SOC and its soluble components [31]. In this research, however, we did not have direct measures to link Rs with crop phenology, e.g., biomass root or SOC observations, and conclude that such measurements would provide more integrated information on Rs dynamics.

4.2. *Effect of Cover Types on Soil Respiration*

Results obtained in this study show a significant difference in Rs among different cover types. Barley had significantly higher mean Rs (19.64 kg CO_2-C ha^{-1} day^{-1}) compared to maize (12.72 kg CO_2-C ha^{-1} day^{-1}), while no significant difference was present between Rs in wheat (14.73 kg CO_2-C ha^{-1} day^{-1}) and barley and between maize and wheat. Non-cropped plots had significantly lower Rs flux than all cover types at −6.75 kg CO_2-C ha^{-1} day^{-1}. The higher Rs recorded in barley in our study is also in agreement with [32], which noted that CO_2 emissions from soils with different crops were similar, with maximum values of about 10 g m^{-2} day^{-1}; however, for barley, CO_2 evolution between 10 and 20 g m^{-2} day^{-1} was recorded. Although some research [30] found wheat to have higher Rs than maize, the present study did not find a significant difference between these two crops, indicating the complexity of factors that regulate Rs. Ref. [1] found differences in Rs in different crop types, i.e., grass, barley, and potato. However, Ref. [33] conducted a study on different crop groups and types, including cereals (oats, barley), rape seed (spring wheat and spring triticale), and row crops (potato, carrot, and parsnip) and found low variation between the compared crops which were inconsistent with seasons and time. Differences in Rs between crops may be due to crop-specific factors, such as photosynthetic potential, root architecture, root biomass, and root exudate plant species, which could differentially influence soil CO_2 efflux [15,21]. While vegetation type clearly contributes to Rs variability, it alone has little influence on Rs [34]. Several studies emphasize the dominant role of environmental factors like temperature, moisture availability, microbial activity, and substrate conditions [15,34]. Factors associated with varietal properties of crops and management practices, such as planting density, can also affect Rs. Barley and wheat have a longer growing season than maize. However, studies that compare barley with other crops are fewer, and the factors that could have caused this relationship remain underexplored. In an earlier study [21], Rs was 23.0% to 36.5% lower in fallow plots than in cropped ones. Similarly, Ref. [35] found that CO_2 fluxes were roughly twice as high under barley as in fallow plots. The rhizosphere respiration contribution to the total Rs is extremely variable among crops. It was 62% to 98% in soybean-planted soil and varied with the crop growth stages [25], and it was comparatively lower for wheat and maize, which were 36% and 29%, respectively [36]. Furthermore, Ref. [21] reported that root respiration comprised 13–29% in different crops. However, it was observed in some cases that non-cropped soils also had greater respiration rates than cropped plots [34]. Nevertheless, Rs is generally higher in cropped sites than in fallow due to the presence of the autotrophic respiration of roots and enhanced microbial activity of the rhizosphere. As supported in the literature, the contribution of roots to the total Rs rates greatly varies with crop types and environmental conditions.

4.3. Implications of Temperature and Moisture Variability on Soil Respiration

In this study, Rs was positively correlated (R^2 = 0.071) to soil temperature and negatively (R^2 = 0.13) to soil moisture. In various studies in the literature, it was found that temperature had a significant role in defining Rs in the seasonal variation of croplands [9,18,23]. Among the several parameters for temperature, it was found that mean weekly air temperature was the best predictor of Rs [9]. A high correlation (R^2 = 0.53–0.86) was found in a 2-year study of winter wheat cultivation [19], and a moderate correlation (R^2 = 0.49) was found in summer-grown maize [23]. Soil temperature stimulates plant growth and microbial activity in the soil and thus increases respiration rates. However, there are not always significant temperature dependencies described in the literature. For instance, a previous study conducted on the same study site had a lower correlation (R^2 = 0.0195) of temperature with Rs [20]. Similarly, Ref. (Akinremi et al.) [35] also reported a low coefficient of determination (R^2 = 0.06 and 0.02) in fallow plots and (R^2 = 0.09–0.24) in barley. The effect of soil moisture on Rs was complex and highly variable in various studies. In the present research, the limited range of soil moisture and lower sampling frequency most likely contributed to its poor negative correlation with Rs. It is supported by other research that shows that in a limited range of soil moisture, its effect on soil CO_2 efflux was negligible [18,31]. Moreover, Rs had different behaviors according to soil types [10]. The Rs rate for peat and clay soils decreased with the increase in soil moisture content (R^2 = 0.39, $p < 0.001$ for peat soil; R^2 = 0.578, $p < 0.001$ for clay soil), whereas only in sandy soil did Rs increase with the increase in moisture (R^2 = 0.29, $p < 0.001$) [10]. In broader ecosystem studies, soil moisture alone did not explain Rs variability [37]. From the same study [37], in cropland ecosystems, Rs displayed poor to no correlation with soil moisture. In line with this, Ref. (Moyano et al.) [38] found no significant effect of moisture on respiration activity, and this suggests that low amounts of aboveground litter can limit microbial substrate supply and thus the moisture response. Ref. [39] found soil depth, nutrient levels and slope positions among the different soil factors greatly influence Rs. These observations showed that the relationship between the temperature and moisture on Rs varies with soil types, biomes, and crop types and can have a differential influence on Rs based on their complex interactions. The complexity of the relation between the two variables can lead to non-linear effects. In this study, when a linear model incorporating the interaction of soil temperature and soil moisture was used, the coefficient of determination was improved (R^2 = 0.16) in line with earlier findings [37,40], where model performance was increased by considering both factors combined. A better frequency of data records might further improve the predictability of temperature and moisture variables with Rs. Rising temperatures and altered precipitation patterns from climate change can influence soil temperature regimes and moisture conditions, thereby altering their role in ecosystem processes. Studies on warming effects on Rs have been found to elevate Rs [12,27], and this was also found when they were coupled with increasing precipitation [41]. However, moisture effects on soil respiration are also a subject of debate in various ecosystems and require attention in light of the possible effects of climate change in the future.

4.4. Limitations of the Study

This study had a limited sampling frequency which did not account for diurnal and spatial variations. Such an approach may have sacrificed important diurnal and short-term variations, particularly under conditions of sharp temperature and humidity changes. Having a single measurement per month could limit our understanding of unknown changes during certain periods of temperature or moisture fluctuations during the season. It has been realized that the soil temperature during the measurements was generally higher and did not correspond well with the mean air temperature, adding potential biases to seasonal Rs measurement. To overcome these limitations, increasing the sampling frequency, particularly at the peak growth seasons and periods of maximum hydrothermal fluctuations, is essential. To enhance the precision of Rs estimations, it is recommended to incorporate diurnal and spatial variations in future studies. Site-specific parameters arising from soil heterogeneity, root distribution, and microclimate conditions also influence Rs. Individual soil samples, representative of different treatment plots, were not considered in this investigation; however, this would provide additional and precise information and is encouraged to keep in consideration during future investigations. Apart from that, there is scarce information on how different crops influence Rs. Based on our results, barley had significantly higher Rs than maize, which suggests crop selection make a difference in Rs behavior and may have climate-smart implications. More comparison studies are required to arrive at a conclusion that barley has a higher Rs than maize. An inter-comparison of Rs in different crops would contribute to better C balance modeling and cropping system approaches towards CO_2 mitigation. Finally, to gain deeper insights into Rs, future studies should integrate additional parameters such as biomass activity and LAI to gain a more comprehensive view of the mechanisms driving Rs variability.

5. Conclusions

We examined the seasonal variation in Rs in three different major crops, maize, wheat, and barley, in continental Croatia and came to the following conclusions:

I. Seasonal variation in crops is governed by phenology and crop growth cycles. Maximum Rs generally corresponded with the peak growth stage of the crop. Rs remained lower at the beginning and end of the crop-growing season. Rs also followed the temperature trend with some exceptions, which are attributed to the interaction effects of moisture and other possible factors.

II. Barley had significantly higher respiration rates compared to maize. This could be relevant in crop selection for climate-smart agriculture. However, additional research under diverse cropping systems and agroclimatic conditions is required to understand these dynamics in more detail. Cropped fields have significantly higher Rs than fallow, indicating the prominent role of autotrophic respiration in cropped fields.

III. No significant correlation was found between Rs and soil temperature and between Rs and soil moisture. Interaction effects play an influential role in masking the individual effect of these factors on Rs. More frequent sampling is helpful to clearly understand the effects of these agroclimatic variables on Rs.

IV. To obtain a better understanding of factors contributing to seasonal Rs dynamics, increasing the sampling frequency of Rs and agroclimatic variables is recommended. Sampling frequency could be increased after heavy rainfalls and during peak growth periods in crops.

V. Seasonal variation in Rs is influenced by both biotic factors, such as crop types and phenology, and abiotic factors, such as temperature and moisture, which can interact in different ways. To understand Rs demands an analysis of its seasonal variation, making it essential to account for these variations when quantifying and modeling Rs.

Author Contributions: Conceptualization: D.B. (Darija Bilandžija); methodology: D.B. (Darija Bilandžija), Z.Z., T.K. and N.B.; software: D.B. (Dija Bhandari) and S.G.; formal analysis: D.B. (Dija Bhandari) and S.G., investigation: D.B. (Darija Bilandžija) and N.B.; resources: T.K. and Z.Z.; writing—original draft preparation: D.B. (Dija Bhandari); writing—review and editing: D.B. (Dija Bhandari) and T.R.P.; visualization: N.B.; supervision: D.B. (Darija Bilandžija), T.K. and T.R.P.; funding acquisition: T.K. All authors have read and agreed to the published version of the manuscript.

Funding: This research was funded by the European Union from the Operational Program Competitiveness and cohesion of European Regional Development Fund via project "Production of food, biocomposites and biofuels from cereals in the circular bioeconomy" (grant number KK.05.1.1.02.0016).

Institutional Review Board Statement: Not applicable.

Informed Consent Statement: Not applicable.

Data Availability Statement: The original contributions presented in this study are included in the article. Further inquiries can be directed to the corresponding author.

Conflicts of Interest: The authors declare no conflicts of interest.

Appendix A

Table A1. Agrotechnical measures in the production of maize, wheat, and barley during the experimental season 2021/2022.

Month	Field Operation	Tools and Equipment	How Was It Conducted?
Wheat and Barley			
October	Primary tillage Secondary tillage	Combined tool Rotary harrow	Up to 15–20 cm depth Up to 5–10 cm depth
October	Fertilization	Mineral fertilizer spreader (Amazone 1500)	Urea 46% (100 kg/ha) NPK 7:20:30 (400 kg/ha)
October	Sowing	Multirow mechanical seeder	Seeding density: Rex (200 kg/ha), Srpanjka (290 kg/ha)
November	Application of rodenticide	33.66	13.61
February	Fertilization	Mineral fertilizer spreader (Amazone 1500)	Top dressing by KAN (100 kg/ha)
March	Fertilization	Mineral fertilizer spreader (Amazone 1500)	Top dressing by KAN (400 kg/ha)
March	Herbicide application		Trimur WG (15 g/ha) + Fluxir (0.5 L/ha)
April	Fungicide application	Mechanical harvester	Impact 25 SC (0.5 L/ha) + Tebusha 25% EW (1 L/ha)
July	Harvest	Mechanical harvester	
Maize			
October April	Primary tillage Secondary tillage	Fendt, 194 kW Fendt, 164 kW	
April	Sowing	Multirow mechanical corn planter	OS 515 Seeding rate: 65,000 plants/ha
April	Herbicide application in corn field	Mechanical sprayer	Dual Gold 960 (1 L/ha) + Koban T (3 L/ha)
May October	Fertilization Harvest of maize	Fendt, 164 kW Mechanical harvester	KAN (250 kg/ha)

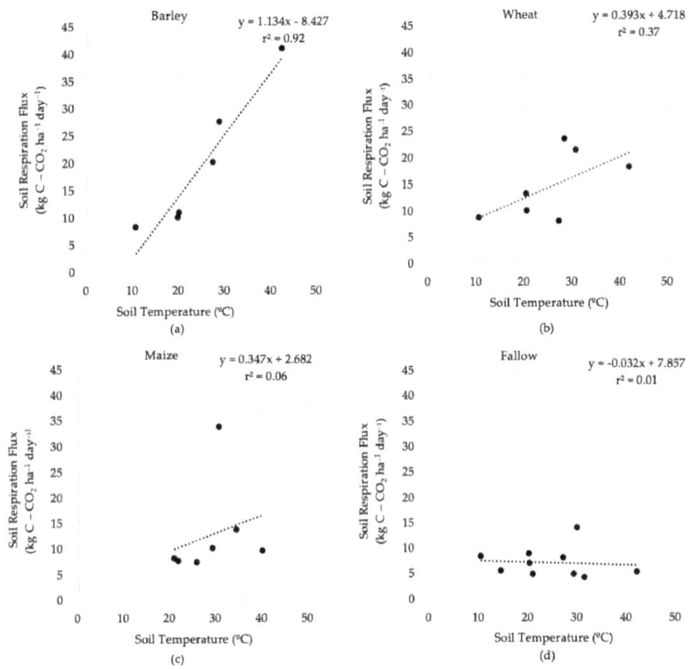

Figure A1. Scatter plot of soil temperature and corresponding measured Rs flux under barley (**a**), wheat (**b**), maize (**c**), and no vegetation (**d**).

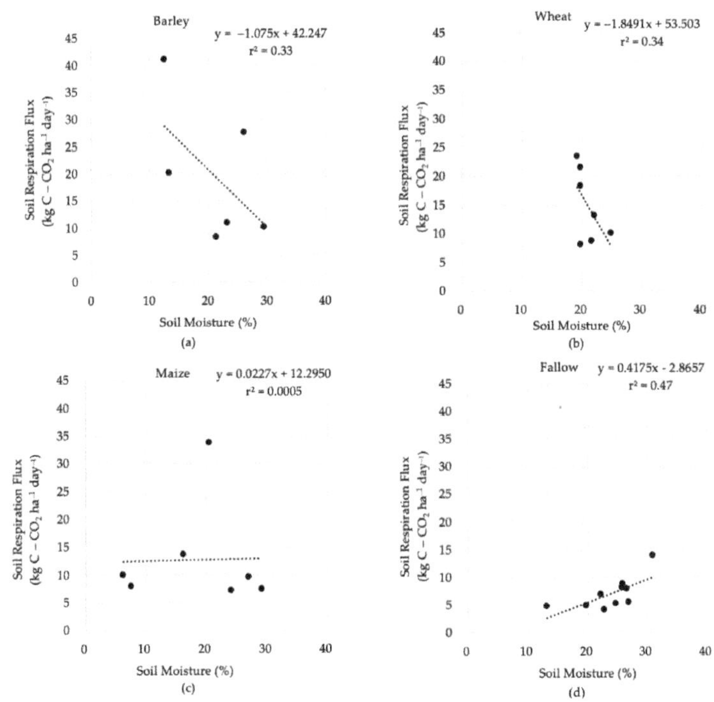

Figure A2. Correlation of soil moisture and corresponding measured Rs flux under barley (**a**), wheat (**b**), maize (**c**), and no vegetation (**d**).

References

1. Lohila, A.; Aurela, M.; Regina, K.; Laurila, T. Soil and total ecosystem respiration in agricultural fields: Effect of soil and crop type. *Plant Soil* **2003**, *251*, 303–317. [CrossRef]
2. European Commission; Environment Directorate-General. *Soil: The Hidden Part of the Climate Cycle*; Publications Office of the European Union: Luxembourg, 2011.
3. Bond-Lamberty, B.; Bailey, V.L.; Chen, M.; Gough, C.M.; Vargas, R. Globally rising soil heterotrophic respiration over recent decades. *Nature* **2018**, *560*, 80–83. [CrossRef]
4. Huang, N.; Wang, L.; Song, X.-P.; Black, T.A.; Jassal, R.S.; Myneni, R.B.; Wu, C.; Wang, L.; Song, W.; Ji, D.; et al. Spatial and Temporal Variations in Global Soil Respiration and Their Relationships with Climate and Land Cover. Available online: https://www.science.org (accessed on 22 March 2024).
5. Zhao, Z.; Peng, C.; Yang, Q.; Meng, F.; Song, X.; Chen, S.; Epule, T.E.; Li, P.; Zhu, Q. Model prediction of biome-specific global soil respiration from 1960 to 2012. *Earth's Future* **2017**, *5*, 715–729. [CrossRef]
6. Wu, Z.; Dijkstra, P.; Koch, G.W.; Peñuelas, J.; Hungate, B.A. Responses of terrestrial ecosystems to temperature and precipitation change: A meta-analysis of experimental manipulation. *Glob. Change Biol.* **2011**, *17*, 927–942. [CrossRef]
7. Hursh, A.; Ballantyne, A.; Cooper, L.; Maneta, M.; Kimball, J.; Watts, J. The sensitivity of soil respiration to soil temperature, moisture, and carbon supply at the global scale. *Glob. Change Biol.* **2017**, *23*, 2090–2103. [CrossRef] [PubMed]
8. Yurak, V.V.; Fedorov, S.A. Review of natural and anthropogenic emissions of carbon dioxide into the earth's atmosphere. *Int. J. Environ. Sci. Technol.* **2024**, *22*, 2719–2736. [CrossRef]
9. Buyanovsky, G.A.; Wagner, G.H.; Gantzer, C.J. Soil Respiration in a Winter Wheat Ecosystem. *Soil Sci. Soc. Am. J.* **1986**, *50*, 338–344. [CrossRef]
10. Koizumi, H.; Kontturi, M.; Mariko, S.; Nakadai, T.; Bekku, Y.; Mela, T. Soil Respiration in Three Soil Types in Agricultural Ecosystems in Finland. *Acta Agric. Scand. Sect. B—Soil Plant Sci.* **1999**, *49*, 65–74. [CrossRef]
11. Bünemann, E.K.; Bongiorno, G.; Bai, Z.; Creamer, R.E.; De Deyn, G.; de Goede, R.; Fleskens, L.; Geissen, V.; Kuyper, T.W.; Mäder, P.; et al. Soil quality—A critical review. *Soil Biol. Biochem.* **2018**, *120*, 105–125. [CrossRef]
12. Zhou, L.; Zhou, X.; Shao, J.; Nie, Y.; He, Y.; Jiang, L.; Wu, Z.; Bai, S.H. Interactive effects of global change factors on soil respiration and its components: A meta-analysis. *Glob. Change Biol.* **2016**, *22*, 3157–3169. [CrossRef]
13. Ryan, M.G.; Law, B.E. Interpreting, measuring, and modeling soil respiration. *Biogeochemistry* **2005**, *73*, 3–27. [CrossRef]
14. Zhang, S.; Hussain, H.A.; Wang, L.; Hussain, S.; Li, B.; Zhou, H.; Luo, H.; Zhang, X.; Ma, Z.; Long, L.; et al. Responses of soil respiration and organic carbon to straw mulching and ridge tillage in maize field of a triple cropping system in the hilly region of southwest China. *Sustainability* **2019**, *11*, 3068. [CrossRef]
15. Yao, X.; Chen, S.; Ding, S.; Zhang, M.; Cui, Z.; Linghu, S.; Xu, J. Temperature, Moisture, Hyperspectral Vegetation Indexes, and Leaf Traits Regulated Soil Respiration in Different Crop Planting Fields. *J. Soil Sci. Plant Nutr.* **2021**, *21*, 3203–3220. [CrossRef]
16. Cvjeticanin, P.; Kanizaj, Z.; Valentic, M.; Grlica, A.; Pejkovic, M.; Dujman, I. *Agricultural Production, 2023*; Statistical Reports; Croatian Bureau of Statistics: Zagreb, Croatia, 2024; ISSN 1333-3518.
17. Bilandžija, D.; Martinčić, S. Agroclimatic Conditions of the Osijek Area During Referent (1961–1990) and Recent (1991–2018) Climate Periods. 2020. Volume 54, pp. 1–154. Available online: http://hrcak.srce.hr/hmc (accessed on 2 May 2024).
18. Shi, P.-L.; Zhang, X.-Z.; Zhong, Z.-M.; Ouyang, H. Diurnal and seasonal variability of soil CO2 efflux in a cropland ecosystem on the Tibetan Plateau. *Agric. For. Meteorol.* **2006**, *137*, 220–233. [CrossRef]
19. Liang, G.; Houssou, A.A.; Wu, H.; Cai, D.; Wu, X.; Gao, L.; Li, J.; Wang, B.; Li, S. Seasonal patterns of soil respiration and related soil biochemical properties under nitrogen addition in winter wheat field. *PLoS ONE* **2015**, *10*, e0144115. [CrossRef]
20. Bilandžija, D.; Zgorelec, Ž.; Bilandžija, N.; Zdunić, Z.; Krička, T. Contribution of winter wheat and barley cultivars to climate change via soil respiration in continental Croatia. *Agronomy* **2021**, *11*, 2127. [CrossRef]
21. Benbi, D.K.; Toor, A.; Brar, K.; Dhall, C. Soil respiration in relation to cropping sequence, nutrient management and environmental variables. *Arch. Agron. Soil Sci.* **2020**, *66*, 1873–1887. [CrossRef]
22. Lamptey, S.; Xie, J.; Li, L.; Coulter, J.A.; Jagadabhi, P.S. Influence of organic amendment on soil respiration and maize productivity in a semi-arid environment. *Agronomy* **2019**, *9*, 611. [CrossRef]
23. Gao, Y.; Chu, Z.; Jiang, L.; Zhou, Y.; Zhao, F.; Gao, F.; Liu, D.; Zhao, H. Soil respiration dynamics and effects of hydrothermal factors and NPP during growing period of maize cropland ecosystem in Songnen Plain, China. *Trop. Ecol.* **2024**, *66*, 55–66. [CrossRef]
24. Cicuéndez, V.; Rodríguez-Rastrero, M.; Recuero, L.; Huesca, M.; Schmid, T.; Inclán, R.; Litago, J.; Sánchez-Girón, V.; Palacios-Orueta, A. First insights on soil respiration prediction across the growth stages of rainfed barley based on simulated MODIS and sentinel-2 spectral indices. *Remote Sens.* **2020**, *12*, 2724. [CrossRef]
25. Yang, L.-F.; Cai, Z.-C. Soil Respiration During a Soybean-Growing Season. *Pedosphere* **2006**, *16*, 192–200. [CrossRef]
26. Bilandžija, D.; Zgorelec, Ž.; Kisić, I. The Influence of Agroclimatic Factors on Soil CO_2 Emissions. *Coll. Antropol.* **2014**, *38*, 77–83.

.7. Chen, S.; Wang, J.; Zhang, T.; Hu, Z.; Zhou, G. Warming and straw application increased soil respiration during the different growing seasons by changing crop biomass and leaf area index in a winter wheat-soybean rotation cropland. *Geoderma* **2021**, *391*, 114985. [CrossRef]

.8. Rochette, P.; Desjardins, R.L.; Pattey, E. Spatial and temporal variability of soil respiration in agricultural fields for personal use only. *Can. J. Soil Sci.* **1991**, *71*, 189–196. [CrossRef]

.9. Han, G.; Zhou, G.; Xu, Z.; Yang, Y.; Liu, J.; Shi, K. Biotic and abiotic factors controlling the spatial and temporal variation of soil respiration in an agricultural ecosystem. *Soil Biol. Biochem.* **2007**, *39*, 418–425. [CrossRef]

0. Bilandžija, D.; Zgorelec, Ž.; Kisić, I. Influence of tillage practices and crop type on soil CO_2 emissions. *Sustainability* **2016**, *8*, 90. [CrossRef]

1. Wang, Y.; Bölter, M.; Chang, Q.; Duttmann, R.; Scheltz, A.; Petersen, J.F.; Wang, Z. Driving factors of temporal variation in agricultural soil respiration. *Acta Agric. Scand. Sect. B—Soil Plant Sci.* **2015**, *65*, 589–604. [CrossRef]

2. Monteith, J.L.; Szeicz, G.; Yabuki, K. Crop Photosynthesis and the Flux of Carbon Dioxide Below the Canopy. *J. Appl. Ecol.* **1964**, *1*, 321. [CrossRef]

3. Norberg, L.; Berglund, Ö.; Berglund, K. Seasonal CO2 emission under different cropping systems on Histosols in southern Sweden. *Geoderma Reg.* **2016**, *7*, 338–345. [CrossRef]

4. Raich, J.W.; Tufekcioglu, A. Vegetation and soil respiration: Correlations and controls. *Biogeochemistry* **2000**, *48*, 71–90. [CrossRef]

5. Akinremi, O.O.; McGinn, S.M.; McLean, H.D.J. Effects of soil temperature and moisture on soil respiration in barley and fallow plots. *Can. J. Soil Sci.* **1999**, *79*, 5–13. [CrossRef]

6. Zhang, Q.; Lei, H.-M.; Yang, D.-W. Seasonal variations in soil respiration, heterotrophic respiration and autotrophic respiration of a wheat and maize rotation cropland in the North China Plain. *Agric. For. Meteorol.* **2013**, *180*, 34–43. [CrossRef]

7. Lai, L.; Zhao, X.; Jiang, L.; Wang, Y.; Luo, L.; Zheng, Y.; Chen, X.; Rimmington, G.M. Soil Respiration in Different Agricultural and Natural Ecosystems in an Arid Region. *PLoS ONE* **2012**, *7*, e48011. [CrossRef] [PubMed]

8. Moyano, F.; Kutsch, W.; Schulze, E. Response of mycorrhizal, rhizosphere and soil basal respiration to temperature and photosynthesis in a barley field. *Soil Biol. Biochem.* **2007**, *39*, 843–853. [CrossRef]

9. Adhikari, K.; Anderson, K.R.; Smith, D.R.; Owens, P.R.; Moore, P.A.; Libohova, Z. Identifying key factors controlling potential soil respiration in agricultural fields. *Agric. Environ. Lett.* **2023**, *8*, e20117. [CrossRef]

0. Wu, R.E.; Gauland, T.T.R.; Bus, R.L. The interdependent effects of soil temperature and water content on soil respiration rate and plant root decomposition in arid grassland soils. *Soil Biol. Biochem.* **1975**, *7*, 373–378.

1. Chang, X.; Zhu, X.; Wang, S.; Luo, C.; Zhang, Z.; Duan, J.; Bai, L.; Wang, W. Temperature and moisture effects on soil respiration in alpine grasslands. *Soil Sci.* **2012**, *177*, 554–560. [CrossRef]

sustainability

MDPI

Article

Analysis of Soil Nutrient Content and Carbon Pool Dynamics Under Different Cropping Systems

Huinan Xin, Caixia Lv, Na Li, Lei Peng *, Mengdi Chang, Yongfu Li, Qinglong Geng, Shuhuang Chen and Ning Lai

Institute of Agricultural Resources and Environment, Xinjiang Academy of Agricultural Sciences, Urumqi 830091, China; xinhuinan@163.com (H.X.); lvcaixia@xaas.ac.cn (C.L.); lina@xaas.ac.cn (N.L.); 18160525063@163.com (M.C.); liyongfu@xaas.ac.cn (Y.L.); gengql2024@163.com (Q.G.); chensh66@163.com (S.C.); lainingde@163.com (N.L.)

* Correspondence: penglei1028@stu.xju.edu.cn; Tel.: +86-130-0968-6236

Abstract: Understanding the effects of agricultural practices on soil nutrient dynamics is critical for optimizing land management in arid regions. This study analyzed spatial patterns, driving factors, and surface stocks (0–20 cm) of soil organic carbon (SOC), total nitrogen (TN), total phosphorus (TP), and their stoichiometric ratios (C:N, C:P, and N:P) across six cropping systems (paddy fields, cotton fields, wheat–maize, orchards, wasteland, and others) in the Aksu region, Northwest China, using 1131 soil samples combined with geostatistical and field survey approaches. Results revealed moderate to low levels of SOC, TN, and TP, and stoichiometric ratios, with moderate spatial autocorrelation for SOC, TN, TP, and C:N but weak dependence for C:P and N:P. Cropping systems significantly influenced soil nutrient distribution: intensive systems (paddy fields and orchards) exhibited the highest SOC (22.31 ± 10.37 t hm^{-2}), TN (2.20 ± 1.07 t hm^{-2}), and TP stocks (peaking at 2.58 t hm^{-2} in orchards), whereas extensive systems (cotton fields and wasteland) showed severe nutrient depletion. Soil pH and elevation were key drivers of SOC and TN variability across all systems. The C:N ratio ranked highest in "other systems" (e.g., diversified rotations), while wheat–maize fields displayed elevated C:P and N:P ratios, likely linked to imbalanced fertilization. These findings highlight that sustainable intensification (e.g., paddy and orchard management) enhances soil carbon and nutrient retention, whereas low-input practices exacerbate degradation in arid landscapes. The study provides actionable insights for tailoring land-use strategies to improve soil health and support ecosystem resilience in water-limited agroecosystems.

Keywords: land use; carbon; nitrogen; phosphorus; semi-variogram; soil carbon pool

1. Introduction

Soil, a core component of terrestrial ecosystems, significantly impacts agricultural sustainability and the global carbon cycle through its nutrient cycling and carbon pool dynamics [1]. Amid intensifying climate change and population pressure, exploring how different cropping systems affect soil nutrient levels and carbon pool evolution has become a key focus in modern soil science and agricultural ecology [2]. Soil nutrient composition, a central indicator of ecosystem material cycling, greatly influences farm productivity, ecological stability, and agricultural management effectiveness [3]. From an eco—stoichiometric perspective, soil element contents (C, N, and P) and their ratios (C:N, C:P, and N:P) are crucial for assessing soil fertility, understanding energy distribution in soil-plant systems, and microbial activity [4]. For instance, a high C:N ratio may indicate nitrogen limitation for crops [5], while a low N:P ratio suggests phosphorus limitation. These ratios affect organic matter decomposition, nutrient mineralization, and root absorption, directly impacting crop yield and quality.

In agriculture, the spatial and temporal heterogeneity of soil nutrients guides precision fertilization [6]. Studies show that optimizing nitrogen and phosphorus fertilizer ratios based on regional soil C:N:P characteristics can reduce nutrient loss by over 20% and improve fertilizer efficiency [7]. Soil nutrient stoichiometric balance also affects soil health. An optimal C:N ratio (10:1–12:1) promotes organic matter accumulation and microbial succession, while a low C:P ratio (<50:1) may increase phosphorus fixation and soil acidification or heavy metal activation risks [8]. Under climate change, soil nutrient stoichiometric features serve as environmental indicators [9]. Long-term nitrogen deposition lowers the soil C:N ratio, while drought stress may raise the C:P ratio by inhibiting decomposition [10,11], providing data for predicting carbon sequestration potential and formulating management strategies. Thus, understanding soil nutrient composition and its regulation is essential for efficient agricultural resource use and addressing food and ecological security challenges.

Analyzing the balance and coupling mechanisms of key elements (C, N, P) in ecosystems is vital for understanding material cycling and energy flow regulation [12,13]. In cropland ecosystems, soil C, N, and P stoichiometric characteristics reflect nutrient supply capacity and limiting factors influenced by both natural factors and human-driven land management [14]. Changes in cropland use, such as altering crop rotation [15], optimizing crop structure, or implementing organic substitution [16], affect external nutrient inputs, crop residue return, and microbial metabolism, reshaping the C:N:P ecological stoichiometric network through soil–plant–microbe interactions [17]. For example, continuous monocropping can imbalance the soil C:N ratio and increase phosphorus fixation, while balanced organic–inorganic fertilization can enhance nutrient use efficiency by adjusting element coupling [18].

Agricultural ecosystems are highly dynamic, multifactorial networks where nutrient cycling (e.g., carbon, nitrogen, and phosphorus), microbial activity, and crop responses are directly governed by tillage practices. For example, no-till farming preserves soil organic carbon by minimizing soil disturbance, while intercropping enhances nutrient use efficiency through root complementarity [19]. Monoculture systems reduce soil microbial diversity by 15–30%, whereas diversified practices (e.g., crop rotation combined with cover crops) restore the abundance of functional microbial communities [20]. Using an extended Lotka–Volterra modeling framework, different tillage systems modulate competitive interactions among crops, weeds, insects, and microbes, thereby influencing nutrient use efficiency. For instance, intercropping systems improve nitrogen availability via root exudates (e.g., nitrogen fixation by legumes), whereas continuous monoculture exacerbates nutrient depletion [21]. A 2024 Nature study revealed that 51–60% of global anthropogenic ammonia emissions originate from rice, wheat, and maize cultivation, and optimized fertilizer management (e.g., deep placement of enhanced-efficiency fertilizers) could reduce ammonia emissions by 38% [22]. Excessive nitrogen fertilization further induces nitrate leaching and eutrophication, posing ecological risks. In this study, we quantify the long-term stability of nutrient cycling under various tillage systems using semi-variogram equations and soil C-N-P stock estimation models. Currently, there is a lack of systematic analysis on soil C, N, and P stoichiometric responses and driving mechanisms under different cropping systems, especially regarding regional—scale management measures to coordinate element balance for soil health and sustainable agriculture. Clarifying the response mechanisms of soil ecological stoichiometric characteristics to land management practices under various cropping systems is crucial for precision fertilization, soil fertility management, and agricultural low-carbon transition.

2. Materials and Methods

2.1. Overview of the Study Area

The Aksu region is located between 78°03′ and 84°07′ east longitude and 39°30′ and 42°41′ north latitude, on the southern slope of the Tianshan Mountains in the Xinjiang Uygur Autonomous Region and the northern edge of the Tarim Basin (Figure 1), in the alluvial plain area of the Aksu River. The terrain of the region is higher in the north and lower in the south, with numerous peaks in the north, the boundless Taklamakan Desert in the south, and the middle part is a mixture of the mountain foot, gravelly fan-shaped land, alluvial plain areas, Gobi, and oases. The region administers 7 counties, 2 cities, 84 townships, and 56 agricultural and forestry farms. Aksu has a warm temperate, arid climate, located on the northern edge of the Tarim Basin, with little precipitation, high evaporation, a dry climate, and abundant light and heat resources. The farmland cropping system can be divided into six types: paddy field, wheat–corn rotation, cotton field, orchard, others (mainly vegetables and some cash crops), and wasteland (Table 1). The main food crops are wheat and corn, with paddy fields in some areas.

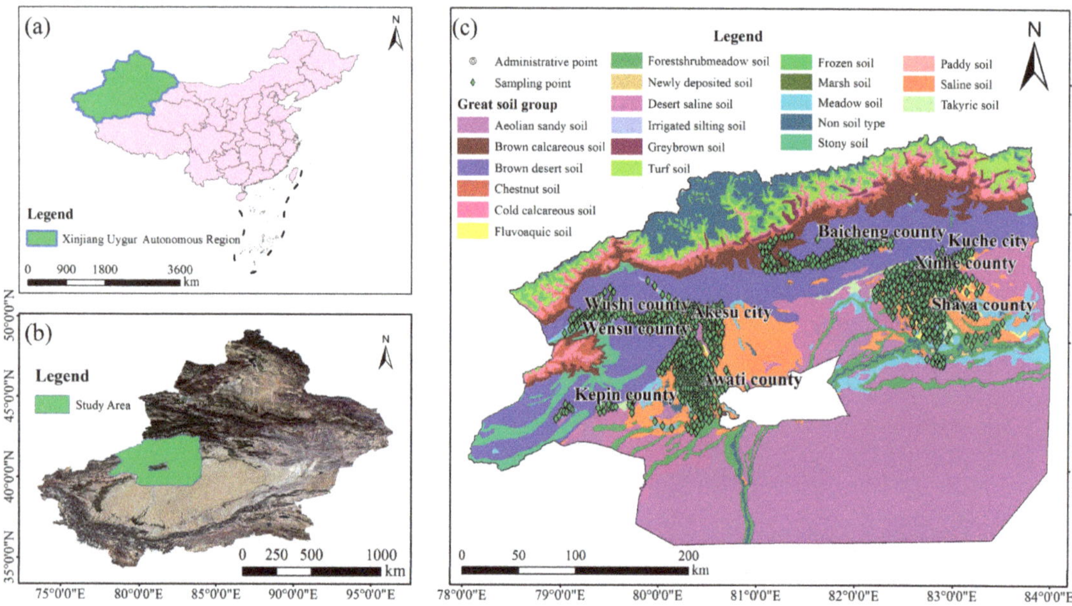

Figure 1. Overview maps of the (**a**) Xinjiang Uygur Autonomous Region and (**b**) study area. (**c**) Data sampling points.

Table 1. Planting pattern type description.

Type	Sample Sites Overview
Orchard	Mainly fruit trees
Cotton	Long-term cotton cultivation
Wheat–corn	Long-term cultivation of wheat or corn and wheat–corn crop rotation
Paddy field	Mainly planted with rice
Wasteland	Newly reclaimed or abandoned plots
Other	Plots planted with vegetables and cash crops

2.2. Study Methods

2.2.1. Sample Collection

The data comes from the Aksu region's arable land quality evaluation database. Soil samples are mainly from farmland areas in various counties and districts of the Aksu region. Sampling follows the Technical Regulations for Arable Land Productivity Survey and Quality Evaluation (NY/T 1634-2008). Samples were collected from October to November 2020. In the lab, multiple soil nutrient indicators were measured, including soil organic carbon (SOC), total nitrogen (TN), and total phosphorus (TP). Soil C:N:P content was determined using the dry burning method, Kjeldahl method, and molybdenum-antimony colorimetric method, respectively; Soil bulk weight (ring knife method); Gravel content (sieve method). Considering both natural and human-related conditions of the region's arable land, we randomly set 1131 sampling points. At each point, the mixed sampling method was used to collect soil samples from 0–20 cm depth. We also gathered information on cropping conditions and the latitude and longitude of each point, converting sample point data into vector data with spatial coordinates.

2.2.2. Semi-Variable Function Models

Spatial variability of soil C, N, and P was quantified using semi-variogram modeling. The experimental semi-variogram was calculated as:

$$\gamma(h) = \frac{1}{2N(h)}\sum_{i=1}^{N(h)} [z(x_i) - z(x_i + h)]^2 \tag{1}$$

where $\gamma(h)$ is the semi-variance at lag distance h, $N(h)$ is the number of data pairs separated by h, and $z(x_i)$ represents the measured value (C, N, or P) at location x_i.

Three theoretical models (spherical, exponential, and Gaussian) were fitted to the experimental semi-variograms to evaluate spatial dependence [23]. Model parameters, including nugget (C_0), sill (C_0 + C), and range (A_0), were optimized by minimizing the residual sum of squares (RSS). The spatial dependence index (SDI) was calculated as $C_0/(C_0 + C) \times 100\%$, with SDI $\leq 25\%$ indicating strong spatial dependence, 25–75% moderate, and $\geq$75% weak.

2.2.3. Model for Estimating Soil C, N and P Stocks

Based on previous studies, the following equations were used to determine C, N, and P stocks in the soil horizons of different land use types [24].

$$S = \sum_{i=1}^{n} c_i \times p_i \times H_i \times \frac{1-\delta_i}{10} \tag{2}$$

In the formula, c represents the carbon, nitrogen, and phosphorus content in the soil surface ($g{\cdot}kg^{-1}$), p represents the soil surface bulk density ($g{\cdot}cm^{-3}$), H represents the soil layer depth (cm), and δ_i represents the volume percentage of gravel larger than 2 mm in the soil (%).

2.2.4. Data Processing

WPS Office 2021 and SPSS 22.0 software were used for data processing and statistical analysis. The K-S test in SPSS 22.0 checked if soil nutrient data followed a normal distribution. Non-normal data were transformed to meet normality. The LSD method tested differences in soil nutrient content and C:N:P ratios across farmland uses. If data had unequal variances, multiple comparisons were done before correlating soil nutrients and C:N:P ratios with their influencing factors.

3. Results

3.1. Statistical Characteristics of Soil Nutrient Content and C:N:P Ratios

Among the 1131 sampling points in the Aksu region, 408 were orchards, 355 cotton fields, 288 wheat–corn rotations, 21 paddy fields, 23 wastelands, and 36 others (see Table 2).

Table 2. Data table of sampling points in the Aksu area.

Mode County	Aksu	Awati	Baicheng	Keping	Kuche	Shaya	Wensu	Wushi	Xinhe	Total
Orchard	103	37	19	1	64	25	83	51	25	408
Wasteland	10	-	-	7	1	1	2	-	2	23
Wheat–corn	18	29	101	3	36	14	24	42	21	288
Cotton	36	92	-	19	57	96	26	-	29	355
Paddy field	2	-	1	-	-	-	17	1	-	21
Other	1	4	19	-	2	2	8	-	-	36
Total	170	162	140	30	160	138	160	94	77	1131

Descriptive statistical results (Figure 2) revealed that across six cropping systems in the Aksu region, the average contents of soil organic carbon (SOC), total nitrogen (TN), and total phosphorus (TP) in the plough layer ranged from 9.37–15.63 g/kg, 0.52–0.85 g/kg, and 0.67–0.75 g/kg, respectively. The stoichiometric ratios of C:N, C:P, and N:P varied between 18.39 and 18.91, 14.13 and 24.00, and 0.78 and 1.31, respectively. Overall, the regional averages for SOC, TN, TP, C:N, C:P, and N:P were 13.25 g/kg, 0.73 g/kg, 0.72 g/kg, 18.52, 19.68, and 1.08, respectively. Additionally, the mean contents of alkali-hydrolyzable nitrogen and available phosphorus in the plough layer were 62.90 mg/kg and 24.80 mg/kg, respectively.

Significant differences in SOC, TN, TP contents, and stoichiometric ratios were observed among cropping systems. The coefficients of variation (CV) for SOC, TN, and TP ranged primarily between 20% and 60%, indicating moderate to high variability, which suggests substantial anthropogenic influences on these nutrients. The remaining nutrient ratios also exhibited considerable variability, ranked as TN > SOC > C:P > C:N, reflecting distinct spatial distribution patterns of soil carbon, nitrogen, and stoichiometric ratios across different agricultural practices in the Aksu region.

3.2. Spatial Autocorrelation of Soil Nutrient Content and C:N:P Ratio

The spatial distribution of soil nutrients exhibits both stochastic and structural features, which can be effectively modeled using semi-variogram analysis. Structural factors primarily include natural conditions such as soil type, parent material, topography, soil texture, and climate, while stochastic factors refer to anthropogenic influences from land use practices and field management. The nugget (C_0) represents spatial heterogeneity caused by random factors, and the sill ($C_0 + C$) indicates total variability within the dataset. The nugget-to-sill ratio ($C_0/(C_0 + C)$) quantifies the proportion of spatial heterogeneity attributable to stochastic factors, with higher values suggesting dominant random effects.

Using GS+ 9.0 geostatistical software, semi-variogram models were fitted for SOC, TN, TP contents, and C:N:P ratios. Optimal models were selected based on spatial heterogeneity and trend analysis (Table 3 and Figure 3). Exponential models best described SOC, TN, and C:N ratios, whereas spherical models were optimal for TP, C:P, and N:P ratios. The coefficients of determination (R^2) for SOC, TN, TP, C:N, C:P, and N:P were 0.389, 0.352, 0.362, 0.639, 0.447, and 0.469, respectively. Nugget-to-sill ratios for SOC, TN, and TP ranged between 50.01–58.14%, indicating moderate spatial autocorrelation driven by both structural and stochastic factors. In contrast, C:N, C:P, and N:P ratios exhibited nugget-to-sill ratios > 80%, suggesting dominant stochastic influences with no clear spatial patterns.

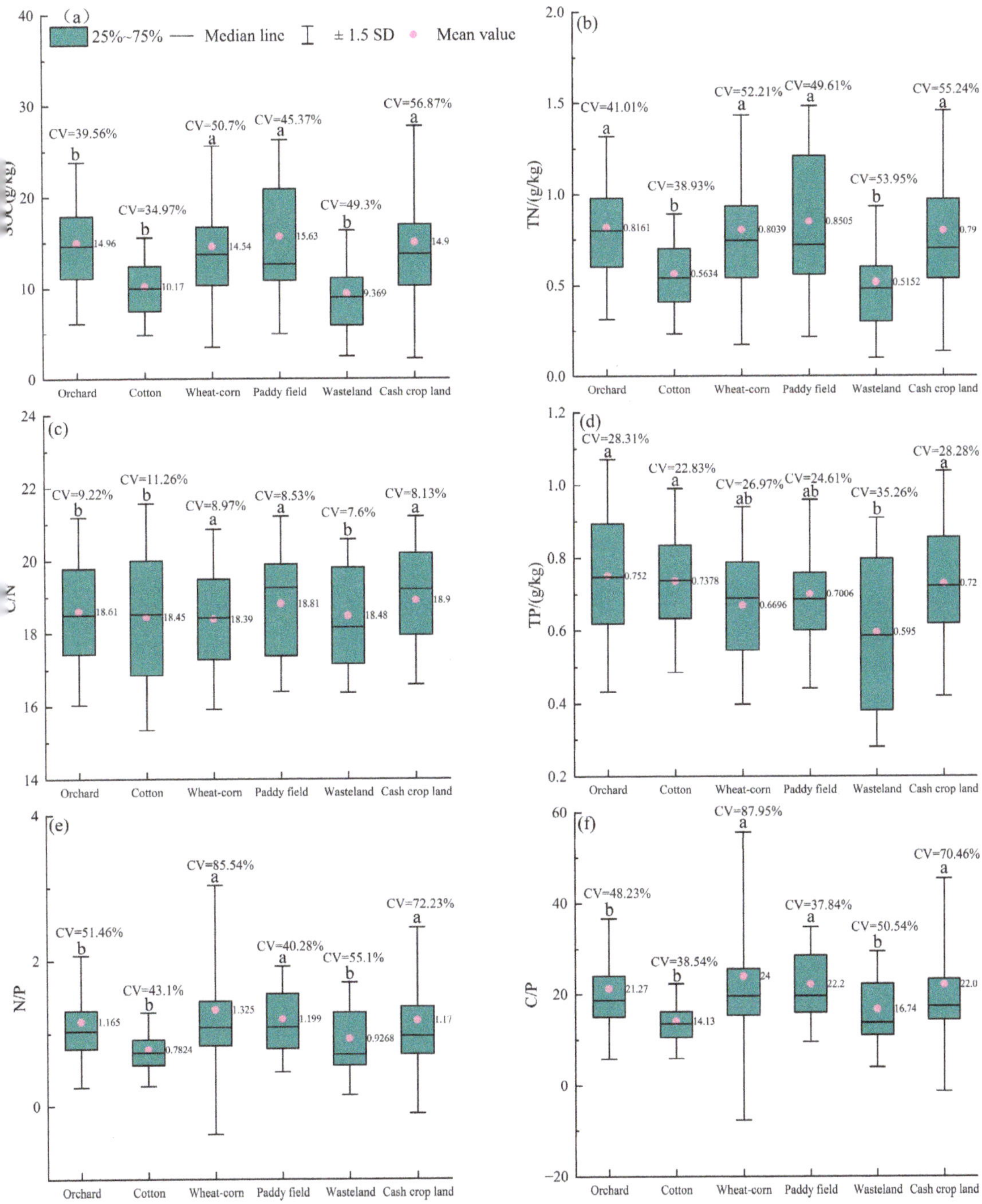

Figure 2. Descriptive statistics of soil C, N, and P, and other indicators and their ecological stoichiometric characteristics ((**a–f**) are soil SOC, TN, C/D, TP, N/P, and C/P mathematical and statistical characteristics, respectively; Numbers a and b represent significant differences).

Table 3. Semi-variance models of the soil SOC, TN, TP, and C:N, C:P, and N:P ratios and their corresponding parameters in the Aksu area.

Plough Layer Nutrient	Fitted Model	Coefficient of Determination (R^2)	Nugget-to-Sill Ratio ($C_0/(C_0 + C)$)
SOC	E	0.389	50.01%
TN	E	0.352	52.55%
TP	S	0.362	58.14%
C:N	E	0.639	83.03%
C:P	S	0.447	81.91%
N:P	S	0.469	85.53%

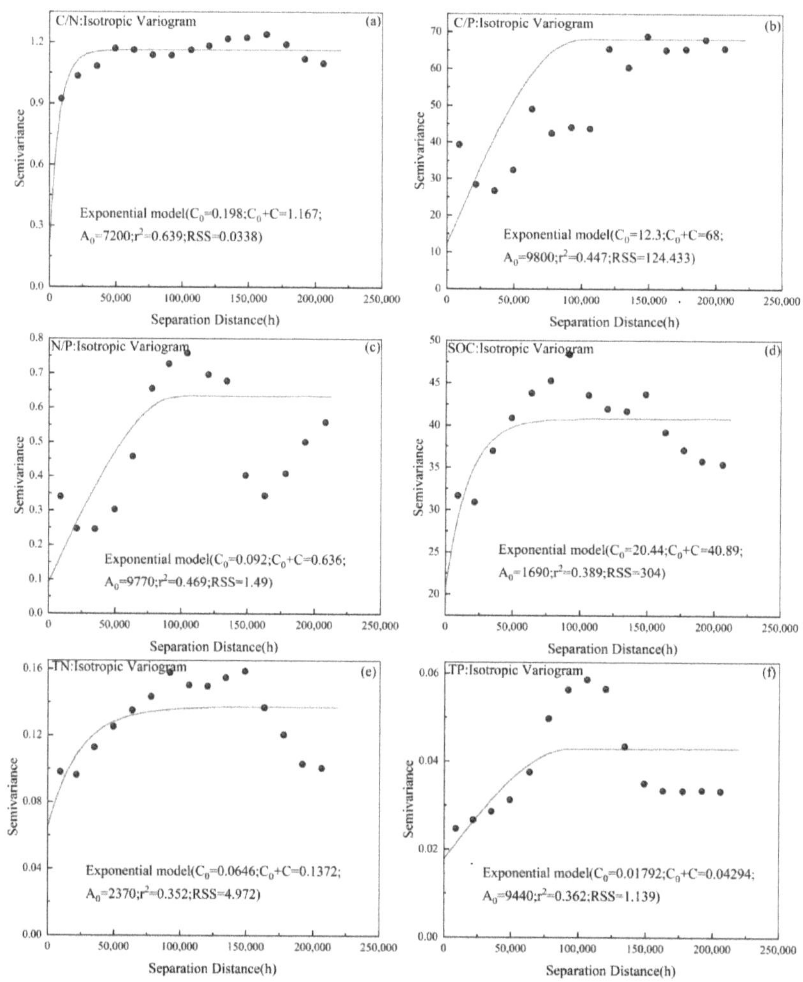

Figure 3. Semi-variogram of the soil C:N ratio (**a**), C:P ratio (**b**), N:P ratio (**c**), and contents of SOC (**d**), TN (**e**), and TP (**f**) in the Aksu region.

The ranges (spatial correlation distances) of SOC, TN, and TP were 50.7 km, 78.1 km, and 94.4 km, respectively. The C:N ratio showed a shorter range (24.6 km), implying higher localized spatial autocorrelation compared to other nutrients. The ranges of C:P (98.0 km) and N:P (97.7 km) ratios were smaller than those of SOC and TP, reflecting weaker spatial dependencies in stoichiometric relationships.

3.3. Spatial Distribution Patterns of Soil Nutrient Contents and C:N:P Ratios

Based on the optimal semi-variogram models for SOC, TN, TP, and stoichiometric ratios, spatial distribution maps of these parameters in the Aksu region were generated using ordinary Kriging interpolation (Figure 4). The results show that SOC and TN contents ranged from 4.22–13.60 g/kg and 0.41–1.21 g/kg, respectively, with similar spatial patterns characterized by higher values in the north and lower in the south and higher values in the eastern part of the oasis compared to the west—a trend attributable to the significant positive correlation between SOC and TN. High SOC and TN concentrations were clustered in Wushi, Wensu, and Baicheng, whereas low values predominated in Xinhe, Shaya County, and Awat.TP content exhibited relatively homogeneous spatial distribution, with moderate to high levels concentrated in Wensu, Aksu, and Kuche. In contrast, the C:N ratio displayed strong spatial randomness, consistent with the semi-variogram analysis results.

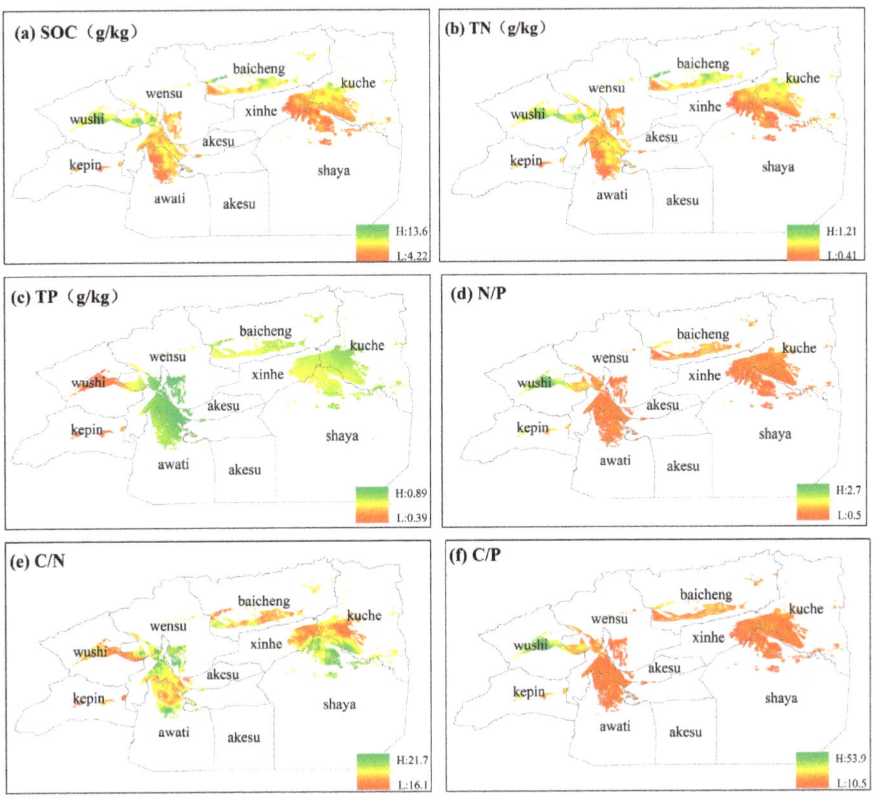

Figure 4. Spatial distribution of SOC, TN, and TP, contents and C:N, C:P, and N:P ratios in the Aksu region.

3.4. Analysis of Factors Influencing Soil SOC, TN, TP, and C:N:P Ratios Across Cropping Systems

The spatial heterogeneity of soil nutrients in the Aksu region arises from both natural environmental factors and anthropogenic activities. Pearson correlation analysis revealed a strong positive correlation between SOC and TN in the plough layer (r = 0.979 **, $p < 0.01$). SOC and TN showed significant negative correlations with a C:N ratio (r = −0.179 ** and

−0.353 **, respectively, $p < 0.01$), indicating that the C:N ratio is influenced by SOC and TN levels. As illustrated in Figure 5, the soil C:N ratio was further modulated by pH, elevation, and groundwater depth, with varying dominant factors across cropping systems.

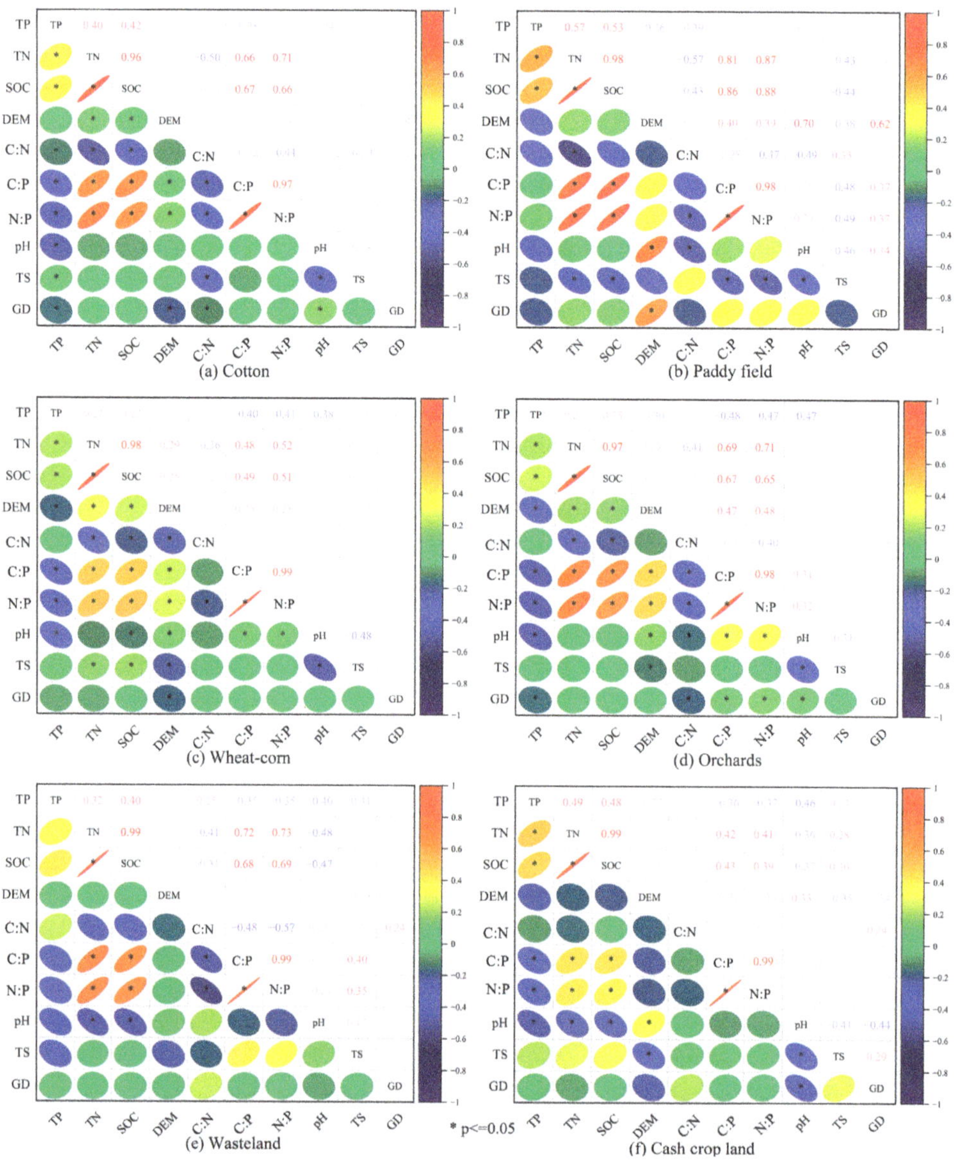

Figure 5. Correlation analysis of influencing factors for soil SOC, TN, and C:N ratio under different cropping systems (GD: ground water; DEM: digital elevation model; TS: total salt).

3.5. Analysis of Environmental Drivers and Interrelationships of Soil Carbon, Nitrogen, and Phosphorus Stocks, and Stoichiometric Ratios in Different Agricultural Systems

Orchards: Elevation was the key factor affecting SOC and TN, while C:P and N:P ratios were closely linked to pH and elevation. TP exhibited significant correlations with pH, elevation, and groundwater depth. Wheat–maize systems: SOC, TN, TP, and stoichiometric ratios (C:N, C:P, and N:P) were all significantly influenced by elevation. TP and C:N ratio showed significant negative correlations with elevation, while only TP was negatively correlated with pH. Cotton fields: SOC, C:N, and C:P ratios showed no significant correlations with elevation, pH, or groundwater depth. TN positively correlated with elevation, whereas TP negatively correlated with pH and groundwater depth. Paddy fields and wasteland: No significant correlations were observed between soil parameters and environmental factors. Other systems: Only TP exhibited a significant correlation with elevation. Comprehensive analysis identified soil pH and elevation as the primary drivers of SOC, TN, and C:N ratio variability across all six cropping systems.

Correlations between C:N:P stocks and stoichiometric ratios are summarized in Table 4: Orchards: C and N stocks positively correlated with C:P and N:P ($p < 0.01$), but negatively with C:N ($p < 0.01$). P stock negatively correlated with C:P and N:P ($p < 0.05$). Cotton and wheat–maize systems: C and N stocks strongly positively correlated with C:P and N:P ($p < 0.01$); P stock negatively correlated with these ratios ($p < 0.01$). C stock negatively correlated with C:N, which is significant in cotton systems. Paddy fields: C and N stocks positively correlated with C:P and N:P ($p < 0.01$), negatively with C:N ($p < 0.01$). P stock showed no correlations. Wasteland: C and N stocks positively correlated with C:P and N:P ($p < 0.01$); no correlations for P stock. Other systems: C stock positively correlated with C:P and N:P ($p < 0.05$); N stock correlated with C:P ($p < 0.01$); P stock weakly correlated with C:P and N:P ($p < 0.05$).

Table 4. Correlation of soil carbon, nitrogen, and phosphorus reserves with C:N, C:P, and N:P.

Planting Patterns	C, N, P Storage Capacity	C:N	C:P	N:P
Orchard	C	−0.206 **	0.655 **	0.631 **
	N	−0.408 **	0.691 **	0.705 **
	P	0.019	−0.477 **	−0.468 **
Cotton	C	−0.260 **	0.677 **	0.663 **
	N	−0.501 **	0.657 **	0.711 **
	P	−0.135 *	−0.282 **	−0.246 **
Wheat–corn	C	−0.304	0.640 **	0.645 **
	N	−0.411	0.721 **	0.735 **
	P	0.254	−0.347	−0.350
Paddy field	C	−0.430	0.837 **	0.864 **
	N	−0.573 **	0.811 **	0.872 **
	P	−0.385	0.054	0.133
Wasteland	C	−0.304	0.640 **	0.645 **
	N	−0.411	0.721 **	0.735 **
	P	0.254	−0.347	−0.350
Other	C	−0.001	0.415 *	0.383 *
	N	−0.159	0.425 **	0.410 *
	P	−0.094	−0.362 *	−0.366 *
Total cultivated land	C	−0.304	0.640 **	0.645 **
	N	−0.411	0.721 **	0.735 **
	P	0.254	−0.347	−0.350

Note: ** indicates $p < 0.01$, * indicates $p < 0.05$.

3.6. Analysis of Surface Soil Carbon, Nitrogen, and Phosphorus Stocks and Their Correlations

Based on soil bulk density data from the Aksu region (2020), average values for orchards, cotton fields, wheat–maize systems, paddy fields, other systems, and wasteland were 1.441, 1.444, 1.44, 1.457, 1.459, and 1.456 g cm^{-3}, respectively. Surface soil C, N, and P stocks were calculated accordingly (Figure 6). The mean surface soil C stock was 22.31 ± 10.37 t hm^{-2}, ranked as paddy fields (26.28 t hm^{-2}) > other systems (25.49 t hm^{-2}) > orchards (24.98 t hm^{-2}) > wheat–maize (24.21 t hm^{-2}) > cotton fields (17.06 t hm^{-2}) > wasteland (15.84 t hm^{-2}), with significant differences ($p < 0.05$) between cotton/other systems and the remaining four. Mean N stock (2.20 ± 1.07 t hm^{-2}) followed paddy fields (2.58 t hm^{-2}) > orchards (2.47 t hm^{-2}) > wheat–maize (2.43 t hm^{-2}) > other systems (2.41 t hm^{-2}) > cotton fields (1.71 t hm^{-2}) > wasteland (1.56 t hm^{-2}), where cotton and wasteland differed significantly ($p < 0.05$) from others. Mean P stock (2.21 ± 0.59 t hm^{-2}) showed orchards (2.58 t hm^{-2}) > cotton fields (2.47 t hm^{-2}) > other systems (2.43 t hm^{-2}) > paddy fields (2.41 t hm^{-2}) > wheat–maize (1.71 t hm^{-2}) > wasteland (1.56 t hm^{-2}), with wheat–maize and wasteland significantly lower ($p < 0.05$). These results highlight pronounced anthropogenic impacts on surface soil nutrient stocks across land-use types.

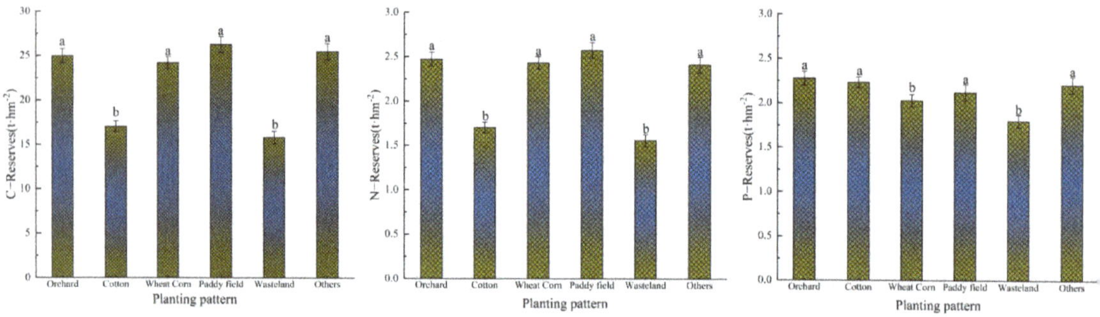

Figure 6. Soil organic carbon (SOC), total nitrogen (TN), and total phosphorus (TP) stocks under different cropping systems. (Numbers a and b represent significant differences).

Soil P is mainly derived from matric weathering or anthropogenic inputs (e.g., fertilization) rather than as a direct product of organic matter decomposition. The accumulation of C and N may be correlated with organic matter inputs, whereas P storage is dominated by physicochemical processes, such as mineral adsorption-desorption and fixation of Fe-Al oxides, and is decoupled from the C/N cycle. The ability of microorganisms to mineralize P is regulated by phosphatase activity, whereas changes in C:N have a small effect on the P cycle, resulting in no significant correlation between C:P and C/N storage.

In most terrestrial ecosystems, P is readily immobilized by soil minerals to form insoluble phosphates (e.g., Fe-P, Al-P), and its effectiveness is strongly influenced by pH and redox conditions. Even if C and N stocks increase, P effectiveness may still be limited by mineral adsorption, resulting in insignificant C:P fluctuations. Plant and microbial demand for P is relatively independent of C/N metabolism. For example, in nitrogen-limited ecosystems, plants may activate P by increasing organic acid secretion from the root system, but this adaptive strategy does not significantly alter the relationship between C and N stocks, and, thus, changes in C:P ratios may not be correlated.

Overall, C and N stocks in Aksu soils exhibited significant positive correlations with C:P and N:P ($p < 0.01$) but no linkage to C:N. P stock showed no consistent relationships with any stoichiometric ratios, suggesting its dynamics are decoupled from C and N cycling.

4. Discussion

4.1. Impacts of Cropping Systems on Soil SOC, TN, TP Contents, and C:N:P Ratios

The diverse cropping systems and heterogeneous management practices in the Aksu region contribute to pronounced spatial variability in soil nutrient dynamics across counties or subregions. Soil carbon (C), nitrogen (N), and phosphorus (P) storage and release are closely linked to cropping systems, which further regulate their contents. Among the six cropping systems, soil organic carbon (SOC) content ranked as paddy fields > other systems > orchards > wheat–maize > cotton fields > wasteland. This pattern arises because prolonged waterlogging in paddy fields suppresses microbial activity and facilitates cation-organic matter complexation, thereby enhancing SOC and total nitrogen (TN) accumulation [25,26]. The "other systems", dominated by vegetable cultivation exhibit high SOC due to substantial organic inputs and intensive irrigation.

Soil TN followed the order: paddy fields > orchards > wheat–maize > other systems > cotton fields > wasteland. Orchards showed elevated TN due to litter accumulation, minimal tillage, and reduced solar radiation, which collectively inhibit N loss. Wasteland soils exhibited the lowest TN due to long-term abandonment and lack of fertilization. In wheat–maize and cotton systems, nitrogen-dominated fertilization (with limited P inputs) combined with crop residue incorporation enhanced soil N retention.

The C:N ratio ranked as other systems > paddy fields > orchards > wasteland > cotton fields > wheat–maize. This reflects variations in hydrothermal conditions, fertilization regimes, and management practices. Cotton and wheat–maize systems, characterized by improved soil aeration and porosity, promote aerobic microbial activity, accelerating organic carbon decomposition and N loss [27]. High C:N ratios in "other systems" result from heavy organic fertilization, while waterlogged paddy fields exhibit slower organic matter decomposition due to low soil and water temperatures, leading to higher C:N.

The C:P ratio followed wheat–maize > paddy fields > other systems > orchards > wasteland > cotton fields. In Xinjiang, N-based fertilization predominates, with P applied secondarily. Wheat–maize systems demonstrate higher P use efficiency than cotton, partially explaining their elevated C:P ratios. Notably, Aksu's farmland soils exhibit higher C:N (18.52) than China's average (12.30), but markedly lower C:P (19.68 vs. 52.64) and N:P (1.08 vs. 4.20). Higher C:N ratios (18.52) may have a dual impact on ecosystem carbon sinks and agricultural sustainability by inhibiting microbial nitrogen conversion efficiency, slowing down organic matter decomposition, and exacerbating nitrogen limitation. In ecosystems, this imbalance may promote carbon sequestration at the expense of productivity, while in agriculture, optimizing the ratio of carbon and nitrogen inputs, regulating microbial function, and improving fertilization techniques are needed to break the nitrogen limitation bottleneck. Future research is needed to quantify the threshold effect of elevated C:N on the coupled carbon and nitrogen cycle by combining long-term positional observations with process modeling. Although soil testing and formulated fertilization have improved nutrient management in recent years, potassium (K) application remains suboptimal, and imbalanced fertilizer ratios persist. Expanding precision fertilization technologies is critical to address these challenges.

4.2. Factors Influencing SOC, TN, TP, and C:N:P Ratios in the Aksu Farmland

Through correlation analysis with soil physicochemical properties and environmental factors, it was found that total salt, soil pH, elevation, and groundwater significantly influence soil carbon, nitrogen, and phosphorus at varying degrees [28,29]. In the cultivated layer soils of the Aksu region, soil organic carbon (SOC) and total nitrogen (TN) exhibited a strong positive correlation ($r = 0.979$, $p < 0.01$). SOC showed a significant negative correlation with the C:N ratio ($r = -0.179$ **), while TN also displayed a significant negative correlation with the C:N ratio ($r = -0.353$ **), both reaching highly significant levels ($p < 0.01$). This indicates that the soil C:N ratio is primarily determined by SOC and TN contents. Beyond these elemental proportions, the C:N ratio is additionally influenced by pH, elevation, and other factors. Variations in SOC, TN, and C:N ratios under different agricultural land-use types also differ in their characteristic patterns and driving factors.

Under orchard systems, elevation emerged as a key factor affecting SOC, TN, and the C:P ratio, while groundwater depth correlated with C:N, C:P, and N:P ratios. In wheat–maize rotation systems, SOC, TN, and the C:N ratio showed significant correlations with elevation and total salt, with the C:N ratio exhibiting a significant negative correlation with soil pH. In cotton cultivation, SOC and TN were significantly associated with elevation. In paddy fields, SOC and TN demonstrated significant correlations with pH. Overall, soil nutrient contents generally displayed positive correlations with elevation: SOC, TN, C:P, and N:P ratios increased with rising elevation, whereas TP and C:N ratios showed negative correlations with elevation. The observed relationships between SOC, TN, and elevation partially diverged from previous studies, which may be attributed to zonal soil characteristics [30]. The negative correlation between the C:N ratio and elevation aligns with earlier findings [31,32].

Total salt exerted substantial effects on nutrients, showing negative correlations with all measured nutrient indices except TP. Soil pH exhibited a significant negative correlation with TP but positive correlations with N:P and C:P ratios. These patterns are consistent with the ecological stoichiometric characteristics observed in farmland ecosystems across Xinjiang.

4.3. Implications of Soil C:N:P Stoichiometry for Nutrient Management

Human activities under different cultivation practices profoundly influence soil carbon (C), nitrogen (N), and phosphorus (P) stocks and their cycling processes. In the Aksu region, the variation patterns of soil C and N stocks across five cultivation systems were similar. Paddy fields exhibited higher C and N stocks compared to the other four systems, while orchard systems showed superior soil P stocks. Thus, in terms of carbon sequestration and nitrogen retention efficiency, paddy fields represent a more effective approach than other systems and serve as primary "carbon sources" and "nitrogen sources" [33,34]. Regarding phosphorus fixation capacity, orchards outperform paddy fields and act as dominant "phosphorus sinks" [35]. Overall, soil C:P and N:P ratios in the Aksu cultivated area demonstrated significant correlations with C and N stocks ($p < 0.01$), indicating that large-scale soil ecological stoichiometric characteristics also provide critical insights into soil carbon and nitrogen storage dynamics.

5. Conclusions

This study analyzed 1131 surface soil samples (0–20 cm) in the Aksu region, revealing moderate-to-low levels of SOC, TN, TP, and stoichiometric ratios (C:N, C:P, and N:P), with weak spatial autocorrelation for C:P and N:P. Intensive systems (paddy fields, orchards) significantly enhanced C and N stocks (22.31 ± 10.37 and 2.20 ± 1.07 t hm^{-2}, respectively), whereas cotton and wasteland showed nutrient depletion. Soil pH and elevation were key drivers of SOC and TN variability. Stock assessments demonstrated that intensive cultivation systems (e.g., paddy fields and orchards) significantly enhanced surface soil carbon (22.31 ± 10.37 t·ha^{-1}) and nitrogen stocks (2.20 ± 1.07 t·ha^{-1}), whereas cotton fields and wastelands, characterized by extensive management practices, exhibited notably lower nutrient stocks. Phosphorus stocks were highest in orchards (2.58 t·ha^{-1}), while wheat–maize systems and wastelands showed insufficient stocks due to soil impoverishment. The observed stoichiometric imbalances (e.g., elevated C:P and N:P ratios in cotton systems) align with the ecological stoichiometry theory, which posits that deviations from optimal elemental ratios disrupt microbial-mediated nutrient cycling and plant productivity. The weak spatial autocorrelation of C:P and N:P further reflects a departure from homeostatic equilibrium—a hallmark of stable ecosystems—suggesting that arid agroecosystems in Aksu are operating under suboptimal nutrient coupling. Long-term monitoring is critical to assess whether current nutrient stocks reflect transient states or irreversible degradation, particularly under climate-induced aridification. The 0–20 cm sampling depth overlooks subsurface nutrient fluxes (e.g., deep-rooted crops accessing deeper P reserves). Future studies should integrate vertical profiling (e.g., 0–100 cm) to refine stock estimates. This work positions arid agroecosystems as critical testbeds for advancing agricultural ecology theory, particularly in reconciling human-driven intensification with biogeochemical resilience. By integrating stoichiometric principles into land management, we contribute to the emergent paradigm of ecological precision agriculture—a discipline that harmonizes data-driven decision-making with ecosystem thresholds. Future efforts should scale these insights through partnerships with regional land-use planners, ensuring that Xinjiang's fragile farmlands serve as a model for sustainable dryland agriculture under global change.

Author Contributions: Conceptualization, L.P. and Q.G.; methodology, M.C.and S.C.; software, N.L. (Na Li); validation, N.L. (Ning Lai); formal analysis, C.L.; investigation, Y.L.; resources, H.X.; data curation, H.X.; writing—original draft preparation, H.X.; writing—review and editing, L.P. All authors have read and agreed to the published version of the manuscript.

Funding: This research was supported by stabilization support from the Xinjiang Academy of Agricultural Sciences (nkyzzkj-014,xjnkywdzc-2023002-5,xjnkywdzc-202401-05-0101) and (XinJiang Agriculture Research System—Wheat (XJARS-01-21)).

Institutional Review Board Statement: Not applicable.

Informed Consent Statement: Not applicable.

Data Availability Statement: The data presented in this study are available on request from the corresponding author.

Acknowledgments: We would like to thank our laboratory colleagues who collected and processed the data. We sincerely thank the editor-in-chief and the two anonymous reviewers for their helpful comments on improving this paper.

Conflicts of Interest: The authors declare no conflicts of interest.

References

1. George, T.S.; Giles, C.D.; Menezes-Blackburn, D.; Condron, L.M.; Gama-Rodrigues, A.C.; Jaisi, D.; Lang, F.; Neal, A.L.; Stutter, M.I.; Almeida, D.S. Organic phosphorus in the terrestrial environment: A perspective on the state of the art and future priorities. *Plant Soil* **2018**, *427*, 191–208. [CrossRef]
2. Abbas, F.; Hammad, H.M.; Ishaq, W.; Farooque, A.A.; Bakhat, H.F.; Zia, Z.; Fahad, S.; Farhad, W.; Cerdà, A. A review of soil carbon dynamics resulting from agricultural practices. *J. Environ. Manag.* **2020**, *268*, 110319. [CrossRef]
3. Sanaullah, M.; Usman, M.; Wakeel, A.; Cheema, S.A.; Ashraf, I.; Farooq, M. Terrestrial ecosystem functioning affected by agricultural management systems: A review. *Soil Tillage Res.* **2020**, *196*, 104464. [CrossRef]
4. Feng, W.-L.; Yang, J.-L.; Xu, L.-G.; Zhang, G.-L. The spatial variations and driving factors of C, N, P stoichiometric characteristics of plant and soil in the terrestrial ecosystem. *Sci. Total Environ.* **2024**, 175543. [CrossRef] [PubMed]
5. Dawson, J.C.; Huggins, D.R.; Jones, S.S. Characterizing nitrogen use efficiency in natural and agricultural ecosystems to improve the performance of cereal crops in low-input and organic agricultural systems. *Field Crops Res.* **2008**, *107*, 89–101. [CrossRef]
6. Tomaz, A.; Martins, I.; Catarino, A.; Mourinha, C.; Dôres, J.; Fabião, M.; Boteta, L.; Coutinho, J.; Patanita, M.; Palma, P. Insights into the spatial and temporal variability of soil attributes in irrigated farm fields and correlations with management practices: A multivariate statistical approach. *Water* **2022**, *14*, 3216. [CrossRef]
7. Zheng, S.; Xia, Y.; Hu, Y.; Chen, X.; Rui, Y.; Gunina, A.; He, X.; Ge, T.; Wu, J.; Su, Y. Stoichiometry of carbon, nitrogen, and phosphorus in soil: Effects of agricultural land use and climate at a continental scale. *Soil Tillage Res.* **2021**, *209*, 104903. [CrossRef]
8. Shen, H.; Huang, Y.; Lin, X.; Dai, Z.; Zhao, H.; Su, W.-Q.; Dahlgren, R.A.; Xu, J. Recoupling of Soil Carbon, Nitrogen, and Phosphorus Cycles along a 30 Year Fire Chronosequence in Boreal Forests of China. *Environ. Sci. Technol.* **2025**, *59*, 4432–4443. [CrossRef]
9. Luo, H.; Li, J.; Li, P.; Rioual, P.; Zou, Y.; Zhang, J.; Yan, Y.; Li, J.; Wang, L. Variation in the seasonal response to climate change during the past 1000 years as inferred from a Maar Lake sediment record, northeast China. *J. Paleolimnol.* **2022**, *68*, 133–154. [CrossRef]

10. Shen, F.; Wu, J.; Fan, H.; Liu, W.; Guo, X.; Duan, H.; Hu, L.; Lei, X.; Wei, X. Soil N/P and C/P ratio regulate the responses of soi microbial community composition and enzyme activities in a long-term nitrogen loaded Chinese fir forest. *Plant Soil* **2019**, *43*, 91–107. [CrossRef]
11. Zhao, X.; Tian, Q.; Huang, L.; Lin, Q.; Wu, J.; Liu, F. Fine-root functional trait response to nitrogen deposition across fores ecosystems: A meta-analysis. *Sci. Total Environ.* **2022**, *844*, 157111. [CrossRef] [PubMed]
12. Sardans, J.; Rivas-Ubach, A.; Penuelas, J. The elemental stoichiometry of aquatic and terrestrial ecosystems and its relationship with organismic lifestyle and ecosystem structure and function: A review and perspectives. *Biogeochemistry* **2012**, *111*, 1–3 [CrossRef]
13. Yuan, Z.; Chen, H.Y. Decoupling of nitrogen and phosphorus in terrestrial plants associated with global changes. *Nat. Clim. Change* **2015**, *5*, 465–469. [CrossRef]
14. Sokol, N.W.; Slessarev, E.; Marschmann, G.L.; Nicolas, A.; Blazewicz, S.J.; Brodie, E.L.; Firestone, M.K.; Foley, M.M.; Hestrin, R Hungate, B.A. Life and death in the soil microbiome: How ecological processes influence biogeochemistry. *Nat. Rev. Microbio* **2022**, *20*, 415–430. [CrossRef] [PubMed]
15. Schönhart, M.; Schmid, E.; Schneider, U.A. CropRota–A crop rotation model to support integrated land use assessments. *Eur. J Agron.* **2011**, *34*, 263–277. [CrossRef]
16. Tian, S.; Xu, Y.; Wang, Q.; Zhang, Y.; Yuan, X.; Ma, Q.; Feng, X.; Ma, H.; Liu, J.; Liu, C. The effect of optimizing chemical fertilizer consumption structure to promote environmental protection, crop yield and reduce greenhouse gases emission in China. *Sci Total Environ.* **2023**, *857*, 159349. [CrossRef]
17. Yuan, S.; Li, R.; Zhang, Y.; Tang, J.; Wang, L.; Ji, H.; Huang, S. Effects of long-term partial substitution of inorganic fertilizer with pig manure and/or straw on nitrogen fractions and microbiological properties in greenhouse vegetable soils. *J. Integr. Agric.* **2024** *23*, 2083–2098. [CrossRef]
18. Zhang, L.; Feng, Y.; Zhao, Z.; Cui, Z.; Baoyin, B.; Wang, H.; Li, Q.; Cui, J. Maize/soybean intercropping with nitrogen supply levels increases maize yield and nitrogen uptake by influencing the rhizosphere bacterial diversity of soil. *Front. Plant Sci.* **2024** *15*, 1437631. [CrossRef]
19. Almagro, M.; Díaz-Pereira, E.; Boix-Fayos, C.; Zornoza, R.; Sánchez-Navarro, V.; Re, P.; Fernández, C.; Martínez-Mena, M The combination of crop diversification and no tillage enhances key soil quality parameters related to soil functioning withou compromising crop yields in a low-input rainfed almond orchard under semiarid Mediterranean conditions. *Agric. Ecosyst Environ.* **2023**, *345*, 108320.
20. Lv, Y.-J.; Zhang, X.-L.; Gong, L.; Huang, S.-B.; Sun, B.-L.; Zheng, J.-Y.; Wang, Y.-J.; Wang, L.-C. Long-term reduced and no tillage increase maize (*Zea mays* L.) grain yield and yield stability in Northeast China. *Eur. J. Agron.* **2024**, *158*, 127217. [CrossRef]
21. Homulle, Z.; George, T.S.; Karley, A.J. Root traits with team benefits: Understanding belowground interactions in intercropping systems. *Plant Soil* **2022**, *471*, 1–26. [CrossRef]
22. Xu, P.; Li, G.; Zheng, Y.; Fung, J.C.; Chen, A.; Zeng, Z.; Shen, H.; Hu, M.; Mao, J.; Zheng, Y. Fertilizer management for global ammonia emission reduction. *Nature* **2024**, *626*, 792–798. [CrossRef] [PubMed]
23. Oliver, M.A.; Webster, R. Semi-variograms for modelling the spatial pattern of landform and soil properties. *Earth Surf. Process Landf.* **1986**, *11*, 491–504. [CrossRef]
24. Tang, X.; Hu, J.; Lu, Y.; Qiu, J.; Dong, Y.; Li, B. Soil C, N, P stocks and stoichiometry as related to land use types and erosion conditions in lateritic red soil region, south China. *Catena* **2022**, *210*, 105888. [CrossRef]
25. Wei, L.; Ge, T.; Zhu, Z.; Luo, Y.; Yang, Y.; Xiao, M.; Yan, Z.; Li, Y.; Wu, J.; Kuzyakov, Y. Comparing carbon and nitrogen stocks in paddy and upland soils: Accumulation, stabilization mechanisms, and environmental drivers. *Geoderma* **2021**, *398*, 115121 [CrossRef]
26. Lin, S.; Wang, W.; Sardans, J.; Lan, X.; Fang, Y.; Singh, B.P.; Xu, X.; Wiesmeier, M.; Tariq, A.; Zeng, F. Effects of slag and biochar amendments on microorganisms and fractions of soil organic carbon during flooding in a paddy field after two years in southeastern China. *Sci. Total Environ.* **2022**, *824*, 153783. [CrossRef] [PubMed]
27. Ghezzehei, T.A.; Sulman, B.; Arnold, C.L.; Bogie, N.A.; Berhe, A.A. On the role of soil water retention characteristic on aerobic microbial respiration. *Biogeosciences* **2019**, *16*, 1187–1209. [CrossRef]
28. Zhang, W.; Zhu, J.; Zhou, X.; Li, F. Effects of shallow groundwater table and fertilization level on soil physico-chemical properties, enzyme activities, and winter wheat yield. *Agric. Water Manag.* **2018**, *208*, 307–317. [CrossRef]
29. Ping, C.L.; Michaelson, G.J.; Stiles, C.A.; González, G. Soil characteristics, carbon stores, and nutrient distribution in eight forest types along an elevation gradient, eastern Puerto Rico. *Ecol. Bull.* **2013**, *54*, 67–86.
30. Wu, X.; Cai, C.; Wang, J.; Wei, Y.; Wang, S. Spatial variations of aggregate stability in relation to sesquioxides for zonal soils, South-central China. *Soil Tillage Res.* **2016**, *157*, 11–22. [CrossRef]
31. Jiang, L.; He, Z.; Liu, J.; Xing, C.; Gu, X.; Wei, C.; Zhu, J.; Wang, X. Elevation gradient altered soil C, N, and P stoichiometry of Pinus taiwanensis forest on Daiyun Mountain. *Forests* **2019**, *10*, 1089. [CrossRef]
32. Ravikovitch, S.; Porath, A. The effect of nutrients on the salt tolerance of crops. *Plant Soil* **1967**, *26*, 49–71. [CrossRef]

3. Chen, S.; Elrys, A.S.; Zhao, C.; Cai, Z.; Zhang, J.; Müller, C. Global patterns and controls of yield and nitrogen use efficiency in rice. *Sci. Total Environ.* **2023**, *898*, 165484. [CrossRef] [PubMed]
4. Jun, X.; Evgenia, B.; Zhang, Y.; Yu, W.; Hu, Q.-J.; Zhang, C.-M.; Jie, W.; Zhang, Y.-Q.; Shi, X.-J. Substituting nitrogen and phosphorus fertilizer with optimal amount of crop straw improves rice grain yield, nutrient use efficiency and soil carbon sequestration. *J. Integr. Agric.* **2022**, *21*, 3345–3355.
5. Solangi, F.; Zhu, X.; Khan, S.; Rais, N.; Majeed, A.; Sabir, M.A.; Iqbal, R.; Ali, S.; Hafeez, A.; Ali, B. The global dilemma of soil legacy phosphorus and its improvement strategies under recent changes in agro-ecosystem sustainability. *ACS Omega* **2023**, *8*, 23271–23282. [CrossRef]

sustainability

MDPI

Article

Effects of Long-Term Multi-Treatment Experiments on Organic Matter and Enzymatic Activity in Sandy Soil

Krystyna Kondratowicz-Maciejewska [1], Joanna Lemanowicz [1,*] and Iwona Jaskulska [2]

[1] Department of Biogeochemistry, Soil Science and Irrigation and Drainage, Faculty of Agriculture and Biotechnology, Bydgoszcz University of Science and Technology, Bernardyńska 6 St., 85-029 Bydgoszcz, Poland; krystyna.kondratowicz-maciejewska@pbs.edu.pl

[2] Department of Agronomy, Faculty of Agriculture and Biotechnology, Bydgoszcz University of Science and Technology, Profesora Sylwestra Kaliskiego 7, 85-796 Bydgoszcz, Poland; iwona.jaskulska@pbs.edu.pl

* Correspondence: joanna.lemanowicz@pbs.edu.pl

Abstract: This study shows an evaluation of the condition of organic matter against enzymatic activity in soil. Long-term static field experiments with fertilisation with manure (FYM), different minerals, and mineral–manure were used for the research. Assays were obtained of the content of total organic carbon (TOC), dissolved fraction (DOC), susceptibility to oxidation (CL1) and (CL), total nitrogen (TN), dissolved nitrogen fraction (DTNT), and available forms of potassium, phosphorus, and magnesium. The activity of enzymes dehydrogenases, catalase, β-glucosidase, proteases, alkaline, and acid phosphatase was determined. We calculated the enzymatic indices and those evaluating the labile organic carbon management (CMI and CPI) in soil. An increase in TOC, up to 8.85 g kg^{-1} and 8.56 g kg^{-1} (FYM, FYM + KN), respectively, as compared with the control (5.67 g kg^{-1}), did not have a significant effect on the content of labile carbon fraction CL for the fertilisation treatments. Only a higher CL content was found in the soil with the FYM + PN and FYM + NPK + Mg treatments (2.07 g kg^{-1} and 2.05 g kg^{-1}). All the fertilisation treatments under study demonstrated a decrease in the value of the carbon management index (CMI). Similar DOC values (on average, 75.14 mg kg^{-1}) were noted. The average percentage share of the DOC fraction accounted for 1.163% TOC, and it was lower as compared with the control variant (1.33% TOC). The mineral fertilisation treatments decreased soil enzyme activities. Multiparametric enzymatic soil fertility indices differed due to soil properties, depending on the fertilisation applied.

Keywords: soil organic matter; fertilisation; oxidation; activity redox and hydrolytic enzymes

1. Introduction

The organic matter of soil is subject to transformations in mineralization and humification processes. Agrotechnical practices affect the course and the rate of its decomposition [1,2]. The long-term dynamic use of soils and frequently unbalanced fertilisation result in many processes leading to, e.g., a decrease in biological activity, acidification, and lowering the content of humus [3,4]. The assessment of these processes should be based not only on determining the current organic carbon content but also on the share of fractions of organic matters related to mineralization processes.

The method that offers such possibilities is the analysis of susceptibility to the chemical oxidation of soil organic matter, with the use of potassium manganate (VII) solution [5,6]. The content of the soil organic matter fraction, which is water soluble or soluble in salt solutions of a pH~7 (DOM), also facilitates evaluating the quality of the organic matter of soils [7]. DOM in soil is expressed as the content of dissolved organic carbon (DOC). That fraction includes mostly low-molecular organic compounds, which can act as growth stimulants and inhibitors in plants. DOC is also an easily available energy material for microorganisms [8]. Manure (FYM) and mineral fertilisers are common fertilisers applied to manage nutrients, enhancing soil fertility and quality, and thus increasing the yields [9,10]. However, an intensive use of mineral fertilisers can lead to soil degradation. For that reason, using FYM as an alternative for mineral fertilisers can be an effective fertilisation strategy to maintain soil health and biodiversity [9]. Nutrients found in FYM are gradually released, which depends on the origin [11] and the ratio of TOC/TN. Fertilisers change the environmental parameters of soil, e.g., TOC and pH; they can change the composition and diversity of soil microorganisms and, as a result, affect biochemical processes [2,12]. The processes are a parameter, showing the biological conditions of soils. The level of the enzymatic activity of soils involves a sensitive soil fertility index and soil productivity index as well as information about the ecological changes in the soil environment [13,14]. Soil enzymes are released by microorganisms and plant roots [15]. They take part in the formation and decomposition of soil organic matter (SOM) and the biogeochemistry of C, N, P, and S [16]. As reported by Li et al. [17], fertilisation with mineral nitrogen has an inconsiderable effect on the biomass of microorganisms and thus on the enzymatic activity of soil. The application of enzymatic soil quality indices, which facilitate specifying the anthropogenic changes, allows the identification of the trends that occur in soil for a longer period [18].

The research hypothesis assumes that the manure and mineral fertilisation are factors that significantly modify the quality of organic matter and differentiate the activity of soil enzymes. With that in mind, the aim of this study was to determine the quantitative and qualitative changes in the organic matter of light soil (Luvisol) under the influence of long-term, varied manure and mineral fertilisation. We also investigated the activity of redox (dehydrogenases and catalase) and hydrolytic enzymes responsible for C (β-glucosidase), N (proteases), and P (alkaline and acid phosphatase) transformations. The values of indices evaluating the labile organic carbon management in soil and the enzymatic soil fertility indices have been demonstrated.

2. Materials and Methods

2.1. Material

The experiment was set up in 1948 in Mochełek, in close vicinity to Bydgoszcz, the Kujawsko-Pomorskie province, Poland (53°13′ N, 17°51′ E, 95 m above sea level) in Luvisol. The grain size composition of topsoil (0–20 cm) was classified as sand 61.1%, silt 33.6%, and clay 5.3%. The experiment was established in the region of the impact of a cold climate with dry seasons and a warm summer; Dfb [19]. The annual mean precipitation is 485 mm, and the temperature—8.1 °C. The aim of the experiment has been an evaluation of the effects of the mineral and natural forms of manure (FYM) fertilisation variants on soil properties and crop productivity. For 76 years, some changes were made in experimental treatments and crop rotation [20]. In 2002, reclamation liming was carried out, and since 2011, for the control, until then, there was no fertilisation; mineral fertilisation was introduced equal to half an NPK rate (nitrogen, phosphorus, and potassium). The crop rotation included pea, winter wheat, winter rapeseed, corn, and spring barley. Soil was sampled in 2022, after the corn harvest, and prior to spring barley sowing, from 12 fertilisation treatments, each treatment had four replications: T0—control (without fertilisation); T1—NPK; T2—FYM (FYM every 4 years); T3—FYM + PK; T4—FYM + KN; T5—FYM + KN + Mg (+magnesium); T6—FYM + PN; T7—FYM + PN + Mg; T8—FYM + NPK; T9—FYM + NPK + Mg; T10—FYM + NPK + Ca (liming once every 4 years); T11—FYM + NPK + Ca + Mg. Over the last two decades the annual mean fertilisation was as follows: nitrogen—96 kg N ha^{-1}, phosphorus—88 kg P_2O_5 ha^{-1}, potassium—120 kg K_2O ha^{-1}, magnesium—20 kg MgO ha^{-1}, FYM 8t ha^{-1}, and liming (CaO) 1.2 t ha^{-1}.

2.2. Methods

2.2.1. Chemical Analysis of the Included Soil

In the air-dried and sieved (<2 mm) soils, the following were assayed:

- pH in 1 M KCl and H_2O by potentiometric method;
- total organic carbon (TOC) and total nitrogen (TN) were assayed with the Vario Max CN analyser (Elementar, Langenselbold, Germany);
- the amount of dissolved organic carbon (DOC) and dissolved total nitrogen (DTN) were measured in the solutions following the extraction with 0.004 M $CaCl_2$. DOC and DTN were assayed with the Muli N/C 3100 Analityk Jena analyzer (Jena, Germany, assay sensitivity of 1 µg L^{-1}) and expressed in mg kg^{-1} d.w. of the soil sample as well as the percentage share in the pool of TOC and TN, respectively;
- humus fractions susceptible to oxidation [5,6]. The method is based on assaying the fractions of organic carbon susceptible to oxidation by acting on the soil sample with a 0.333 M solution (CL labile carbon) and (CNL non-labile carbon) as well as a 0.0333 M $KMnO_4$ solution (CL1 labile carbon) in a neutral environment.

The results were used to calculate soil organic carbon management indices:

- carbon pool size index CPI:

$$CPI = \frac{TOC\ sample}{TOC\ reference} \quad (1)$$

- carbon management index CMI:

$$CMI = CPI \times LI \times 100 \quad (2)$$

where the lability index

$$LI = \frac{L\ sample}{L\ reference},$$

and L is calculated as follows:

$$L = \frac{CL}{CNL},$$

where the L–lability of C (C in fraction oxidised by $KMnO_4$; C remaining unoxidized by $KMnO_4$).

In the soil, the content of available macroelements was also assayed:

- the content of available phosphorus (P) PN-R-04023 [21] and potassium (K) PN-R-04022 [22] were assayed with the Egner–Riehm method (DL) [23], and the content of available magnesium (Mg) PN-R-04020 [24] was assayed, according to the Schachtschabel method [25].

2.2.2. The Activity of Enzymes in Soil

The activity of selected enzymes representing the oxidoreductases (dehydrogenases and catalase) and hydrolases (alkaline and acid phosphatase, β-glucosidase, and proteases) was measured for fresh sieved (<2 mm) soils. The soils were stored at 4 °C (for two weeks). The activity of the following enzymes in soil was investigated:

- the activity of dehydrogenases (DEH) was assayed with the Thalmann [26] method after sample incubation with 2,3,5-triphenyltetrazolium chloride, and the measurement of absorbance of triphenylformazan (TPF) at 546 nm was expressed in mg TPF kg^{-1} 24 h^{-1}.

- the activity of catalase (CAT) was assayed with the Johnson and Temple [27] method with a 0.3% solution of hydrogen peroxide as a substrate. The other H_2O_2 was determined with a titration of 0.02 M $KMnO_4$ in acid conditions.
- the activities of alkaline phosphatase (AlP) and acid phosphatase (AcP) were measured from the detection of p-nitrophenol (pNP) released after incubation (37 °C, 1 h) for a pH~6.5 for acid phosphatase and for a pH~11.0 for alkaline phosphatase [28].
- the activity of β-glucosidase (BG) was measured with the Eivazi and Tabatabai [29] method, applying p-nitrophenyl-β-D-glucopyranoside as a substrate. The concentrations of p-nitrophenol were assayed with an immediate readout of the sample at 400 nm after alkalization with the buffer Tris/NaOH (pH 10.0) and $CaCl_2$.
- the activity of proteases (PRO) was assayed with the Ladd and Butlera [30] method, where the concentration of the amino acid tyrosine (Tyr) was assayed in the soil samples after incubation with sodium caseinate. Absorbance was measured with the spectrophotometer at a rem $\lambda = 680$.

The results of the activity of the enzymes under study were used to calculate the soil indices:

- enzymatic index of the soil pH from the activity of alkaline (AlP) and acid (AcP) phosphatase [31]:

$$AlP/AcP \tag{3}$$

- the geometric mean GMea [32]:

$$GMea = \sqrt[6]{DEH \times CAT \times AlP \times AcP \times BG \times PRO} \tag{4}$$

- to evaluate the total activity of soil enzymes (TEI) (total enzyme activity index), the following was calculated [33]:

$$TEI = \sum \frac{Xi}{\overline{Xi}} \tag{5}$$

where Xi is the activity of soil enzyme i, and $\overline{Xi}$ is the mean activity of enzyme i in all the samples.

- the results of the metabolic activity index (MAI) [34] for the total soil activity are also presented:

$$MAI = \sum \frac{Pij}{Pcij} \tag{6}$$

where $P_{ij} = \frac{A_{ij}}{Ref_j}$, $Pc_{ij} = \frac{Ac_{ij}}{Refc_j}$ and A_{ij} is the value of activity of each enzyme; Ref_j is the reference parameter—TOC; Ac_{ij} is the value of the activity of each enzyme in the control soil; Ref_{cj} is the reference parameter in the control soil.

2.3. Statistical Analyses

The data received were exposed to the statistical analysis performed in MS Excel, Statistica 13.3. The normality of the distribution of the parameters was verified with the Shapiro–Wilk test of normality. To determine the dependence among the basic soil properties, total organic carbon (TOC), total nitrogen (TN), dissolved organic carbon and dissolved nitrogen (DOC, DTN), fractions of labile carbon (CL1), content of available P, K, and Mg forms, the activity of selected soil enzymes, and the coefficients of correlation were calculated using PAST 4.13 [35]. Statistically significant values of correlation coefficients are presented as the correlogram. For the results, a single-factor analysis of variance was performed with the post hoc HSD Tukey test for a level of significance of $p = 0.05$. The results were expressed as the arithmetic mean. To account for the diversification of soil in terms of physical, chemical, and biochemical parameters, we used a multidimensional exploration technique, principal component analysis (PCA), for the first two components.

3. Results and Discussion

3.1. Physicochemical Properties of the Soil

The soils studied in the surface horizon demonstrated a decrease in the active acidity of pH of H_2O (by 5.96 on average), and exchangeable acidity in the pH of KCl (by 4.68 on average), as compared with the T0 (Table 1). The fertilisation variants which included liming were the only ones for which the values of those parameters corresponded to the T0 level (5.3; 6.3). Research by Jaskulska et al. [2] showed that over 70 years of applying manure and mineral fertilisation caused strong acidification of the soil. The soil reaction affects the solubility of the minerals and the content of their forms that is available to plants [36]. It is also the basic factor regulating many biological processes in soil [37]. Soil acidity has an unfavourable effect on soil conditions, and it has already become a serious problem in intensifying agricultural systems all across the world. Unfortunately, such changes were observed in long-term field experiments, especially in light soils. In extreme cases, soil degradation was noted [38].

Table 1. Values of pH and the organic matter properties.

Fertilisation	pH KCl	pH H_2O	TOC g kg^{-1}	TN g kg^{-1}	TOC/TN	DOC mg kg^{-1}	DTN mg kg^{-1}	DOC %TOC	DTN %TN
T0	**5.3**	**6.3**	**5.67 ±0.005**	**0.53 ±0.005**	**10.7**	**75.65 ±0.10**	**7.28 ±0.005**	**1.33**	**1.37**
T1	4.5	6.0	5.07 ±0.005	0.94 ±0.005	5.4	64.55 ±0.15	6.30 ±0.005	1.27	0.70
T2	4.4	5.8	8.85 ±0.005	0.68 ±0.005	13.0	67.95 ±0.05	6.41 ±0.01	0.80	0.94
T3	4.2	5.6	6.75 ±0.005	0.55 ±0.005	12.3	78.20 ±0.00	6.60 ±0.005	1.16	1.20
T4	4.2	5.6	8.56 ±0.005	0.50 ±0.005	17.1	80.05 ±0.05	11.41 ±0.01	0.92	2.28
T5	4.4	5.6	5.57 ±0.005	0.46 ±0.005	12.1	79.50 ±0.10	8.02 ±0.005	1.43	1.74
T6	4.4	5.7	6.52 ±0.01	0.57 ±0.005	11.4	78.40 ±0.20	7.58 ±0.005	1.20	1.33
T7	4.5	5.9	6.16 ±0.005	0.54 ±0.005	11.4	80.20 ±0.10	8.03 ±0.005	1.30	1.50
T8	4.6	5.9	6.82 ±0.01	0.58 ±0.005	11.7	82.10 ±0.10	10.87 ±0.005	1.20	1.88
T9	4.9	6.2	6.36 ±0.01	0.56 ±0.005	11.3	70.25 ±0.05	7.59 ±0.02	1.10	1.35
T10	5.4	6.5	6.72 ±0.005	0.58 ±0.000	11.6	67.10 ±0.00	8.08 ±0.005	1.00	1.39
T11	5.4	6.5	6.23 ±0.005	0.53 ±0.005	11.7	77.70 ±0.10	7.70 ±0.005	1.25	1.45
Mean HSD			**6.607** 0.039	**0.585** 0.028	**11.642** n.s.	**75.137** 0.516	**7.993** 0.063	**1.163** 0.062	**1.428** n.s.

Abbreviations: T0—control (without fertilisation); T1—NPK; T2—FYM (FYM every 4 years); T3—FYM + PK; T4—FYM + KN; T5—FYM + KN + Mg (+magnesium); T6—FYM + PN; T7—FYM + PN + Mg; T8—FYM + NPK; T9—FYM + NPK + Mg; T10—FYM + NPK + Ca (liming once every 4 years); T11—FYM + NPK + Ca + Mg. TOC—total organic carbon; TN—nitrogen total; DOC—dissolved organic carbon; DTN—dissolved total nitrogen; DOC%—percentage share in the pool TOC, DTN%—percentage share in the pool TN; HSD LSD—Honestly Significant Difference± Standard Deviation; n.s.—not statistically significant.

Soil organic carbon (SOC) determines the persistence of ecosystems and environmental processes, including the quality of soil. An inadequate long-term use of agrotechnical practices usually leads to SOC exhaustion, and it affects the content of soil organic carbon. The average TOC in the soils analysed was 6.607 g kg^{-1}, and it was higher than the T0 (5.67 g kg^{-1}). The treatment with only mineral fertilisation was the only one with a decrease in TOC to 5.07 g kg^{-1} in soil, which is also reported by Nardi et al. [3]. The content of the labile fractions of organic carbon and the value of indices evaluating the management of that parameter in soil are equally important [39,40]. The analysis of the condition of organic matter in the soils under study involved the use of the organic carbon fraction susceptible to oxidation with $KMnO_4$ solution with a concentration of 0.0333 M dm^{-3}, namely the fraction most susceptible to oxidation (CL1), and a concentration of 0.3333 M dm^{-3} (CL). The content of CL1 for the control variant was 0.321 g kg^{-1}, whereas the average content for the treatments studied was 0.264 g kg^{-1} (Table 2). Important information is provided by the percentage share of that fraction in the TOC pool. However, also in that case, the CL1 values (% TOC) for all the fertilisation variants were lower (from 2.3% TOC for FYM to 5.2% TOC for FYM + NPK + Ca + Mg), as compared with the control: 5.7% TOC. The indices of carbon management (CMI) and carbon level (CPI) facilitate evaluating the effects of varied mineral and FYM fertilisation on the quality of organic matter (Figure 1). The results of this study demonstrate that an increase in TOC, up to 8.85 g kg^{-1} and 8.56 g kg^{-1} (FYM, FYM + KN), respectively, as compared with the control (5.67 g kg^{-1}), did not have a significant impact on the content of the labile fraction (CL) in respective fertilisation treatments. We noted only a higher content of that fraction for FYM + PN and FYM + NPK + Mg treatments (2.07 g kg^{-1} and 2.05 g kg^{-1}). Unfortunately, all the fertilisation treatments demonstrated a decrease in the value of the carbon management index (CMI) (Table 2).

Table 2. The impact of different fertilisations on labile and non-labile organic carbon fractions.

Fertilisation	CL1 g kg^{-1}	CL1 %TOC	CL g kg^{-1}	CNL g kg^{-1}
T0 *	0.321 ± 0.001	5.7	1.07 ± 0.005	4.60
T1	0.191 ± 0.000	3.8	0.64 ± 0.005	4.43
T2	0.200 ± 0.000	2.3	0.61 ± 0.000	8.24
T3	0.266 ± 0.000	3.9	0.64 ± 0.005	6.11
T4	0.211 ± 0.000	2.5	0.56 ± 0.005	8.00
T5	0.253 ± 0.003	4.5	0.65 ± 0.010	4.92
T6	0.243 ± 0.003	3.7	2.07 ± 0.010	4.45
T7	0.267 ± 0.003	4.3	0.64 ± 0.010	5.52
T8	0.274 ± 0.004	4.0	0.80 ± 0.010	6.02
T9	0.311 ± 0.001	4.9	2.05 ± 0.005	4.31
T10	0.302 ± 0.002	4.5	1.05 ± 0.005	5.67
T11	0.323 ± 0.003	5.2	1.07 ± 0.010	5.16
Mean	**0.264**	**4.108**	**0.998**	**5.428**
HSD	n.s.	n.s.	n.s.	3.791

Abbreviations: *; ±—see Table 1. CL1—labile carbon (fraction oxidised 0.0333 M $KMnO_4$), CL—labile carbon (fraction oxidised 0.333 M $KMnO_4$), CNL—non-labile carbon (C remaining unoxidized by 0.333 M $KMnO_4$), n.s.—not statistically significant

The lowest decrease in the CMI value (down to 78.4% and to 65.3% of the reference sample) was recorded for the treatments with the highest CL fraction content. According to Blair et al. [41], introducing FYM and mineral fertilisation increases the content of total organic carbon and labile fractions. Unfortunately, the results of the analyses of soils in respective fertilisation treatments do not confirm the working hypothesis presented by the authors. Only the application of FYM + PN and FYM + NPK + Mg increased the labile fraction of organic carbon, as compared with the control. The soil carbon accumulation index points to an unfavourable condition of soil organic matter. It is, therefore, worth stressing that the agrotechnical treatments in the present field experiment did not enhance the quality of organic matter despite higher contents of organic carbon in the fertilisation treatments [42]. It was found that the average content of total nitrogen (TN) in the treatments analysed was 0.583 g kg^{-1} (Table 1). As compared with the TN (0.53 g kg^{-1}) in the control soil, a decrease in the content of that parameter was recorded for FYM + KN (0.50 g kg^{-1}) and FYM + KN + Mg (0.46 g kg^{-1}) treatments. However, the highest content of TN was recorded for the treatment with mineral fertilisation only; NPK (0.94 g kg^{-1}). The TOC and TN results were used to calculate the values of the ratio of TOC/TN, which is an indicator of the degree of soil organic matter decomposition. In general, it is assumed that FYM fertilisation results in a decrease in the value of that ratio, which is due to a greater accumulation of the content of nitrogen than carbon [43]. The TOC/TN value for the control was 10.7, and the average value for the fertilisation treatments analysed was 11.6 (Table 1). Importantly, higher TOC/TN values, from 11.3 to 17.1, were recorded for almost all the fertilisation treatments, except for the NPK variant (5.4), which is due to the lowest TOC (5.07 g kg^{-1}) and the highest TN (0.94 g kg^{-1}). A high value of the ratio of TOC/TN (17.10) for the FYM + KN variant, on the other hand, can point to a mineralization process slowdown [42].

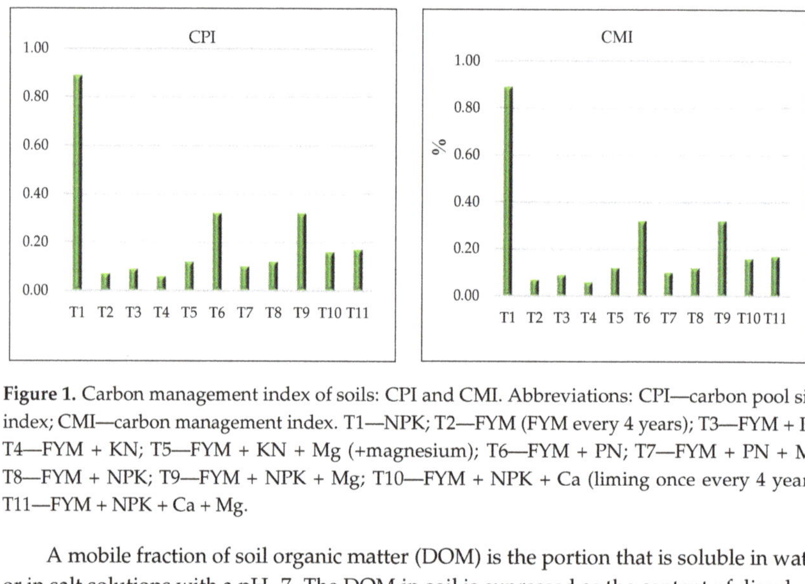

Figure 1. Carbon management index of soils: CPI and CMI. Abbreviations: CPI—carbon pool size index; CMI—carbon management index. T1—NPK; T2—FYM (FYM every 4 years); T3—FYM + PK; T4—FYM + KN; T5—FYM + KN + Mg (+magnesium); T6—FYM + PN; T7—FYM + PN + Mg; T8—FYM + NPK; T9—FYM + NPK + Mg; T10—FYM + NPK + Ca (liming once every 4 years); T11—FYM + NPK + Ca + Mg.

A mobile fraction of soil organic matter (DOM) is the portion that is soluble in water or in salt solutions with a pH~7. The DOM in soil is expressed as the content of dissolved organic carbon (DOC). It is the most mobile and the most active soil component, acting as an easily accessible source of nutrients and energy for microorganisms and other living organisms. It determines the processes of soil structure formation, and it is responsible for nutrient transport [8]. Both the agrotechnical practices and the organic material introduced into soil have an unquestionable effect on the content of dissolved organic matter in soils under agricultural use [7]. In the soil under study, for the fertilisation variants analysed, there were noted similar DOC contents (on average, 75.14 mg kg^{-1}) (Table 1). The significantly highest DOC content, as compared with the control variant (75.65 mg kg^{-1}), was identified in the soil from FYM + NPK (82.10 mg kg^{-1}) and FYM + PN + Mg (80.20 mg kg^{-1}) treatments. One should also note a decrease in the content of the fraction of organic carbon in such fertilisation variants such as NPK, FYM, FYM + NPK + Mg, and FYM + NPK + Ca.

However, the DOC contents in the soils studied did not affect the percentage share of that carbon fraction in soil humus. The average percentage share of the DOC fraction in the total organic carbon accounted for 1.16% TOC, and it was lower than in the control variant (1.33% TOC). A lack of significant differences in the DOC expressed in % TOC can suggest, according to Gonet et al. [7], that it depends on the absolute amount of soil organic matter. For the dissolved total nitrogen (DTN), the average content was 7.99 mg kg^{-1}, and it was higher than the contents in the control variant (7.28 mg kg^{-1}). Interestingly, however, for the NPK, FYM, and FYM + PK fertilisation variants, we recorded a decrease in the content of that fraction (6.30; 6.41 and 6.60 mg kg^{-1}, respectively). The fertilisation variants applied in the long-term field experiment did not have a significant effect on the percentage share of that fraction in the total nitrogen.

3.2. *The Content of Available Macronutrients*

The ANOVA demonstrated a significant effect of fertilisation on changes in the content of available P, K, and Mg (Table 3). The significantly highest content of available P (168.9 mg kg^{-1}) was recorded in the soil T10. There was 130% more compared with the control (73.52 mg kg^{-1}). The lowest P content was recorded in the soil T4 (89.3 mg kg^{-1}) and T5 (92.7 mg kg^{-1}). Menšík et al. [44] reported the highest content of N, P, and K in the topsoil for the NPK, FYM, cattle slurry and straw, and cattle slurry fertilisation treatments and the lowest for the control. FYM is an important source of organic and mineral P; it is mostly mineral P which is uptaken by the plants and, in general, it accounts for 45% to 90% of P in FYM. The soil from all the variants, except for the control, according to PN-R-04023 [21], can be considered class I with a very high P content. In such a case, phosphorus fertilisation is considered redundant. P is an element that is hardly mobile, available to plants only in the immediate vicinity of roots, and P uptake depends a lot on soil reaction and temperature. An excess of phosphorus is not harmful to plants; however, too much of that element in soil can impair the absorption of some microelements, e.g., potassium, copper, iron, and zinc. An intensive phosphorus fertilisation to increase the yield and its quality poses a greater risk of dispersion of that nutrient to the environment and penetration into waters. Filipek and Skowron [45] showed that 20–30% of the excessive phosphorus introduced into soil with fertiliser, as compared with phosphorus uptake with the yield, is accumulated in soil in an available form of phosphorus extractable with the Egner–Riehm solution. Such an amount is a potential source of phosphorus loss due to leaching.

The significantly highest content of available K (211.5 mg kg^{-1}) was recorded for T11, which accounted for 254% more as compared with the T0 (59.8 mg kg^{-1}) (Table 3). According to PN-R-04022 [22], soil from that treatment is considered class I with a very high K content. We found no significant differences in the K content across T1, T2, T5, T8, and T9 (168.2 mg kg^{-1}, 172.9 mg kg^{-1}, 175.9 mg kg^{-1}, 172.4 mg kg^{-1}, and 179.2 mg kg^{-1}, respectively). Reports by Arbačauskas et al. [46] demonstrated that the application of only potassium resulted in an increase in available P, contrary to nitrogen and phosphorus fertilisation. The T6 and T7 (without K) fertilisation variants had a significantly decreased K content in soil (78.5 mg kg^{-1} and 67.9 mg kg^{-1}, respectively). According to PN-R-04022 [22], those soils are considered class IV with a low content of that macronutrient. A long-term field experiment carried out by Balik et al. [47] showed that the content of available K decreased over 21 years in non-fertilised plots, which must have been due to a release of potassium from the unexchangeable form. Those changes, however, do not always reflect the K equilibrium.

Table 3. The content of available phosphorus, potassium, and magnesium.

Fertilisation	P	K	Mg
		mg kg^{-1}	
T0 *	73.52 ± 1.26	59.80 ± 3.58	10.32 ± 1.98
T1	118.4 ± 3.59	168.2 ± 9.12	11.91 ± 2.59
T2	128.7 ± 7.25	172.9 ± 7.56	12.11 ± 2.14
T3	154.2 ± 9.37	195.4 ± 8.23	13.29 ± 1.89
T4	89.31 ± 2.53	141.2 ± 6.12	15.04 ± 2.56
T5	92.70 ± 3.57	175.9 ± 12.11	22.98 ± 3.45
T6	142.7 ± 8.21	78.50 ± 5.73	19.63 ± 4.11
T7	149.7 ± 7.11	67.90 ± 6.32	23.17 ± 4.96
T8	163.8 ± 11.89	172.4 ± 9.63	21.75 ± 4.87
T9	159.7 ± 10.46	179.2 ± 10.45	24.89 ± 3.28
T10	168.9 ± 8.26	183.7 ± 9.58	17.56 ± 3.56
T11	142.9 ± 7.63	211.5 ± 13.44	26.97 ± 2.31
Mean	**132.0**	**150.6**	**18.30**
$HSD_{0.05}$	4.170	11.62	3.151

Abbreviations: *, ±—see Table 1. P—available phosphorus; K—available potassium; Mg—available magnesium.

The content of available Mg was significantly modified by long-term mineral, natural, and natural and mineral fertilisation (Table 3). The highest content (26.97 mg kg^{-1}, 24.89 mg kg^{-1}, and 23.17 mg kg^{-1}) was noted in the soil samples with Mg fertilisation (T11, T9, and T7, respectively). However, we recorded no significant differences in the Mg content in the soils sampled from those treatments. These soils, according to PN-R-04020 [24], can be qualified as soils with a low content (class IV) of available Mg. The significantly lowest Mg content was found in soil T0 (10.32 mg kg^{-1}), T1 (11.91 mg kg^{-1}), and T2 (12.11 mg kg^{-1}). Introducing combined organic and mineral fertilisation into the soil resulted in a significant increase in the content of that macroelement. Similar results were recorded by Sienkiewicz et al. [48]. They found that FYM with mineral fertilisers increases the Mg content as compared with only mineral fertilisation. FYM enhances the sorption of the soil complex, which retains cations due to physicochemical sorption. It limits the process of element leaching, including Mg.

3.3. *The Activity of Enzymes*

The activity of the enzymes is presented in Table 4. The results demonstrate that FYM and mineral fertilisers significantly changed the enzymatic activity of soil. T2 and T4 fertilisation increased the activity of dehydrogenases (0.489 mg TPF kg^{-1} 24 h^{-1} and 0.471 mg TPF kg^{-1} 24 h^{-1}, respectively) and catalase (0.265 g H_2O_2 kg^{-1} h^{-1} 0.251 g H_2O_2 kg^{-1} h^{-1}), respectively. For DEH, it was about 73%, and for CAT, it was around 300% higher activity, as compared with the control. The application of NPK (T1) also increased the activity of redox enzymes: DEH—6% only; CAT—41%. Some mineral ions, e.g., Ca, Mg, and Fe, are cofactors that activate enzymes [49]. Their availability in soil determines the level of activity of the enzyme. The activity of dehydrogenases is considered an indicator of oxidative metabolism in soil. Catalase, on the other hand, is an antioxidating enzyme that protects against oxidative stress and catalyses the decomposition of hydrogen peroxide to water and oxygen. The activity of those two enzymes is used to acquire information on the microbiological activity in soil. The AlP and AcP activities were significantly highest in T10 (0.648 mM pNP kg^{-1} h^{-1} and 1.203 mM pNP kg^{-1} h^{-1}, respectively) and T11 (0.499 mM pNP kg^{-1} h^{-1} and 1.099 mM pNP kg^{-1} h^{-1}, respectively) (Table 4). Applying only FYM (T2) significantly increased the activity of AlP (0.351 mM pNP kg^{-1} h^{-1}) and AcP (0.868 mM pNP kg^{-1} h^{-1}) (by 75.5% and 34%), as compared with the control (0.200 mM pNP kg^{-1} h^{-1} and 0.649 mM pNP kg^{-1} h^{-1}), respectively. The NPK (T1) application also resulted in a significant increase in the AlP activity (0.237 mM pNP kg^{-1} h^{-1}) and AcP (0.709 mM pNP kg^{-1} h^{-1}), however, only by 18.5% and 9%.

Table 4. The activity of enzymes in soil.

Fertilisation	DEH	CAT	AlP	AcP	BG	PRO
T0 *	0.271 ± 0.08	0.063 ± 0.01	0.200 ± 0.09	0.649 ± 0.05	0.529 ± 0.08	16.23 ± 1.23
T1	0.289 ± 0.09	0.089 ± 0.02	0.237 ± 0.07	0.709 ± 0.07	0.612 ± 0.07	37.57 ± 2.35
T2	0.489 ± 0.08	0.265 ± 0.09	0.351 ± 0.07	0.868 ± 0.01	1.351 ± 0.11	22.43 ± 1.98
T3	0.358 ± 0.04	0.183 ± 0.08	0.362 ± 0.08	0.875 ± 0.10	1.107 ± 0.09	18.56 ± 1.23
T4	0.471 ± 0.04	0.251 ± 0.08	0.291 ± 0.06	0.826 ± 0.05	1.309 ± 0.12	19.65 ± 1.58
T5	0.321 ± 0.05	0.138 ± 0.07	0.252 ± 0.07	0.753 ± 0.07	0.531 ± 0.11	15.22 ± 1.66
T6	0.334 ± 0.04	0.169 ± 0.06	0.286 ± 0.09	0.806 ± 0.03	1.122 ± 0.10	19.56 ± 2.08
T7	0.317 ± 0.03	0.156 ± 0.05	0.267 ± 0.11	0.767 ± 0.08	0.961 ± 0.09	17.72 ± 1.83
T8	0.365 ± 0.05	0.198 ± 0.06	0.309 ± 0.11	0.861 ± 0.08	1.128 ± 0.11	20.08 ± 1.98
T9	0.343 ± 0.06	0.172 ± 0.06	0.271 ± 0.09	0.784 ± 0.12	1.139 ± 0.12	19.05 ± 1.76
T10	0.361 ± 0.04	0.181 ± 0.07	0.648 ± 0.12	1.203 ± 0.11	1.186 ± 0.11	20.19 ± 2.08
T11	0.328 ± 0.03	0.161 ± 0.05	0.499 ± 0.11	1.099 ± 0.09	0.984 ± 0.09	17.96 ± 1.81
Mean	**0.354**	**0.169**	**0.331**	**0.850**	**0.997**	**20.35**
$HSD_{0.05}$	0.022	0.015	0.044	0.029	0.131	2.161

Abbreviations: *; ±—see Table 1. DEH—dehydrogenases (mg TPF kg^{-1} 24 h^{-1}); CAT—catalase (mg H_2O_2 kg^{-1} h^{-1}); AlP—alkaline phosphatase (mM pNP kg^{-1} h^{-1}); AcP—acid phosphatase (mM pNP kg^{-1} h^{-1}); BG—β-glukosidase (mM pNP kg^{-1} h^{-1}); PRO—protease (mg TYR kg^{-1} h^{-1}).

We recorded the significantly highest increase in BG activity after the application of FYM (T2) (1.351 mM pNP kg^{-1} h^{-1}) and FYM + KN (T4) (1.309 mM pNP kg^{-1} h^{-1}) (Table 4). BG activity was significantly higher for T2 and T4 (155% and 147%, respectively) than in the control (0.529 mM pNP kg^{-1} h^{-1}). We found no significant differences in BG activity between the T0 and T1 applications (0.612 mM pNP kg^{-1} h^{-1}). As for the negative effects of chemical fertilisation, Dincă et al. [50] noted a decrease in the enzymatic activity. It was noticeable especially in the soil after applying higher rates of mineral fertilisers, which coincides with organic matter losses. Mineral fertilisation only (T1) increased the PRO activity significantly (37.57 mg TYR kg^{-1} h^{-1}) (Table 4). It was 131% more than for the control (16.23 mg TYR kg^{-1} h^{-1}). Also, T2 resulted in a significant PRO increase (by 38%). The significantly lowest PRO activity (15.22 mg TYR kg^{-1} h^{-1}) was recorded with T5. We found no significant differences in the PRO activity due to the application of T3, T4, T6, T8, T9, and T10. A higher SOC content can be a possible reason for a higher activity of enzymes for FYM as compared with mineral fertilisers [9]. Yang et al. [51] demonstrated a significant impact of magnesium fertilisation in the form of $MgSO_4$ on hydrolytic enzymes (urease, phosphatase, invertase, and protease). Many enzymes depend on Mg, which can bond the magnesium substrate complex with a poor interaction with Mg or Mg, which can bond directly with the enzyme and change its structure, serving a catalytic function [52]. According to Keeler et al. [53], N added into soil increased the activity of acid phosphatase by an average of 13%, cellobiohydrolase by 17%, and a β-1,4-N-acetylglucosaminidase by 18%. The authors suggest that adding N enhances the count of soil microorganisms and increases the demand for P and C. It leads to an increase in the activity of enzymes taking part in the biogeochemistry of those macroelements. We found, however, no effect of N on the activity of redox enzymes degrading lignin. To evaluate the global model of AlP and AcP activity in soil due to the application of N and/or P, Zheng et al. [54] performed a meta-analysis. It demonstrated that the N application activated AcP by 10.1% (±2.9%); however, it showed a minimal effect on AlP, whereas FYM P fertiliser decreased the AcP activity by 7.7% ± 2.6%, yet it increased the AlP activity slightly. The report by Wang et al. [55] identified that the activities of β-glucosidase and alkaline phosphatase were highest after the application of organic fertilisation and after combined NPK with organic fertilisation (NPKMg). That activity was 161–171% and 75–91% higher, respectively, as compared with the control (without fertilisation). A study of 21 years by Choudhary et al. [56] demonstrated that the activity of enzymes (dehydrogenases, β-glucosidase, invertase, alkaline and acid phosphatase, arylsulfatase, and urease) depended significantly on the FYM + NPK fertilisation, as compared with the other treatments (control, N 120, NPK, FYM, and FYM + N). The application of inorganic fertilisers diversifies the activity of soil enzymes. The activity of some enzymes increases with the concentration of inorganic nutrients. Mineral nutrients applied for fertilisation are an immediate food for soil microorganisms. Their count increases and, as a result, the enzymatic activity increases. Yang and Norton [57] report on the application of organic fertilisers (compost) increasing the activity of soil enzymes (protease, chitinase, urease, and arginase), as compared with the inorganic N fertiliser (ammonium sulphate). Usually, organic additives essentially increase the activity of enzymes [58], which is probably due to a stimulation of the growth of microorganisms and a correlated increase in the activity of extracellular enzymatic complexes.

A significant correlation was recorded between TOC and DEH activity (r = 0.98), CAT (r = 0.98), and BG (r = 0.85) (Figure 2). The activity of those enzymes depended on TOC, 96%, 96%, and 72%, respectively. Dehydrogenases are respiratory enzymes that oxidize organic compounds, allocating their two atoms of hydrogen to the acceptors of electrons, producing energy [59]. Those enzymes play an essential role in the biological oxidation of the organic matter of soil by transferring hydrogen from organic substrates to inorganic acceptors [60]. β-glucosidase is an enzyme catalysing the final stage of cellulose hydrolysis, decomposing disaccharides and releasing the glucose available to soil microorganisms [18]. A higher content of organic carbon increases the rate of mineralization by microorganisms in soil. It results in an increase in enzymatic activity. SOC determines an increase in the concentration of the substrate available in soil. An increased C content in soil can often lead to increased water retention, which can increase substrate and enzyme diffusion [61]. Also, Gautam et al. [9] found that the long-term use of organic fertilisers (16 years) enhanced the content of SOC, which is the main substrate of the activity of enzymes in soil.

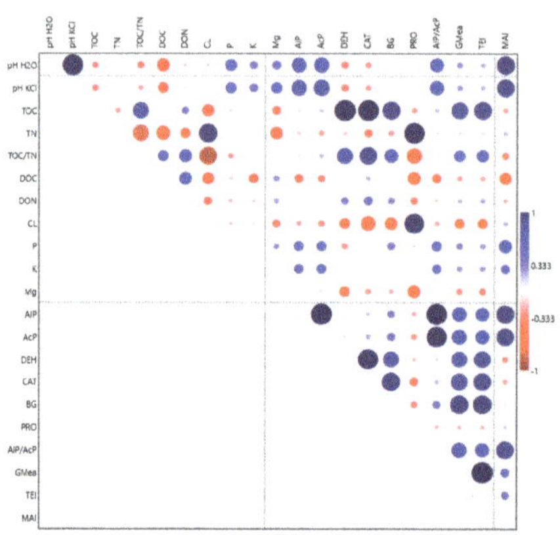

Figure 2. Correlogram of the soil variables. Abbreviations: TOC–total organic carbon; TN–nitrogen total; DOC–dissolved organic carbon; DON–dissolved organic nitrogen; CL—labile carbon; P—available phosphorus; K—available potassium; Mg—available magnesium; AlP—alkaline phosphatase; AcP—acid phosphatase; DEH—dehydrogenases; CAT—catalase; BG—β-glucosidse; PRO—proteases; AlP/AcP—enzymatic pH index; GMea—geometric mean; TEI—total enzyme activity index; MAI—metabolic activity index.

3.4. Enzymatic Indicators of Soil Quality

Chemical compounds present in the soil environment can activate or inhibit the effect of a single enzyme. For that reason, many multiparametric indicators differentiating soil due to the effect of, e.g., the use or type of vegetation have been developed [62]. According to Paz-Ferreiro and Fu [63], effective indicators should provide early warning of the upcoming changes in the soil environment due to the effect of biotic and abiotic factors. From the activity of soil enzymes determined with varied organic and mineral fertilisation, we calculated the enzymatic indicators of the soil quality (AlP/AcP, GMea, TEI, MAI) (Figure 3). The enzymatic pH indicator (AlP/AcP) was calculated with the results of AlP and AcP activity for varied fertilisation. The AlP/AcP value ranged from 0.31 to 0.54, depending on the fertilisation applied. According to Dick et al. [31], for optimal plant growth and development, such a soil pH value can be considered where AlP/AcP is about 0.50. In the present experiment, the AlP/AcP value exceeded 0.5 only for the T10 application (0.54). The results have been confirmed with excessive soil pH in H_2O and 1 M KCl (Table 1). The highest value of the GMea index was observed in the soil following T2 and T10 applications. According to Jat et al. [14], the GMea of the enzymes studied is related to the physicochemical and biological soil properties. For that reason, it serves as a soil quality indicator. Higher GMea values stand for a higher soil quality, and they can describe qualitative changes in soil, disregarding the physicochemical properties [18]. The MAI value ranged from 7.91 (T5) to 10.82 (T11). According to Picariello et al. [34], the MAI values can range from +1 to $+\infty$; however, the values increase with an increase in the metabolic activity of soil. MAI calculated from the enzymatic activity also provides information on the functional stability of microbial communities. Analysing the soils with varied long-term mineral and organic fertilisation, no differences were observed in the MAI values across T2, T3, T4, T6, T7, T8, and T9. To evaluate the total enzymatic activity of soil, the total enzyme activity index (TEI) was used. Its value varied depending on the fertilisation applied. The highest TEI value was recorded following the application of FYM (T2) (13.04) as well as FYM + NPK + Ca (T10) (12.78). Changes in the TEI value depending on fertilisation were the same as for GMea. Similar results were reported by Zhang et al. [64], who suggest that, indeed, those two indicators reflect changes in the soil environment most and that they are reliable for a comprehensive evaluation of the enzymes' reactions to biotic and abiotic factors. Wojewódzki et al. [65], on the other hand, demonstrated that MAI and TEI effectively differentiated the soil properties depending on the biocarbon applied. The analysis of correlation between the physical and chemical parameters and enzymatic soil indicators showed a significant positive correlation between TOC and GMea ($r = 0.79$) and TEI ($r = 0.83$). These indicators accounted for 62% and 69%, respectively, depending on TOC. The indicator AlP/AcP and MAI depended significantly on the pH of H_2O ($r = 0.66$ and $r = 0.87$, respectively) and the pH of KCl ($r = 0.68$ and $r = 0.84$, respectively) (Figure 2).

We identified a significant positive correlation between NT and PRO activity ($r = 0.98$) (Figure 2). Enzymes are compounds that are rich in N, and so their production is closely regulated with N availability. Proteases take part in nitrogen mineralization in soil. They catalyse the hydrolysis of peptide bonds in proteins and peptides to amino acids [66]. We recorded a positive significant correlation between TOC/TN and DTN ($r = 0.61$), DEH ($r = 0.74$), CAT ($r = 0.83$), and BG ($r = 0.63$) and a negative one with PRO ($r = -0.70$) and CL1 ($r = -0.83$). We also confirmed a significant negative correlation between DOC and CL1 ($r = -0.55$) and between TOC and CL1 ($r = -0.55$). As reported by Błońska et al. [67], a high activity of enzymes is closely connected with a higher OM content, with a good rate of decomposition, and with a low TOC/TN ratio. We recorded a significant negative correlation between DOC and the H_2O pH (r = −0.61) and KCl pH (r = 0.50). Cincotta et al. [68] confirm that acidity determines the stability or release of dissolved organic matter.

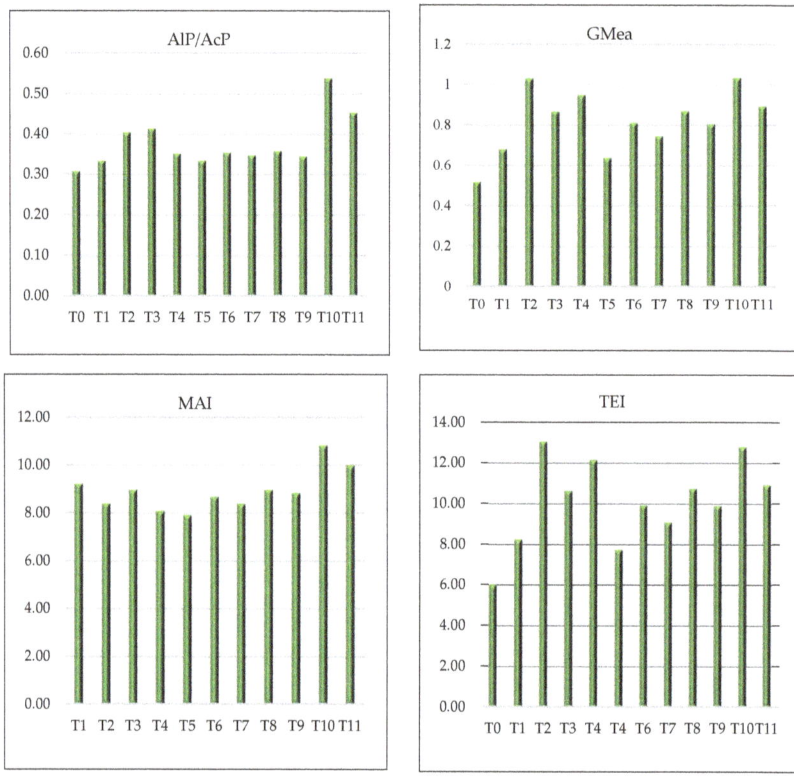

Figure 3. Index of soil enzymes: AlP/AcP, GMea, MAI, and TEI. Abbreviations: AlP/AcP—enzymatic pH index; GMea—geometric mean; TEI—total enzyme activity index; MAI—metabolic activity index. T0–T11—see Figure 1.

A significant correlation was also noted between the activity of AlP and H_2O pH (r = 0.71) and KCl pH (r = 0.75) as well as AcP and H_2O pH (r = 0.72) and KCl pH (r = 0.76). According to Dick et al. [31], phosphatases are enzymes that are most sensitive to changes in pH, which is one of the significant factors affecting the rate of chemical reactions [65]. pH affects the level of ionisation of the enzyme and substrate and changes the conditions of an enzyme–substrate complex. Zheng et al. [54] demonstrated that a change in soil pH was the key factor accounting for varied AlP and AcP activities. A significant positive correlation was recorded between the content of available P and the activity of AlP (r = 0.46) and AcP (r = 0.45). Those enzymes were the right parameter for the soils analysed in terms of the content of available phosphorus. However, the activity of AlP and AcP depended only slightly (21% and 20%, respectively) on P. Phosphatases are enzymes that play a key role in the process of the biochemical mineralization of organic phosphorus bonds [69]. To a large extent, the knowledge of the values of those two parameters should facilitate estimating the content of phosphorus available to plants, which can refer to phosphorus assayed with the Egner–Riehm method [70]. A negative significant correlation was identified between the content of CL1 and the activity of CAT (r = −0.68), BG (r = −0.57), and a positive correlation with TN (r = 0.88) and PRO (r = 0.93). We recorded a significant positive correlation between P and pH in H_2O (r = 0.56) and pH in KCl (r = 0.51). Similar dependencies were reported by Lemanowicz et al. [18]. The soil reaction affects the solubility of mineral nutrients and thus their availability to plants. As for the change in the value of soil pH below optimal for a given element, a fast decrease in yield occurs [36]. We identified a significant correlation between TOC and DEH activity (r = 0.98), CAT (r = 0.98), and BG (r = 0.85) (Figure 2). The activity of those enzymes depended on TOC (by 96%, 96%, and 72%, respectively). Dehydrogenases are respiratory enzymes that oxidise organic compounds, allocating two of their hydrogen atoms to the acceptors of electrons and producing energy [59]. These enzymes play an essential role in biological soil organic matter oxidation by transferring hydrogen from organic substrates to inorganic acceptors [60]. β-glucosidase is an enzyme catalysing the final cellulose hydrolysis stage by decomposing disaccharides, releasing glucose available to soil microorganisms [18]. A higher content of organic carbon increases the rate of mineralization by microorganisms in soil. It results in an increase in the enzymatic activity. SOC determines an increase in the concentration of substrate available in soil. An increased C content in soil can often lead to increased water retention, which can increase substrate and enzyme diffusion [61]. Also, Gautam et al. [9] found that the long-term application of organic fertilisers (16 years) enhanced the content of SOC, which is the main substrate of enzyme activity in soil.

The principal component analysis (PCA) was made to identify the potential factors affecting the parameters in soil. We transformed 17 variables into three orthogonal components, which together account for 76.37% of the total variance, two components of which, PC1 and PC2, are presented in Figure 4. The first two components account for 57.94%. The first component (PC 1), accounting for 35.33% of the variance, was most strongly negatively correlated with TOC (−0.873), TOC/TN (−0.876), DON (−0.532), DEH (−0.849), CAT (−0.937), and BG (−0.847) and positively correlated with TN (0.485), CL (0.758), and PRO (0.475). The second component (PC2), accounting for 22.61%, was most strongly negatively correlated with DOC (−0.517), and positively with H_2O pH (0.848), KCl pH (0.863), P (0.475), Mg (0.473), AlP, (0.790), and AcP (0.789). The loading values >0.75, 0.75–0.5, and 0.5–0.3 are referred to as "strong", "moderate", and "poor", respectively [71].

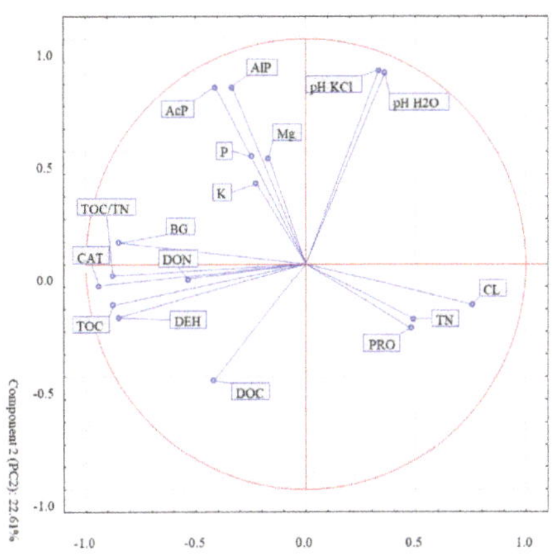

Figure 4. Projection of soil parameters on the factor plane. Abbreviations: TOC—total organic carbon, TN–nitrogen total, DOC—dissolved organic carbon, DON—dissolved organic nitrogen, and CL—labile carbon. P—available phosphorus, K—available potassium, Mg—available magnesium, AlP—alkaline phosphatase, AcP—acid phosphatase, DEH—dehydrogenases, CAT—catalase, BG—β-glucosidse, PRO—proteases, AlP/AcP—enzymatic pH index, GMea—geometric mean, TEI—total enzyme activity index, and MAI—metabolic activity index.

4. Conclusions

The natural fertilisation applied in the fertilisation variants increased the contents of organic carbon in soil. Unfortunately, that state did not determine the values of carbon management (CMI and CPI) significantly. It is evident from a low content of dissolved organic carbon fraction (DOC), as well as from the fractions undergoing oxidation (CL and CL1). The agrotechnical treatments applied in a long-term field experiment did not improve the state of soil organic matter in the fertilisation variants in terms of the DOM availability to microorganisms. It is also worth noting that the soil reaction in the fertilisation variants was slightly acidic, and the FYM + NPK + Ca and FYM + NPK + Ca + Mg treatments were the only ones for which the soil reaction corresponded to that of the control.

A regular use of integrated natural fertilisation in the form of FYM and mineral fertilisation increased the content of the available forms of nutrients.

The lack of manure decreased the activity of catalase, dehydrogenases, alkaline and acid phosphatases, β-glucosidase, and proteases in soil.

The activity of enzymes taking part in C, N, and P biogeochemistry was much higher in soil following the application of manure together with mineral fertilisers and liming as compared with fertilisation with only manure or mineral fertiliser. However, there were observed differences in the activity of the respective enzymes depending on the fertilisation applied. These resulted from their different susceptibility and resistance to environmental factors and from the soil content of substrates specific for enzymatic reactions.

The multiparametric enzymatic soil fertility indices (AlP/AcP, GMea, MAI, and TEI) differentiate the soil properties depending on the fertilisation applied. Therefore, they can be used as indicators of soil fertility. The GMea and TEI indices were mostly correlated with total organic carbon.

Author Contributions: Conceptualization, K.K.-M. and J.L.; methodology, K.K.-M., J.L. and I.J.; software, K.K.-M. and J.L.; validation, K.K.-M., J.L. and I.J.; formal analysis, K.K.-M. and J.L.; investigation, K.K.-M. and J.L.; resources, K.K.-M. and J.L.; data curation, K.K.-M. and J.L.; writing—original draft preparation, K.K.-M., J.L. and I.J.; writing—review and editing, K.K.-M., J.L. and I.J.; visualisation, J.L. All authors have read and agreed to the published version of the manuscript.

Funding: Bydgoszcz University of Science and Technology under Grant BN-WRiB-1/2022, BN-WRiB-2/2022, BN-WRiB-8/2022.

Institutional Review Board Statement: Not applicable.

Informed Consent Statement: Not applicable.

Data Availability Statement: The data presented in this study are available on request from the corresponding author.

Conflicts of Interest: The authors declare no conflicts of interest.

References

1. Gonet, S.S. Ochrona zasobów materii organicznej. [W:] Rola materii organicznej w środowisku. In *Antropogeniczne Przekształcenia Pokrywy Glebowej Brodnickiego Parku Krajobrazowego*; Gonet, S.S., Markiewicz, M., Eds.; PTSH: Wrocław, Poland, 2007; pp. 7–29.
2. Jaskulska, I.; Lemanowicz, J.; Breza-Boruta, B.; Siwik-Ziomek, A.; Radziemska, M.; Dariusz, J.; Białek, M. Chemical and biological properties of sandy loam soil in response to long-term organic–mineral fertilisation in a warm-summer humid continental climate *Agronomy* **2020**, *10*, 1610. [CrossRef]
3. Nardi, S.; Morari, F.; Berti, A.; Tosoni, M.; Giardani, L. Soil organic matter properties after 40years of different use of organic and mineral fertilizers. *Eur. J. Agron.* **2004**, *21*, 357–367. [CrossRef]
4. Mercik, S.; Stępień, M.; Stępień, W.; Sosulski, T. Dynamic of organic carbon content in soil depending on long-term fertilization and crop rotation. *Soil Sci. Ann.* **2005**, *56*, 53–60.
5. Blair, G.J.; Lefroy, R.D.B.; Lisle, L. Soil carbon fractions based on their degree of oxidation, and the development of a carbon management index for agricultural systems. *Aust. J. Agric. Res.* **1995**, *46*, 1459–1466. [CrossRef]

6. Łoginow, W.; Wiśniewski, W.; Gonet, S.S.; Ciećcińska, B. Testowa metoda oceny podatności na utlenianie materii organicznej gleb. *Zesz. Probl. Post. Nauk Rol.* **1993**, *411*, 207–212. (In Polish)
7. Gonet, S.S.; Debska, B. Dissolved organic carbon and dissolved nitrogen in soil under different fertilization treatments. *Plant Soil Environ.* **2006**, *52*, 55–63. [CrossRef]
8. Khalid, M.; Soleman, N.; Jones, D.L. Grassland plants affect dissolved organic carbon and nitrogen dynamics in soil. *Soil Biol. Biochem.* **2007**, *39*, 378–381. [CrossRef]
9. Gautam, A.; Sekaran, U.; Guzman, J.; Kovács, P.; Hernandez, J.L.G.; Kumar, S. Responses of soil microbial community structure and enzymatic activities to long-term application of mineral fertilizer and beef manure. *Environ. Sustain. Indic.* **2020**, *8*, 100073. [CrossRef]
10. Bartkowiak, A.; Lemanowicz, J.; Rydlewska, M.; Drabińska, O.; Ewert, K. Enzymatic activity of soil after applications distillery stillage. *Agriculture* **2022**, *12*, 652. [CrossRef]
11. Yang, Z.; Ha, L. Analysis and comparison of nutrient contents in different animal manures from Beijing suburbs. *Agric. Sci.* **2013**, *4*, 50–55. [CrossRef]
12. Li, C.-X.; Ma, S.-C.; Shao, Y.; Ma, S.-T.; Zhang, L.-L. Effects of long-term organic fertilization on soil microbiologic characteristics, yield and sustainable production of winter wheat. *J. Integr. Agric.* **2018**, *17*, 210–219. [CrossRef]
13. Lemanowicz, J.; Bartkowiak, A.; Zielińska, A.; Jaskulska, I.; Rydlewska, M.; Klunek, K.; Polkowska, M. The Effect of enzyme activity on carbon sequestration and the cycle of available macro- (P, K, Mg) and microelements (Zn, Cu) in Phaeozems. *Agriculture* **2023**, *13*, 172. [CrossRef]
14. Jat, H.S.; Datta, A.; Choudhary, M.; Sharma, P.C.; Dixit, B.; Jat, M.L. Soil enzymes activity: Effect of climate smart agriculture on rhizosphere and bulk soil under cereal based systems of north-west India. *Eur. J. Soil Biol.* **2021**, *103*, 103292. [CrossRef]
15. Jaskulska, I.; Lemanowicz, J.; Debska, B.; Jaskulski, D.; Breza-Boruta, B. Changes in soil organic matter and biological parameters as a result of long-term strip-till cultivation. *Agriculture* **2023**, *13*, 2188. [CrossRef]
16. Burns, R.G.; DeForest, J.L.; Marxsen, J.; Sinsabaugh, R.L.; Stromberger, M.E.; Wallenstein, M.D.; Weintraub, M.N.; Zoppini, A. Soil enzymes in a changing environment: Current knowledge and future directions. *Soil Biol. Biochem.* **2013**, *58*, 216–234. [CrossRef]
17. Li, X.G.; Jia, B.; Lv, J.; Ma, Q.; Kuzyakov, Y.; Li, F. Nitrogen fertilization decreases the decomposition of soil organic matter and plant residues in planted soils. *Soil Biol. Biochem.* **2017**, *112*, 47–55. [CrossRef]
18. Lemanowicz, J.; Haddad, S.A.; Bartkowiak, A.; Lamparski, R.; Wojewódzki, P. The role of an urban park's tree stand in shaping the enzymatic activity, glomalin content and physicochemical properties of soil. *Sci. Total Environ.* **2020**, *741*, 140446. [CrossRef]
19. Peel, M.C.; Finlayson, B.L.; McMahon, T.A. Updated world map of the Köppen-Geiger climate classification. *Hydrol. Earth Syst. Sci.* **2007**, *11*, 1633–1644. [CrossRef]
20. Marks, M.; Jastrzębska, M.; Kostrzewska, M.K. (Eds.) *Long-Term Experiments in Agricultural Studies in Poland*; Wydawnictwo Uniwersytetu Warmińsko-Mazurskiego w Olsztynie: Olsztyn, Poland, 2018; p. 280.
21. *PN-R-04023;* Chemical and Agricultural Analysis—Determination of the Content of Available Phosphorus in Mineral Soils. Polish Standards Committee: Warszawa, Poland, 1996.
22. *PN-R-04022;* Chemical and Agricultural Analysis—Determination of the Content Available Potassium in Mineral Soils. Polish Standards Committee: Warszawa, Poland, 1996.
23. Egnér, H.; Riehm, H.; Domingo, W.R. Untersuchungen uber die chemische Bodenanalyse als Grundlage fur die Beurteilung des Nährstoffzustandes der Böden. II. Chemische Extraktionsmethoden zur Phosphor-und Kaliumbestimmung. *K. Lantbrukshögskolans Ann.* **1960**, *26*, 199–215.
24. *PN-R-04020;* Chemical and Agricultural Analysis. Determination of the Content Available Magnesium. Polish Standards Committee: Warszawa, Poland, 1994.
25. Schachtschabel, P. Das pflanzenverfügbare Magnesium des Boden und seine Bestimmung. *J. Plant. Nutr. Soil Sci.* **1954**, *67*, 9–23. [CrossRef]
26. Thalmann, A. Zur methodic derestimung der Dehydrogenaseaktivität und Boden mittels Triphenyltetrazoliumchlorid (TTC). *Landwirtsch. Forsch* **1968**, *21*, 249–258.
27. Johnson, J.I.; Temple, K.L. Some variables affecting the measurements of catalase activity in soil. *Soil Sci. Soc. Am.* **1964**, *28*, 207–209. [CrossRef]
28. Tabatabai, M.A.; Bremner, J.M. Use of p–nitrophenol phosphate for assay of soil phosphatase activity. *Soil Biol. Biochem.* **1969**, *1*, 301–307. [CrossRef]
29. Eivazi, F.; Tabatabai, M.A. Glucosidases and galactosidases in soils. *Soil Biol. Biochem.* **1988**, *20*, 601–606. [CrossRef]
30. Ladd, J.N.; Butler, J.H.A. Short-term assays of soil proteolytic enzyme activities using proteins and peptide derivates as substrates. *Soil Biol. Biochem.* **1972**, *4*, 19–30. [CrossRef]
31. Dick, W.A.; Cheng, L.; Wang, P. Soil acid alkaline phosphatase activity as pH adjustment indicators. *Soil Biol. Biochem.* **2000**, *32*, 1915–1919. [CrossRef]

32. Hinojosa, M.B.; Garcia-Ruiz, R.; Viñegla, B.; Carreira, J.A. Microbiological rates and enzyme activities as indicators of functionality in soils affected by the Aznalcóllar toxic spill. *Soil Biol. Biochem.* **2004**, *36*, 1637–1644. [CrossRef]
33. Tan, X.; Xie, B.; Wang, J.; He, W.; Wang, X.; Wei, G. County-scale spatial distribution of soil enzyme activities and enzyme activity indices in agricultural land: Implications for soil quality assessment. *Sci. World J.* **2014**, *2014*, 535768. [CrossRef]
34. Picariello, E.; Baldantoni, D.; Muniategui-Lorenzo, S.; Concha-Granã, E.; De Nicola, F. A synthetic quality index to evaluate the functional stability of soil microbial communities after perturbations. *Ecol. Indic.* **2021**, *128*, 107844. [CrossRef]
35. Hammer, Ø.; Harper, D.A.; Ryan, P.D. Past: Paleontological statistics software package for education and data analysis. *Palaeontol. Electron.* **2001**, *4*, 9.
36. Guo, X.; Li, H.; Yu, H.; Li, W.; Ye, Y.; Biswas, A. Drivers of spatio-temporal changes in paddy soil pH in Jiangxi Province, China from 1980 to 2010. *Sci. Rep.* **2018**, *8*, 2702. [CrossRef] [PubMed]
37. Wu, Y.; Zeng, J.; Zhu, Q.; Zhang, Z.; Lin, X. pH is the primary determinant of the bacterial community structure in agricultural soils impacted by polycyclic aromatic hydrocarbon pollution. *Sci Rep.* **2017**, *7*, 40093. [CrossRef]
38. Minasny, B.; Hong, S.Y.; Hartemink, A.E.; Kim, Y.H.; Kang, S.S. Soil pH increase under paddy in South Korea between 2000 and 2012. *Agric. Ecosyst. Environ.* **2016**, *221*, 205–213. [CrossRef]
39. Ciric, V.; Belic, M.; Nesic, L.; Seremesic, S.; Pejic, B.; Bezdan, A.; Manojlovic, M. The sensitivity of water extractable soil organic carbon fractions to land use in three soil types. *Arch. Agron. Soil Sci.* **2016**, *62*, 1654–1664. [CrossRef]
40. Morelli, R.; Bertoldi, D.; Baldantoni, D.; Zanzotti, R. Labile, recalcitrant and stable soil organic carbon: Comparison of agronomic management in a vineyard of Trentino (Italy). *BIO Web Conf.* **2022**, *44*, 02007. [CrossRef]
41. Blair, N.; Faulkner, R.D.; Till, A.R.; Poulton, P.R. Long-term management impacts on soil C, N and physical fertility Part I: Broadbalk experiment. *Soil Tillage Res.* **2006**, *91*, 30–38. [CrossRef]
42. Huang, S.; Rui, W.; Peng, X.; Huang, Q.; Zhang, W. Organic carbon fractions affected by long-term fertilization in a subtropical paddy soil. *Nutr. Cycl. Agroecosyst.* **2010**, *86*, 153–160. [CrossRef]
43. Mando, A.; Bonzi, M.; Woperejs, M.C.S.; Lompo, F.; Stroosnijder, L. Long-term effects of mineral and organic fertilization on soil organic matter fractions and sorghum yield under Sudano-Sahelian conditions. *Soil Use Manag.* **2006**, *21*, 396–401. [CrossRef]
44. Menšík, L.; Hlisnikovský, L.; Pospíšilová, L.; Kunzowa, E. The effect of application of organic manures and mineral fertilizers on the state of soil organic matter and nutrients in the long-term field experiment. *J. Soils Sediments* **2018**, *18*, 2813–2822. [CrossRef]
45. Filipek, T.; Skowron, P. New proposals for phosphorus management in Polish agriculture. W: Scientific basis to mitigate the nutrient dispersion into the environment. *Pr. Zbior. Red. A Sapek. Falenty. Wydaw. IMUZ* **2000**, 151–159.
46. Arbačauskas, J.; Vaišvila, Z.J.; Staugaitis, G.; Žičkienė, L.; Masevičienė, A.; Šumskis, D. The influence of mineral npk fertiliser rates on potassium dynamics in soil: Data from a long-term agricultural plant fertilisation experiment. *Plants* **2023**, *12*, 3700. [CrossRef]
47. Balik, J.; Černy, I.; Kulhanek, M.; Sedlar, O.; Suran, P. Balance of potassium in two long-term field experiments with different fertilization treatments. *Plant Soil. Environ.* **2019**, *65*, 225–232. [CrossRef]
48. Sienkiewicz, S.; Krzebietke, S.; Wojnowska, T.; Żarczyński, P.; Omilian, M. Effect of long-term differentiated fertilization with farmyard manure and mineral fertilizers on the content of available forms of P, K and Mg in soil. *J. Elem.* **2009**, *14*, 779–786. [CrossRef]
49. Jabborova, D.; Sulaymanov, K.; Sayyed, R.Z.; Alotaibi, S.H.; Enakiev, Y.; Azimov, A.; Jabbarov, Z.; Ansari, M.J.; Fahad, S.; Danish, S.; et al. Mineral fertilizers improves the quality of turmeric and soil. *Sustainability* **2021**, *13*, 9437. [CrossRef]
50. Dincă, L.C.; Grenni, P.; Onet, C.; Onet, A. Fertilization and soil microbial community: A review. *Appl. Sci.* **2022**, *12*, 1198. [CrossRef]
51. Yang, W.; Zhang, X.; Wu, L.; Rensinga, C.; Xing, S. Short-term application of magnesium fertilizer affected soil microbial biomass, activity, and community structure. *J. Soil Sci. Plant Nutr.* **2021**, *21*, 675–689. [CrossRef]
52. Cowan, J.A. Structural and catalytic chemistry of magnesium-dependent enzymes. *Biometals* **2002**, *15*, 225–235. [PubMed]
53. Keeler, B.L.; Hobbie, S.E.; Kellogg, L.E. Effects of long-term nitrogen addition on microbial enzyme activity in eight forested and grassland sites: Implications for litter and soil organic matter decomposition. *Ecosystems* **2009**, *12*, 1–15. [CrossRef]
54. Zheng, M.M.; Wang, C.; Li, W.X.; Guo, L.; Cai, Z.J.; Wang, B.R.; Chen, J.; Shen, R.F. Changes of acid and alkaline phosphatase activities in long-term chemical fertilization are driven by the similar soil properties and associated microbial community composition in acidic soil. *Eur. J. Soil Sci.* **2021**, *104*, 103312. [CrossRef]
55. Wang, L.; Wang, J.; Tang, Z.; Wang, J.; Zhang, Y. Long-term organic fertilization reshapes the communities of bacteria and fungi and enhances the activities of C- and P-cycling enzymes in calcareous alluvial soil. *Appl. Soil Ecol.* **2024**, *194*, 105204. [CrossRef]
56. Choudhary, M.; Meena, V.S.; Panday, S.C.; Mondal, T.; Yadav, R.P.; Mishra, P.K.; Bisht, J.K.; Pattanayak, A. Long-term effects of organic manure and inorganic fertilization on biological soil quality indicators of soybean-wheat rotation in the Indian mid-Himalaya. *Appl. Soil Ecol.* **2021**, *157*, 103754. [CrossRef]
57. Yang, O.; Norton, J.M. Short-term nitrogen fertilization affects microbial community composition and nitrogen mineralization functions in an agricultural soil. *Appl. Environ. Microbiol.* **2020**, *86*, e02278-19. [CrossRef]

8. Bowles, T.M.; Acosta-Martínez, V.; Calderón, F.; Jackson, L.E. Soil enzyme activities, microbial communities, and carbon and nitrogen availability in organic agroecosystems across an intensively-managed agricultural landscape. *Soil Biol. Biochem.* **2014**, *68*, 252–262. [CrossRef]
9. Schloter, M.; Nannipieri, P.; Sørensen, S.J.; van Elsas, J.D. Microbial indicators for soil quality. *Biol. Fertil. Soils* **2018**, *54*, 1–10. [CrossRef]
0. Zhang, N.; He, X.; Gao, Y.; Li, Y.; Wang, H.; Ma, D.; Zhang, R.; Yang, S. Pedogenic carbonate and soil dehydrogenase activity in response to soil organic matter in artemisia ordosica Community. *Pedosphere* **2010**, *20*, 229–235. [CrossRef]
1. Datta, A.; Gujre, N.; Gupta, D.; Agnihotri, R.; Mitra, S. Application of enzymes as a diagnostic tool for soils as affected by municipal solid wastes. *J. Environ. Manag.* **2021**, *286*, 112169. [CrossRef]
2. Bastida, F.; Zsolnay, A.; Hernández, T.; García, C. Past, present and future of soil quality indices: A biological perspective. *Geoderma* **2008**, *147*, 159–171. [CrossRef]
3. Paz-Ferreiro, J.; Fu, S. Biological indices for soil quality evaluation: Perspectives and limitations. *Land Degrad. Dev.* **2013**, *27*, 14–25. [CrossRef]
4. Zhang, Y.; Wu, C.; Deng, S.; Zhang, J.; Hou, J.; Wang, C.; Fu, Z. Effect of different washing solutions on soil enzyme activity and microbial community in agricultural soil severely contaminated with cadmium. *Environ. Sci. Pollut. Res.* **2022**, *29*, 54641–54651. [CrossRef]
5. Wojewódzki, P.; Lemanowicz, J.; Debska, B.; Haddad, S.A.; Tobiasova, E. The application of biochar from waste biomass to improve soil fertility and soil enzyme activity and increase carbon sequestration. *Energies* **2023**, *16*, 380. [CrossRef]
6. Kamal, S.; Rehman, S.; Iqbal, H.M.N. Biotechnological valorization of proteases: From hyperproduction to industrial exploitation—A review. *Environ. Prog. Sustain. Energy* **2016**, *36*, 511–522. [CrossRef]
7. Błońska, E.; Lasota, J.; Zwydak, M. The relationship between soil properties, enzyme activity and land use. *Forest Res. Papers* **2017**, *78*, 39–44. [CrossRef]
8. Cincotta, M.M.; Perdrial, J.N.; Shavitz, A.; Libenson, A.; Landsman-Gerjoi, M.; Perdrial, N.; Armfield, J.; Adler, T.; Shanley, J.B. Soil aggregates as a source of dissolved organic carbon to streams: An experimental study on the effect of solution chemistry on water extractable carbon. *Front. Environ. Sci.* **2019**, *7*, 172. [CrossRef]
9. Nannipieri, P.; Giagnoni, L.; Landi, L.; Renella, G. Role of phosphatase enzymes in soil. In *Phosphorus in Action*; Bünemann, E.K., Oberson, A., Frossard, E., Eds.; Series: Soil Biology; Springer: Berlin/Heidelberg, Germany, 2011; Volume 26, pp. 215–243. [CrossRef]
0. Lemanowicz, J.; Bartkowiak, A.; Lamparski, R.; Wojewódzki, P.; Pobereżny, J.; Wszelaczyńska, E.; Szczepanek, M. Physicochemical and enzymatic soil properties influenced by cropping of primary wheat under organic and conventional farming systems. *Agronomy* **2020**, *10*, 1652. [CrossRef]
1. Liu, C.W.; Lin, K.H.; Kuo, Y.M. Application of factor analysis in the assessment of groundwater quality in a blackfoot disease area in Taiwan. *Sci. Total Environ.* **2003**, *313*, 77–89. [CrossRef]

sustainability

MDPI

Article

Tillage System as a Practice Affecting the Quality of Soils and Its Sustainable Management

Joanna Lemanowicz [1], Erika Balontayová [2], Bożena Dębska [1,*], Agata Bartkowiak [1] and Piotr Wasilewski [3]

[1] Department of Biogeochemistry, Soil Science and Irrigation and Drainage, Bydgoszcz University of Science and Technology in Bydgoszcz, 6/8 Bernardyńska Street, 85-029 Bydgoszcz, Poland; joanna.lemanowicz@pbs.edu.pl (J.L.); agata.bartkowiak@pbs.edu.pl (A.B.)

[2] Institute of Agronomic Sciences, Slovak University of Agriculture in Nitra, Tr. A. Hlinku 2 St., 949 76 Nitra, Slovakia

[3] Department of Agronomy and Food Processing, Bydgoszcz University of Science and Technology in Bydgoszcz, 7 Prof. S. Kaliskiego Street, 85-796 Bydgoszcz, Poland; piotr.wasilewski@pbs.edu.pl

* Correspondence: bozena.debska@pbs.edu.pl

Abstract: Sustainable soil management through the use of an appropriate tillage system can positively change the edaphic parameters. The aim of the present study was to compare the effects that reduced tillage (RT) and conventional tillage (CT) systems have on changes in selected physical and chemical properties and enzymatic activity in various soil types. The study included the following soil types: *Eutric Fluvisol*, *Mollic Fluvisol*, *Haplic Chernozem*, *Haplic Luvisol*, *Eutric Regosol*, *Eutric Gleysol*, and *Stagnic Planosol*. Soil samples were collected in the Danubian Lowland and Eastern Slovak Lowland. The following parameters were determined in the soil samples: soil texture, pH, hydrolytic acidity and the sum of basic exchangeable cations, the contents of carbon (TOC), nitrogen (TN), and dissolved organic carbon (DOC), and the activities of dehydrogenases (DEH), catalase (CAT), peroxidases (PER), alkaline phosphatase (AlP), acid phosphatase (AcP), proteases, and β-glucosidase (BG). The reaction of the analysed soils, in the RT and CT cultivations alike, ranged from acidic to neutral, and the sorption properties differed between individual soil types. The TOC ranged from 16.53 to 42.07 g kg^{-1} for conventional cultivation and from 15.51 to 38.90 g kg^{-1} for reduced tillage. The values of enzymatic soil quality indices values correlated with TOC, DOC, and TN, as well as with pH, the sum of exchangeable base cations, cation exchange capacity, and degree of base saturation of the sorption complex. The tillage system determined changes in the activity of the studied enzymes, but the intensity and direction of these changes depended on the soil type. Based on the enzyme activity results, soil quality indices such as GMea and TEI were calculated. TEI proved to be a more sensitive indicator than GMea. It was shown that, of all studied soil types and regardless of the cultivation system, *Eutric Gleyosols* had the most variable properties. For conventional tillage, Haplic *Luvisol* and *Eutric Regosol* were characterised by the greatest uniformity. In general, the edaphic properties of soils under conventional tillage differed from those of soils under simplified tillage.

Keywords: enzyme activity; organic matter; soil types; soil quality indicators; sorption properties

1. Introduction

The production of food in sufficient quantities requires intensive agriculture. The cultivation practices used in intensive agriculture cause the deterioration of the physical, chemical, and biological parameters of soils, thus increasing the rate of mineralisation of organic matter [1,2]. As follows from the main policy objectives of the European Union, the management of an agricultural production space should involve taking actions aimed at protecting soil organic matter and its biodiversity [3].

Agricultural practices can modify the soil environment and provide good conditions for the growth and development of crops. At the same time, they influence nutrient cycling and soil biological activity [4–6]. Therefore, there is a demand for agricultural systems that will not increase soil degradation, while still producing an adequate quantity of agricultural products [7]. A tillage system entails processes that change the edaphic parameters of soil [8,9]. Inappropriate agricultural practices degrade soil quality by, among other things, disturbing the circulation of elements, acidification, and reducing the level of biological activity, including enzymatic activity [10,11]. Excessive mineral fertilisation often brings about far-reaching changes in the soil sorption complex, which may have a negative impact on crop yields and quality [12]. Introducing various tillage systems may also bring about changes in soil pH. Agrotechnical treatments that help accelerate the mineralisation of organic matter and fertilisation with nitrogen fertilisers may decrease pH [13]. According to Aye et al. [14] and Debska et al. [15], long-term liming may lead to a reduction in the TOC content. However, the authors showed that TOC losses caused by liming can be compensated by introducing exogenous organic matter (EOM) into the soil. As reported by, among others, Mensik et al. [16], fertilisation with NPK alone causes a drop in the content and quality of OM (lower contents of TOC and humic substances, and dominance of FA over HA). Therefore, the goal of proper soil tillage is to develop optimal conditions for the growth and development of crops and, consequently, to increase yields.

In recent years, within the concept of sustainable agricultural development, "conservation tillage" has become more popular [17,18]. Conservation tillage excludes tillage operations that turn the soil and bury crop residues. It involves reducing the depth of tillage occasionally or constantly. It recommends using shallower tillage using other tools and/or reducing the intensity of seedbed preparation. Conservation tillage is a broad concept which includes operations such as no tillage, reduced tillage, minimum tillage, or mulching. Some researchers have emphasised that [19] leaving crop residues on the soil surface also reduces evaporation, improves infiltration, and inhibits weed growth. Debska et al. [15] drew attention to the beneficial effect of mulching on TOC content.

Depending on the method applied, tillage can positively or negatively affect soil properties, including soil organic matter [20]. Reduced tillage is probably the most researched agricultural practice [21]. A meta-analysis conducted by Allam et al. [22] has shown that a reduced tillage system using only organic fertilisation can increase the amount of legume grains. However, the combination of organic and inorganic fertilisation was shown to have a positive effect on increasing cereal crop yields. Abandoning plough tillage and introducing reduced tillage causes organic carbon and nutrients to accumulate in the topsoil [23]. According to Debska et al. [23], the use of ploughless tillage and strip-tilling significantly reduces the leaching of C and N from the surface layer to the 30–50-cm layer. As reported in other literature [24,25], reduced tillage has only a minor direct effect on carbon sequestration. Those authors drew attention to the impact of this tillage method by emphasising the need to select appropriate plants in a rotation and in the number of rotations. Han et al. [26], Yue et al. [27], and Armas-Herrera et al. [28] also emphasised the importance of organic fertilisation and post-harvest residues in shaping TOC resources in the soil. According to Balesdent et al. [11], the cessation of ploughing causes changes in soil functioning, i.e., in the availability of nutrients, biodiversity, or soil water-retention capacity. Those authors emphasised that modelling and forecasting TOC sources in soils requires knowledge of the impact that various agricultural practices have in different soils and under different climatic conditions.

Szostek et al. [29] showed that simplifications in soil cultivation contribute to improving soil parameters and enzyme activity through the application of organic fertilisers and limiting the impact on the topsoil. Similarly, Blanco-Canqui and Ruis [30] found that improving soil structure and reducing soil compaction through less intense tillage prevent organic matter and essential macro- and microelements from being lost from a soil ecosystem. However, the reduction in cultivation intensity depends on both biotic and abiotic factors, such as soil type and texture, the tillage system used, and the hydro-thermal conditions [31]. According to Derpsch et al. [32], reducing or completely abandoning soil tillage is currently a popular approach to sustainable agriculture. As reported by Cenini et al. [33] and Rahmati et al. [34], extracellular enzymes produced by soil microorganisms catalyse chemical reactions associated with the decomposition of soil organic carbon. Therefore, an evaluation of soil enzymatic responses to tillage systems may suggest mechanisms for differences in carbon fraction characteristics.

According to Tobiašová et al. [35], each soil type has its own specific profile resulting from its genesis, the place where it is formed, and the method of soil management. Natural classifications divide soils according to genetic criteria, i.e., according to the way a given soil is formed and the soil-forming processes that create it. Soil type is the basic classification unit that categorises soils according to shared configurations of horizons through a soil profile. Soils of such types share similar physical, chemical, and biological properties, as well as a similar type of humus and degree of nutrient richness. According to the IUSS Working Group WRB, [36] soils of various groups differ not only in the structure of the soil profile but also in their properties. Many studies have shown that a soil's biological activity is determined primarily by its physicochemical properties [37]. It is important to understand how soils that differ in properties variously affect the soil parameters that are shaped by various tillage systems [35,38].

Based on the literature on the subject, it can be hypothesised that reduced tillage has a beneficial effect on endogenous soil parameters and increases carbon sequestration. The aim of our study was to assess how soil type determines the impact that two cultivation systems (reduced and conventional tillage) have on soil fertility by assessing:

- the contents of organic carbon, nitrogen, and soluble organic matter;
- sorption properties;
- the activities of selected enzymes involved in the biogeochemistry of C, N, and P in soil.

2. Materials and Methods

2.1. Materials

The localities included in the study are situated in the Danubian Lowland (Nové Zámky, Šal'a, Vráble, Piešťany) and Eastern Slovak Lowland (Trebišov, Michalovce, Sobrance). All fields represent real farms with their management systems, including tillage systems. The geological substrates of the lowlands are Neogene marine sediments clays, sands, and gravels, which are covered with loess and loess loam in some areas. Fluvial sediments are found along the Vah and Laborec rivers. The relief is monotonous, mostly wavy, and covered with loess and loess loam. In some places, Neogene rafts of clays, sands, or gravels are found above the surface [35]. The study areas are located in "slightly warm" to "warm" climatic regions. The average altitude is 100–160 m above sea level, with a range of average temperature of 9.0–10.5 °C and average annual rainfall of 565–610 mm [39].

The study included seven soil types, i.e., *Eutric Fluvisol, Mollic Fluvisol, Haplic Chernozem, Haplic Luvisol, Eutric Regosol, Eutric Gleysol,* and *Stagnic Planosol* [36], each in three repetitions (various crop rotations) of agricultural land in Slovakia. Soil samples were taken in spring 2023. Cereals were the main crops in all of the fields. The organic carbon balance, according to method Jurčová and Bielek [40] used for arable land in SR, fluctuated in a range of 7–16 t ha^{-1} for the last 8 years. Soil samples were collected from a depth of 0–30 cm under simplified tillage (only discing was used; RT) and conventional cultivation (CT). Detailed information on soil types is provided in Table 1

Table 1. Specifics of soil types in relation to their genesis and general properties.

EF *	Shallow humus horizon; huge heterogeneity of soil substrate; natural vegetation in flooded forests; higher underground water before water flow regulation; the composition of the microbial community is strongly influenced by anthropogenic cultivation; the initial process of organic carbon accumulation is dominant
MF	Deep humus horizon; specific hydromorphic regime; rich in a clay content and organic substances of high quality with dominance of humic acids; originally humus accumulation under hydrophilic vegetation with humification as a dominant process; high abundance and diversity of soil organisms
HC	Deep humus horizon with intensive humification; originally rich grass vegetation supported the creation and accumulation of humus substances of high quality; the presence of areas with a longer period of dryness results in the dominance of organisms that adapted to these conditions; ammonium can briefly release and nitrates can accumulate
HL	Shallow humus horizon mixed with genetic horizon in the arable land; strong risk of soil erosion; accumulation of clay in genetic illuvial luvic horizon with dominant process of illimerisation, which results in soil compaction and worse soil structure; biological activity is relatively high with dominance of bacteria
ER	Shallow humus horizon influenced by soil erosion; unsuitable textural composition, usually coarse soils; high aeration and dominance of non-capillary pores; small amount of organic sources for heterotrophic microbial community; high intensity of oxidation processes; risk of nutrient leaching
EG	Unsuitable textural composition with clay dominance; high level of underground water and poor aeration; reduction processes are dominant; accumulation of organic sources is high but with dominant production of low molecular substances and creation of fulvic acids; usually soils with a lower pH and unsuitable soil structure
SP	Soil of the areas with the accumulation of water with unsuitable hydromorphic regimes; frequent fluctuation of water level, changes in oxidation, and reduction conditions in the genetic horizon, mobilisation of Fe, Al, Mn; low diversity of microbial community; special hydrogenetic development of humus-forming process with FA dominance

* EF—Eutric Fluvisol; MF—Mollic Fluvisol; HC—Haplic Chernozem; HL—Haplic Luvisol; ER—Eutric Regosol; EG—Eutric Gleysol; SP—Stagnic Planosol.

2.2. Methods

2.2.1. Physicochemical Properties of Soil

The physical and chemical properties of the soil material were determined as follows:

- Soil texture was determined according to the pipette method [41]. Soil was treated to remove carbonates and organic matter. After dissolution of $CaCO_3$ with 2 M HCl dm^{-3} and oxidation of the organic carbon with 30% H_2O_2, there followed repeated washing, and, finally, the samples were dispersed using $Na(PO_3)_6$. Silt, sand, and clay fractions were determined. The fraction content is expressed in %.
- pH was determined in 1M KCl potentiometrically [42];
- Hydrolytic acidity (Hh) and total exchangeable base cations (TEB) were assessed by the Kappen method [43];

- Cation exchange capacity (CEC) was calculated based on TEB and Hh, and the degree of base saturation (BS) of the sorption complex was calculated from CEC and TEB.

2.2.2. Content of Organic Carbon, Total Nitrogen, and Dissolved Organic Matter

In the air-dried soil samples, the contents of total organic carbon (TOC) and total nitrogen (TN) were determined with a Vario Max CN analyser provided by Elementar (Langenselbold, Germany). In the soil samples, dissolved organic carbon (DOC) and dissolved nitrogen (DTN) were assayed in extracts of 0.004 M $CaCl_2$ (dissolved organic matter, DOM) using a Multi N/C 3100 analyser (Analytik Jena, Jena, Germany). The detailed DOM extraction method was described in a previous work Debska et al. [23].

2.2.3. Activity of Enzymes

The activities of oxidoreductive enzymes (dehydrogenases, catalase, and peroxidases) and hydrolytic enzymes (alkaline and acid phosphatase, β-glucosidase, and proteases) were measured on fresh-sieved (<2 mm) soils.

- The activity of dehydrogenases (DEH) was determined by the Thalmann [44] method after incubating the sample with 2,3,5-triphenyltetrazolium chloride and measuring the absorbance of triphenylformazan (TPF) at 546 nm; results are expressed in mg TPF kg^{-1} 24 h^{-1}.
- Catalase (CAT) was determined by the method of Johnson and Temple [45] using 0.3% hydrogen peroxide solution as the substrate. The remaining H_2O_2 was determined by titration with 0.02 M $KMnO_4$ under acidic conditions.
- The activity of peroxidases (PER) was determined by the method of Barth and Bordeleau [46] by measuring the amount of purpurogallin (PPG) formed by the oxidation of pyrogallol in the presence of H_2O_2.
- The activities of alkaline phosphatase (AlP) and acid phosphatase (AcP) were determined based on the detection of p-nitrophenol (pNP) released after incubation (37 °C, 1 h) at pH ~6.5 for acid phosphatase and pH ~11.0 for alkaline [47].
- β-glucosidase (BG) was measured by the method of Eivazi and Tabatabai [48], using p-nitrophenyl-β-D-glucopyranoside as a substrate. Concentrations of p-Nitrophenol were determined by direct reading of the sample at 400 nm after alkalisation with Tris/NaOH buffer (pH 10.0) and $CaCl_2$.
- The activity of proteases was determined using the method of Ladd and Butler [49], where the concentration of the amino acid tyrosine (Tyr) was determined in soil samples after incubation with sodium caseinate. Absorbance was measured with a spectrophotometer at $\lambda = 680$.

Based on the obtained enzyme activity results, enzymatic soil quality indices were calculated as follows:

- the geometric mean of enzyme activities (GMea) [50]:

$$GMea = \sqrt[7]{DEH \times CAT \times PER \times AlP \times AcP \times BG \times PRO} \quad (1)$$

- total enzyme activity index (TEI) [51]:

$$TEI = \sum \frac{Xi}{\overline{\overline{Xi}}} \quad (2)$$

2.3. *Statistical Analyses*

The research results were statistically analysed in the Statistica software. The normality of distributions of observed features was verified using the Shapiro–Wilk normality test [52] and expressed as the arithmetic mean ±standard deviation (SD). The obtained data were analysed using one-way ANOVA with the tillage system (reduced and conventional tillage) as the factor. Prior to the analysis of variance, in the distribution of each variable assuming hypothesis H0, the variables showing a normal distribution were investigated. The evaluation was made using the Shapiro–Wilk test. This test is based on the so-called studentised range, which makes it possible to group means and allows for comparisons of all possible pairs of groups in a way that controls the risk of making a type I error (falsely rejecting the null hypothesis). In the case that the tillage system was shown to have a statistically significant effect on the intensity of the study parameters, the mean values for tillage system were compared using the Tukey HSD test for $p = 0.05$ and, on this basis, homogeneous groups were designated and marked with the same letters in the tables below.

The paper also presents correlation coefficients between granulometric composition, pH, selected sorption properties, TOC and TN content, DOC and DTN content, and enzyme activity (DEH, CAT, PER, AlP, AcP, BG and PRO), which were calculated using the PAST 4.13 program [53]. A multivariate statistical data analysis also involved cluster analysis (CA). This analysis can include groups that share similar characteristics to one another and, at the same time, possess different from the elements from those in other groups. In a given group, the smaller the Euclidean distance, the more similar the objects. Data clustering was performed with the Ward method [54]. The analysis was performed after data standardisation.

3. Results and Discussion

3.1. Selected Physico-Chemical Properties

Granulometric composition analysis showed that all soil samples shared a similar granulometric composition. Among all soil fractions, the silt fraction dominated (Table 2). The reaction of the analysed soils under RT and CT tillage alike ranged from acidic to neutral (Table 2). The pH values of soils are mainly related to their mineralogical composition (the acidic or alkaline nature of parent rocks), transformations and content of organic matter, and climatic conditions that determine the leaching of components. However, based on studies [55], it was found that the method of cultivation significantly influences the physical and chemical condition of the soil environment. Among the many scientific studies devoted to changes in soil chemical properties affected by tillage, one may find results that indicate both increases and decreases in pH as a result of the introduction of various cultivation systems. However, it is most often assumed that soils in RT systems become slightly acidic [13,56,57]. In our own studies, a decrease in pH was observed only in some soils and only under the RT system.

Table 2. Texture and pH soil.

Soil Types	Sand%		Silt%		Clay%		pH KCl	
	Tillage Systems							
	RT	CT	RT	CT	RT	CT	RT	CT
EF	2.33 ± 0.07	7.33 ± 0.07	64.27 ± 1.41	62.04 ± 0.47	33.4 ± 0.41	30.66 ± 0.06	4.6 ± 0.14	6.29 ± 0.12
MF	19.70 ± 0.37	14.90 ± 0.33	52.44 ± 0.04	45.74 ± 0.86	27.86 ± 0.76	39.36 ± 0.10	7.16 ± 0.21	6.99 ± 0.12
HC	14.54 ± 0.16	13.77 ± 0.15	61.26 ± 0.01	59.29 ± 0.19	24.20 ± 0.70	26.94 ± 0.05	5.57 ± 0.18	6.24 ± 0.05
HL	27.07 ± 0.34	9.75 ± 0.30	51.47 ± 0.05	66.07 ± 0.46	21.46 ± 0.41	24.18 ± 0.10	6.91 ± 0.10	5.27 ± 0.10
ER	10.48 ± 0.76	9.43 ± 0.67	65.40 ± 0.1	61.92 ± 0.69	24.12 ± 0.55	28.65 ± 0.12	7.02 ± 0.19	5.57 ± 0.14
EG	7.17 ± 0.08	16.58 ± 0.08	58.05 ± 0.02	45.30 ± 0.29	4.78 ± 0.23	38.12 ± 0.07	5.55 ± 0.06	5.70 ± 0.10
SP	19.04 ± 0.21	10.45 ± 1.09	64.96 ± 0.11	63.20 ± 0.25	16.00 ± 0.69	26.35 ± 0.46	4.67 ± 0.08	6.05 ± 0.09

EF—*Eutric Fluvisol*; MF—*Mollic Fluvisol*; HC—*Haplic Chernozem*; HL—*Haplic Luvisol*; ER—*Eutric Regosol*; EG—*Eutric* Gleysol; SP—*Stagnic Planosol*; RT—reduced tillage; CT—conventional tillage; ± Standard Deviation.

The sorption properties of the tested soils varied between soil types (Table 3). The hydrolytic acidity of the tested soils ranged from 1.3 to 3.9 cmol(+) kg^{-1} for RT and from 0.37 to 2.25 cmol(+) kg^{-1} for CT. Statistical analysis showed significantly higher values of the discussed parameter in soils of types EF, MF, HC, EG, and SP under RT. The degree of saturation of the sorption complex with basic cations ranged from 1.6 to 5.4 cmol(+) kg^{-1} in soils where RT was used and from 2.50 to 4.80 cmol(+) kg^{-1} in CT soils. The value of this parameter was highest in *Molic Fluvisol* under both RT and CT. *Molic Fluvisol* also showed the highest CEC (5.70 cmol(+) kg^{-1}) and BS (93.59%) values for RT. Statistical analysis showed that the Hh, TEB, and CEC results for *Eutric Fluvisol* and *Mollic Fluvisol* were significantly higher under reduced tillage than under conventional tillage. The sorption capacity of the soil, which includes all the parameters discussed, is considered to be one of the most important factors influencing soil fertility. This soil characteristic is related to the structure of the sorption complex. The quantity and quality of humic compounds in combination with mineral colloids determines the sorption capacity and the composition of adsorbed exchangeable cations [58–60]. A correlation analysis showed highly significant relationships between total organic carbon and TEB (r = 0.780), CEC (r = 0.580), and BS (r = 0.500) (Figure 1). As pH increases, the amount of variable negative charge in organic soil increases and CEC also increases [60,61]. In addition to organic matter content, soil pH also influences cation exchange capacity. In general, acidic soils (low pH) reduce cation exchange capacity, while alkaline soils (high pH) can increase cation exchange capacity. The correlation analysis confirmed this relationship (Figure 1). The literature shows that in arable soils, the soil sorption capacity can also be changed by different soil management practices [62–64]. In our own research, statistical analysis showed that in most of the analysed soils, reduced tillage significantly increased the values of hydrolytic acidity and cation exchange capacity.

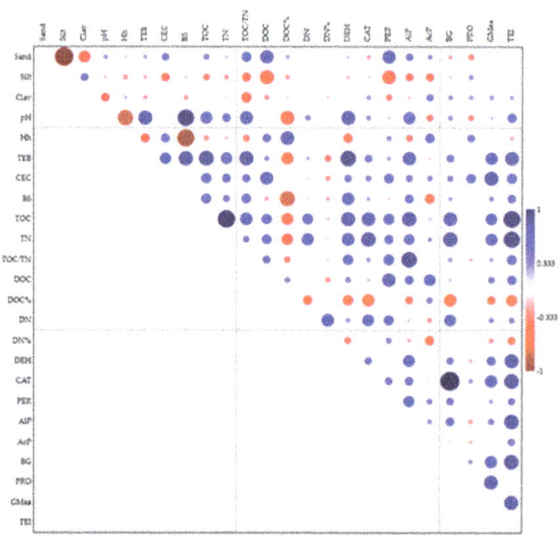

Figure 1. Correlogram of the physicochemical parameters and the activity of enzymes in soil for reduced tillage (RT) and conventional tillage CT. Abbreviations: Hh—hydrolytic acidity; TEB—total exchangeable bases; CEC—cation exchange capacity; BS—base saturation; TOC—total organic carbon, TN—total nitrogen, DOC—dissolved organic carbon; DEH—dehydrogenases, CAT—catalase, PER—peroxidases; AlP—alkaline phosphatase, AcP—acid phosphatase, BG—β-glucosidase; PRO—protease, GMea—the geometric mean of enzyme activities; TEI—total enzyme activity index.

Table 3. Sorption properties of soils.

Soil * Types	Hh (cmol(+) kg^{-1})		TEB (cmol(+) kg^{-1})		CEC (cmol(+) kg^{-1})		BS (%)	
	Tillage Systems							
	RT	CT	RT	CT	RT	CT	RT	CT
EF	3.90 a ± 0.8	0.75 b ± 0.3	3.30 a ± 0.6	2.50 b ± 1.4	7.20 a ± 2.2	3.25 b ± 0.8	45.83 b ± 2.3	76.92 a ± 3.2
MF	1.30 a ± 0.2	0.37 b ± 0.1	5.40 a ± 0.5	4.80 b ± 2.2	6.70 a ± 1.1	5.17 b ± 1.1	80.60 b ± 3.1	92.84 a ± 3.0
HC	2.92 a ± 0.8	0.90 b ± 0.2	2.40 b ± 1.3	3.40 a ± 1.3	5.32 a ± 2.1	4.30 b ± 1.0	45.11 b ± 1.8	79.07 a ± 2.4
HL	1.45 b ± 0.1	2.25 a ± 0.8	4.30 a ± 4.1	2.90 b ± 1.1	5.75 a ± 2.2	5.15 a ± 1.2	74.78 a ± 2.6	56.31 b ± 2.0
ER	1.37 b ± 0.0	1.87 a ± 0.4	3.60 a ± 2.5	3.30 a ± 0.8	4.97 b ± 1.4	5.17 a ± 1.1	72.43 a ± 3.4	63.83 b ± 1.5
EG	1.87 a ± 0.5	1.87 a ± 0.4	1.90 b ± 1.5	2.90 a ± 15.1	3.77 b ± 1.1	4.67 a ± 1.3	50.40 b ± 2.8	59.96 a ± 4.1
SP	2.85 a ± 0.5	0.98 b ± 0.1	1.60 b ± 0.15	3.20 a ± 0.3	4.45 a ± 1.1	4.18 a ± 1.1	35.96 b ± 1.5	76.56 a ± 1.4

Hh—hydrolityc acidity; TEB—total exchangeable bases; CEC—cation exchange capacity; BS—basic saturation; EF—*Eutric Fluvisol*; MF—*Mollic Fluvisol*; HC—*Haplic Chernozem*; HL—*Haplic Luvisol*; ER—*Eutric Regosol*; EG—*Eutric* Gleysol; SP—*Stagnic Planosol*; RT—reduced tillage; CT—conventional tillage; small letters (a, b) indicate a differences on between the soil tillage systems at $p = 0.05$; ± Standard Deviation.

3.2. Content of Organic Carbon, Total Nitrogen and Dissolved Organic Matter

The TOC content (Table 4) in soil samples collected from fields where conventional tillage was used ranged from 16.53 (HL) to 42.07 g kg^{-1} (MF). For soil samples from reduced tillage, the TOC content ranged from 15.51 (SP) to 38.90 g kg^{-1} (HF). For EF, HC, HL, and ER, the TOC content was higher in samples from reduced tillage than from conventional tillage; the inverse relationship was noted for MF, EG, and SP. The difference in TOC content in favour of reduced tillage was highest for *Haplic Luvisols* (92% increase) and lowest for *Eutric Fluvisols* (34% increase). In reduced tillage soils with lower TOC content, the differences in carbon content between cultivation methods were 7.5% (MF), 26.04% (SP), and 36.5% (EG). According to the results of Francaviglia et al. [18], the use of reduced tillage significantly promotes carbon sequestration. Studies conducted by Friedrich et al. [17], Busari et al. [20], Laufer et al. [65], and Powlson et al. [66], among others, showed that reduced tillage systems can lead to carbon sequestration, but it should be remembered that the degree of change in TOC content may vary by soil type (soil properties), crop rotation, and the amount of post-harvest residues left. Kumar et al. [67] believe that, compared to intensive systems, no-till cultivation preserves post-harvest residues on the field surface, thereby increasing or maintaining soil organic carbon resources at a certain level. The relationships obtained in the cited work allow us to conclude that the determinants of the effects of RT may include the physicochemical parameters of the soil.

Table 4. Content of total organic carbon and total nitrogen.

Soil * Types	TOC (g·kg^{-1})		TN (g·kg^{-1});		TOC/TN	
	Tillage Systems					
	RT *	CT	RT	CT	RT	CT
EF *	25.06 a ± 0.35	18.73 b ± 0.22	2.49 a ± 0.05	1.86 b ± 0.04	10.13 a	10.10 a
MF	38.90 b ± 0.50	42.07 a ± 0.55	2.86 b ± 0.06	3.73 a ± 0.08	13.64 a	11.30 b
HC	27.63 a± 0.22	18.13 b ± 0.30	2.87 a ± 0.03	1.81 b ± 0.02	9.63 a	10.02 a
HL	31.67 a ± 0.32	16.53 b ± 0.20	2.51 a ± 0.03	1.57 b ± 0.04	12.71 a	10.53 b
ER	24.20 a ± 0.25	17.75 b ± 0.35	2.12 a ± 0.04	1.91 a ± 0.03	11.41 a	9.32 b
EG	18.63 b ± 0.35	29.36 a ± 0.35	1.84 b ± 0.03	2.89 a ± 0.04	10.13 a	10.17 a
SP	15.51 b ± 0.20	20.97 a ± 0.30	1.49 b ± 0.03	1.92 a ± 0.03	10.41 a	10.95 a

TOC—total organic carbon; TN—total nitrogen; *—see Table 1; a, b—see Table 3.

The TN content for soils ranged from 1.57 (HL) to 3.73 g kg^{-1} (MF) under conventional tillage and, for reduced tillage, from 1.49 to 2.86 (MF) and 2.87 g kg^{-1} (HC). For MF, EG, and SP soils, the nitrogen content was higher under conventional cultivation. The MF,

EG, and SP soils under reduced tillage were characterised by, respectively, 23.4, 36.22, and 22.19% lower TN content compared to conventional tillage (Table 4). In EF, HC, and HL soils, the TN content was, respectively, 34.0, 58.6, and 59.6% higher compared to conventional cultivation. The values of TOC and TN contents in the soil were used to calculate the TOC/TN ratio values (Table 4). The values of this ratio under conventional tillage ranged from 9.32 (ER) to 11.30 (MF), and under simplified tillage ranged from 9.63 (HC) to 13.64 (MF). Only for MF, HL, and ER was a significant difference noted between conventional and reduced tillage, with lower values of this ratio being recorded.

One of the components of organic matter that is very sensitive to agrotechnical treatments is dissolved organic matter (DOM) [68]. DOM content is determined on the basis of dissolved organic carbon content (DOC)—and sometimes on the basis of dissolved nitrogen content (DTN). The DOC content in soils ranged from 141.9 (EF) to 275.1 mg kg^{-1} (EG) under conventional tillage and from 141.2 (EG) to 294.9 mg kg^{-1} (HC) under reduced tillage. In general, soils under reduced tillage (except EG and SP) were characterised by higher DOC contents (Table 5). It should be emphasised that soils under reduced tillage were also characterised by a higher share of DOC (except for HL and EG) in comparison to soils under conventional tillage. The share of DOC ranged from 0.42 (MF) to 1.03% (SP) for conventional tillage and 0.72 (MF) to 1.41% (SP) for reduced tillage. Similarly, Wright et al. [69] and Liu et al. [70] obtained higher DOM contents in soils under reduced tillage compared to conventional tillage. Leinweber et al. [71], however, indicated that ploughing may stimulate the microbiological decomposition of post-harvest residues, which may lead to increased DOM.

Table 5. Content of dissolved organic carbon and dissolved nitrogen.

Soil * Types	DOC (mg kg^{-1})		DOC (%)		DTN (mg kg^{-1})		DTN(%)	
	Tillage Systems							
	RT *	CT	RT	CT	RT	CT	RT	CT
EF *	271.3 a ± 10.5	141.9 b ± 5.5	1.08 a ± 0.05	0.76 b ± 0.03	54.4 a ± 3.0	47.2 b ± 2.5	2.19 a ± 0.05	2.54 a ± 0.07
MF	278.4 a ± 9.3	177.9 b ± 8.5	0.72 a ± 0.03	0.42 b ± 0.01	43.7 b ± 2.1	85.5 a ± 8.8	1.53 b ± 0.03	2.30 a ± 0.09
HC	294.9 a ± 8.5	150.6 b ± 5.8	1.07 a ± 0.06	0.83 b ± 0.05	66.7 a ± 3.3	43.7 b ± 2.5	2.32 a ± 0.03	2.41 a ± 0.11
HL	280.1 a ± 10.2	172.5 b ± 8.5	0.88 b ± 0.05	1.04 a ± 0.07	45.2 a ± 1.6	22.7 b ± 1.8	1.80 a ± 0.05	1.45 a ± 0.08
ER	232.6 a ± 5.5	169.7 b ± 8.5	0.96 a ± 0.05	0.96 a ± 0.08	59.5 a ± 3.8	24.2 b ± 2.0	2.81 a ± 0.08	1.27 b ± 0.05
EG	141.2 b ± 7.5	275.1 a ± 11.0	0.76 b ± 0.04	0.94 a ± 0.08	54.0 a ± 3.0	32.7 b ± 2.2	2.93 a ± 0.10	1.13 b ± 0.04
SP	218.0 a ± 10.8	215.1 a ± 8.9	1.41 a ± 0.09	1.03 b ± 0.07	41.3 a ± 2.5	19.9 b ± 1.9	2.77 a ± 0.09	1.04 b ± 0.05

DOC—dissolved organic carbon; DTN—dissolved nitrogen; *—see Table 1; a, b—see Table 3.

The content and share of DNT was significantly higher in soils under reduced tillage (except MF) compared to conventional tillage. DTN ranged from 1.04 (SP) to 2.54% (EF) under conventional tillage and from 1.53 (MF) to 2.93% (EG) under reduced tillage.

3.3. The Activity of Enzymes in Soil

In order to determine how tillage systems in different soil types might affect levels of enzymatic activity, the activity of selected redox enzymes, i.e., dehydrogenases (DEH), catalase (CAT), and peroxidase (PER) (Table 6), and hydrolytic enzymes, i.e., alkaline (AlP) and acidphosphatase (AcP), β-glucosidase (BG), and proteases (PR), was determined (Table 7). DEH activity was significantly higher under reduced tillage in EF soils (3.42 mg TPF kg^{-1} 24 h^{-1}), HL soils (7.23 mg TPF kg^{-1} 24 h^{-1}), and ER soils (9.73 mg TPF kg^{-1} 24 h^{-1}) as compared to conventional tillage CT (Table 5). DEH activity was 2%, 33%, and 17% higher, respectively. The highest activity of the tested oxidoreductive enzymes (excluding PER under CT) was obtained in MF. *Mollic Fluvisols* are a soil composed of fluvial mud and are characterised by high content and quality of organic matter [72]. According to

Furtak et al. [73], *Fluvisols* are among the most fertile soils. There were no statistically significant differences in DEH between MF and HC, nor in CAT among MF, HL, ER, EG, and SP. Significantly, the highest CAT activity in the RT system was obtained in EF (1.76 mg H_2O_2 $kg^{-1}h^{-1}$) and in HC (1.48 mg H_2O_2 $kg^{-1}h^{-1}$). Soils with high biomass show high CAT activity. The tillage system was the greatest determinant of variation in PER activity. Statistically significant higher PER activity was found in RT soils, with the exception of EG.

Table 6. Activity of oxidoreductive enzymes.

Soil * Types	DEH (mg TPF kg^{-1} 24 h^{-1})		CAT (mg H_2O_2 kg^{-1} h^{-1})		PER (mM PPG kg^{-1} h^{-1})	
	Tillage Systems					
	RT *	CT	RT	CT	RT	CT
EF *	3.42 a ± 0.009	2.36 b ± 0.008	1.76 a ± 0.002	1.52 b ± 0.009	2.08 a ± 0.002	1.86 b ± 0.004
MF	10.85 a ± 0.021	11.28 a ± 0.01	1.93 a ± 0.009	1.89 a ± 0.011	2.20 a ± 0.002	1.92 b ± 0.002
HC	6.54 a ± 0.042	6.12 a ± 0.053	1.48 a ± 0.005	0.96 b ± 0.005	2.11 a ± 0.005	1.81 b ± 0.002
HL	7.23 a ± 0.056	4.87 b ± 0.006	0.89 a ± 0.003	0.81 a ± 0.004	2.16 a ± 0.003	1.72 b ± 0.003
ER	9.73 a ± 0.012	8.09 b ± 0.072	1.06 a ± 0.009	0.86 a ± 0.002	1.95 a ± 0.001	1.62 b ± 0.002
EG	2.63 b ± 0.005	7.23 a ± 0.068	1.72 a ± 0.009	1.69 a ± 0.0011	1.98 a ± 0.002	1.94 a ± 0.004
SP	2.05 b ± 0.004	6.32 a ± 0.009	0.72 a ± 0.004	0.69 a ± 0.008	2.01 a ± 0.004	1.75 b ± 0.001

DEH—dehydrogenases; CAT—catalase; PER—peroxidase; *—see Table 1; a, b—see Table 3.

The analysis of the results obtained for most of the soils showed statistically significant differences in the activity of selected hydrolytic enzymes between tillage systems (Table 7). AlP activity under RT was significantly higher in MF (3.90 mM pNP kg^{-1} kg^{-1}), HL (2.14 mM pNP kg^{-1} kg^{-1}), and ER (1.70 mM pNP kg^{-1} kg^{-1}) soils. Meanwhile, in EG, AlP activity (2.15 mM pNP kg^{-1} kg^{-1}) was significantly highest under CT. No significant differences in AlP were found between the two tillage systems in EF, HC, and SP. A study by Spiegel et al. [74] showed that alkaline phosphatase activity was significantly higher in soils under minimum tillage conditions compared to conventional tillage. According to Nannipieri et al. [75], no-till systems usually have higher enzyme activity in surface soils than in tilled soils due to the increased organic matter content. BG activity was significantly affected by tillage system in all soils except HL. The activity of this enzyme under RT was higher in EF (2.80 mM pNP $kg^{-1}h^{-1}$), HC (2.55 mM pNP $kg^{-1}h^{-1}$), ER (2.24 mM pNP $kg^{-1}h^{-1}$), and EG (2.84 mM pNP $kg^{-1}h^{-1}$). In MF and SP, BG activity was significantly higher under CT (2.23 mM pNP $kg^{-1}h^{-1}$) (Table 6) and tended to be higher under reduced tillage compared to no-till or conventional tillage. Research by Panettieri et al. [76] indicated that reduced tillage had a positive effect on BG activity compared to conventional tillage. Studies by Fernandez-Ortega et al. [77] showed that dehydrogenase activity was higher in the no-tillage system compared to conventional tillage. Also, significant differences found in β-glucosidase activity were attributed to, among other things, the cultivation system. The no-tillage system caused a 53% increase in β-glucosidase activity. PRO activity under CT was significantly higher in HC, HL, ER, EG, and SP soils compared to EF. An absence of differences between RT and CT was found only in MF. Mirzavand et al. [78] showed that replacing conventional tillage with conservation tillage (especially reduced tillage) can increase soil enzyme activity in the short term. The highest acid phosphatase enzyme was obtained in reduced tillage in wheat and maize rotation. However, alkaline phosphatase and urease activity was the highest in no-till soil. Phosphatases catalyse the conversion of organic phosphorus compounds into inorganic phosphates [79]. Research conducted by Piazza et al. [80] showed that minimum tillage together with N fertilisation increased the activity of enzymes (β-cellobiohydrolases, N-acetyl-β-glucosaminidase, β-glucosidase, α-glucosidase, β-xylosidase, acid phosphatase, arylsulphatase, and leucine aminopeptidase). A comprehensive meta-analysis by Wen et al. [81] showed that conservation tillage has a positive effect on soil enzymatic activity. However, the results depend on the type of enzyme and other conservation practices (e.g., no till, straw return). The results of a meta-analysis by Zhu et al. [82] showed that conservation tillage significantly increased SOC and the activities of enzymes related to carbon (C), nitrogen (N), phosphorus (P), and sulphur (S). A meta-analysis conducted by Li et al. [83] showed a significant increase in the activity of enzymes related to C and N cycling and oxidative enzymes under the influence of no-till. The inclusion of legumes in the cultivation significantly increased only enzymes related to P cycling in the soil. However, Woźniak [84] found higher activity of dehydrogenases and phosphatases in soil under CT compared to under RT. The activity of proteases and ureases was higher in soils under RT.

Table 7. The activity of hydrolytic enzymes.

Soil * Types	AlP (mM pNP $kg^{-1}h^{-1}$)		AcP (mM pNP $kg^{-1}h^{-1}$)		BG (mM pNP $kg^{-1}h^{-1}$)		PRO (mg TYR $kg^{-1}h^{-1}$)	
	Tillage Systems							
	RT *	CT	RT	CT	RT	CT	RT	CT
EF *	1.05 a ± 0.002	1.31 a ± 0.001	3.40 a ± 0.014	1.93 b ± 0.085	2.80 a ± 0.014	2.30 b ± 0.021	38.26 a ± 0.081	22.79 b ± 0.065
MF	3.90 a ± 0.008	2.14 b ± 0.005	4.65 a ± 0.055	2.21 b ± 0.091	2.99 b ± 0.021	3.37 a ± 0.008	18.06 a ± 0.058	19.87 a ± 0.041
HC	1.10 a ± 0.001	1.12 a ± 0.031	5.18 a ± 0.061	2.26 b ± 0.047	2.55 a ± 0.009	1.57 b ± 0.009	15.04 b ± 0.035	19.96 a ± 0.022
HL	2.14 a ± 0.002	0.58 b ± 0.008	3.07 a ± 0.034	3.50 a ± 0.009	1.94 a ± 0.008	1.87 a ± 0.006	17.42 b ± 0.009	20.88 a ± 0.012
ER	1.70 a ± 0.001	1.17 b ± 0.012	1.85 b ± 0.019	3.58 a ± 0.014	2.24 a ± 0.012	1.98 b ± 0.011	18.64 b ± 0.017	20.76 a ± 0.061
EG	1.23 b ± 0.001	2.15 a ± 0.038	3.36 b ± 0.022	5.91 a ± 0.052	2.84 a ± 0.008	2.50 b ± 0.025	17.51 b ± 0.012	18.35 a ± 0.016
SP	1.55 a ± 0.002	1.42 a ± 0.018	3.81 a ± 0.020	3.17 a ± 0.017	1.58 b ± 0.006	2.23 a ± 0.019	13.57 b ± 0.008	15.79 a ± 0.014

AlP—alkaline phosphatase (mM pNP $kg^{-1}h^{-1}$); AcP—acid phosphatase (mM pNP $kg^{-1}h^{-1}$); BG—β-glucosidase (mM pNP $kg^{-1}h^{-1}$); PRO—proteases (mg TYR $kg^{-1}h^{-1}$); *—see Table 1; a, b—see Table 3.

Based on the enzyme activity results, enzymatic soil quality indices such as GMea and TEI were calculated (Figure 2). Soil enzyme activity indices TEI and GMea were used to analyse different soil types under two different tillage systems. GMea can directly show the variability of total enzyme activity, whereas TEI values can be used to make easy comparisons of total enzyme activity and soil quality between samples. As compared to GMea, the TEI index showed more varied soil quality results under both tillage systems in the different soil types. Significant differences in GMea were found between tillage systems only in EF, EG, and SP. Under CT, a significantly higher GMea value was obtained in EG and SP soils. The TEI value was similar in these two soils. In the remaining soils, except for ER, the TEI value was significantly higher under RT. Similarly, studies by Tan et al. [44] and Jaskulska et al. [10] showed that TEI was a more sensitive enzymatic indicator of soil quality compared to GMea.

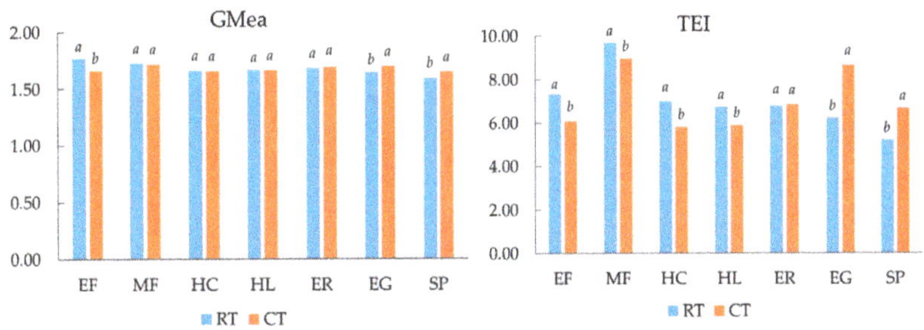

Figure 2. Index of soil enzymes: GMea and TEI (EF—*Eutric Fluvisol*; MF—*Mollic Fluvisol*; HC—*Haplic Chernozem*; HL—*Haplic Luvisol*; ER—*Eutric Regosol*; EG—*Eutric Gleysol*; SP—*Stagnic Planosol*; RT—reduced tillage; CT—conventional tillage).

A correlation analysis showed a significant positive correlation between pH and DEH (r = 0.74) and AlP (r = 0.63) activity (Figure 1). Phosphatases are the enzymes most sensitive to changes in soil pH [79], which also controls phosphorus availability in the soil. A positive significant correlation was also obtained between Hh and AcP (r = 0.48). Soil reaction is a factor that stimulates the activity of most soil enzymes, which is related to the appropriate/related state of ionisation of the active site of protein enzymes. In some soils, pH is masked by the content of organic matter, which affects enzyme activity [85]. Soil enzymes mediate soil biochemical processes and are closely involved in decomposing organic matter, nutrient cycling and environmental quality [86].

The strongest correlation coefficients were found to be between enzyme activity and TOC content (r = 0.74 with DEH; r = 0.72 with CAT; r = 0.57 with PER; r = 0.75 with AlP and r = 0.72 with BG). The coefficient of determination (R^2) showed that as much as 54.7% of the variability of DEH, 51.8% of CAT, 32.5% of PER, 56.3% of AlP, and 51.8% was related to the TOC content in soil. Dehydrogenases are enzymes that play an important role in the biological oxidation of soil organic matter by transferring protons and electrons from substrates to acceptors [87]. Normally, organic matter stabilises and protects extracellular enzymes in the soil, slowing their degradation [88]. The process of OM degradation is related to the activity of BG, as it involves an enzyme catalysing the hydrolysis of cellulose to glucose as a source of energy for soil microorganisms [89]. De Almeida et al. [90] showed that soils enriched with OM with a high C/N ratio are characterised by lower BG activity and slow OM decomposition. Higher soil moisture levels with the no-tillage system, which reduces soil disturbance, led to increased dehydrogenase and β-glucosidase activity. This contributed to greater stabilisation of soil organic carbon [91]. Significant correlations were observed between the activity of PER, AlP, and AcP and the content of DOC, whose fraction contains carbon compounds that are soluble in aqueous solutions and, thus, is the most available in the soil system. Extracellular oxidative and hydrolytic enzymes are responsible for converting organic matter in DOC from high- to low-molecular-weight compounds [92]. Typically, an increased DOC content stimulates the production of a substrate for microbial metabolism and enzyme synthesis. Similarly to Shao et al. [86], no significant correlation was found herein between BG activity and DOC content. BG catalyses the hydrolysis of oligosaccharides to glucose, which can be rapidly utilised by microorganisms. This led to a decrease in DOC content in the soil [86]. Enzymes have their own specific substrates and abilities to catalyse specific biochemical reactions [93]. Differences in substrate availability sources and composition may lead to altered enzyme activity behaviour. Furthermore, soil enzyme activity was significantly and positively correlated with TN content (r = 0.66 with DEH; r = 0.0.76 with CAT; r = 0.55 with AlP; and r = 0.76 with BG). According to Uwituze et al. [94], soil pH was positively correlated with glycosidase activity, which increased with TOC content. Increased N content in the soil caused an increase in the number of microorganisms associated with the C cycle, which affects the production of glucosidases. However, no significant correlations were found between TN and PRO activity. This was confirmed by the meta-analysis conducted by Chen et al. [95]. However, the opposite results were also obtained by Chen et al. [96]. Proteases are enzymes that catalyse the hydrolysis of proteins by catalysing the cleavage of peptide bonds to produce peptides and/or amino acids [97].

Also noteworthy were the high correlations of TEI with TOC values (r = 0.89), DOC (r = 0.47), TN (r = 0.86), pH (r = 0.47), TEB (r = 0.72), CEC (r = 0.52), and BS (r = 0.47). Similarly, studies by Mierzwa-Hersztek et al. [98] and Jaskulska et al. [10] found that TEI is usually positively correlated with TOC and TN content. It was also found that GMea was significantly positively correlated with TEB (r = 0.64), CEC (r = 0.74), TOC (r = 0.60), and TN (r = 0.63) (Figure 1).

4. Summary and Conclusions

The dendograms presented in Figure 3a,b are based on the discussed edaphic parameters of the tested soils. Figure 3a shows the dendogram obtained for soils under reduced tillage and Figure 3b for those under conventional tillage. For reduced tillage, two groups were distinguished, one containing ER and SP, the other HC, HL, and EF. MF and EG lay outside these groups. As the relationships presented in Figure 3a,b show, *Eutric Gleyosols* were the most divergent in properties—regardless of tillage system. Under conventional tillage, ER and HL soils were the two most similar to one another. In general, Euclidean distances were slightly smaller between the studied soil parameters from conventional tillage compared to reduced tillage.

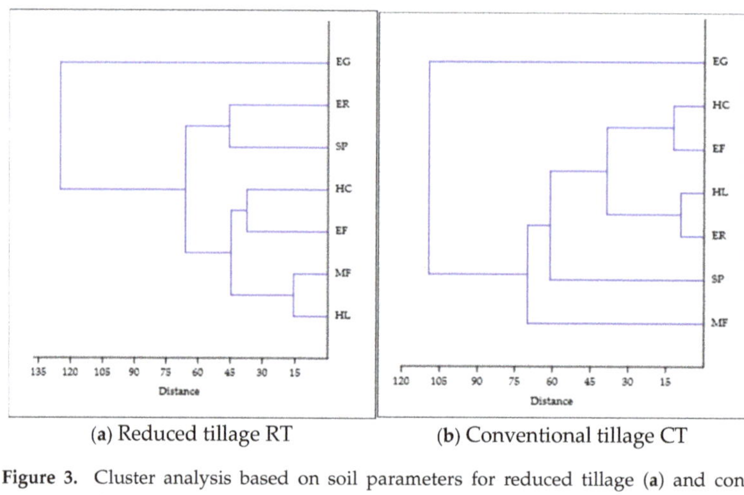

(**a**) Reduced tillage RT (**b**) Conventional tillage CT

Figure 3. Cluster analysis based on soil parameters for reduced tillage (**a**) and conventional tillage (**b**). The distance described on the x-axis represents the Euclidean distance; (EF—*Eutric Fluvisol*; MF—*Mollic Fluvisol*; HC—*Haplic Chernozem*; HL—*Haplic Luvisol*; ER—*Eutric Regosol*; EG—*Eutric Gleysol*; SP—*Stagnic Planosol*).

The identified relationships confirm, among other things, the relationships obtained by Banach-Szott et al. [99], i.e., that organic matter characteristics are a distinctive feature of a given soil type, but that these can nonetheless be modified by the tillage method. Jaskulska et al. [10] clearly demonstrated that the properties of soil under conventional tillage differ from those under reduced tillage and strip-till. The tillage system determined changes in the activity of the studied enzymes, but the intensity and direction of these changes depended on the soil type. Soil type should thus be the key factor in selecting a tillage system.

As can be seen from the data presented in the paper, the intensity and direction of changes in soil quality parameters depend on its type. The TOC decreased in the following order:

For RT: MF > HL > HC > EF = ER > EG > SP; for CT: MF > EG > SP > EF = HC > ER > HR.

The CEC parameter decreased in the following order:

For RT: EF > MF > HL = HC > ER > SP > EG; for CT: MF > ER = HL > EG > HC > SP > EF.

On the basis of the TEI parameter—the enzymatic activity index—the tested soils can be ranked as follows:

For RT: MF > EF + HC > ER + HL > EG > SP for CT: MF = EG > ER > SP > EF > HL = HC.

The total enzyme activity index proved to be a more sensitive enzymatic indicator of soil quality than the geometric mean of enzyme activity indices. However, the diverse properties of soils prevented the precise selection of an enzyme or enzymes for use as soil quality indicators. Therefore, further research is necessary in a direction that will take into account the processes resulting from changes in the use of different types of soil.

Author Contributions: Conceptualisation, E.B. and J.L.; methodology, E.B., J.L., B.D. and A.B.; validation, J.L. and P.W.; formal analysis, E.B., J.L., B.D. and A.B.; investigation, E.B., J.L., B.D. and A.B.; resources, E.B.; writing—original draft preparation, J.L., B.D., A.B. and E.B. writing—review and editing, J.L. and P.W.; visualisation, J.L.; supervision, J.L., B.D. and A.B.; project administration, E.B. All authors have read and agreed to the published version of the manuscript.

Funding: The research was part of the project KEGA 005SPU-4/2022: "Incorporation of contemporary environmental topics into the teaching of soil-related subjects". Bydgoszcz University of Science and Technology under Grant BN-WRiB-1/2022 and BN-WRiB-2/2022.

Institutional Review Board Statement: Not applicable.

Informed Consent Statement: Not applicable.

Data Availability Statement: Data are contained within the article.

Conflicts of Interest: The authors declare no conflict of interest.

References

1. Hartmann, M.; Frey, B.; Mayer, J.; Mäder, P.; Widmer, F. Distinct soil microbial diversity under long-term organic and conventional farming. *ISME J.* **2015**, *9*, 1177–1194. [CrossRef]
2. Wiesmeier, M.; Urbanski, L.; Hobley, E.; Lang, B.; von Lutzow, B.; Marin-Spiotta, E.; van Wesemael, B.; Rabot, E.; Ließ, M.; Garcia-Franco, N.; et al. Soil organic carbon storage as a key function of soils—A review of drivers and indicators at various scales. *Geoderma* **2019**, *333*, 149–162. [CrossRef]
3. European Commision. Communication from the Commission to the European Parliament, the Council, the European Economic and Social Committee and the Committee of the Regions—The European Green Deal, Brussels. 2019. Available online: https://eur-lex.europa.eu/resource.html?uri=cellar:b828d165-1c22-11ea-8c1f-01aa75ed71a1.0016.02/DOC_1&format=PDF (accessed on 11 December 2019).
4. Vittori Antisari, L.; Trenti, W.; De Feudis, M.; Bianchini, G.; Falsone, G. Soil quality and organic matter pools in a temperate climate (Northern Italy) under different land uses. *Agronomy* **2021**, *11*, 1815. [CrossRef]
5. Simon, L.M.; Augustine, K.O.; Holman, J.D.; Roozeboom, K.L. Long-term cover crop management effects on soil properties in dryland cropping systems. *Agric. Ecosyst. Environ.* **2022**, *328*, 107852. [CrossRef]
6. Karaca, A.; Cetin, S.C.; Turgay, O.C.; Kizilkaya, R. Soil enzymes as indication of soil quality. In *Soil Enzymology*; Shukla, G., Varma, A., Eds.; Springer: Berlin/Heidelberg, Germany, 2010; Volume 22. [CrossRef]

7. Lori, M.; Symnaczik, S.; Mäder, P.; De Deyn, G.; Gattinger, A. Organic farming enhances soil microbial abundance and activity—A meta-analysis and meta-regression. *PLoS ONE* **2017**, *12*, e0180442. [CrossRef]
8. Jaskulska, I.; Romaneckas, K.; Jaskulski, D.; Gałęzewski, L.; Breza-Boruta, B.; Dębska, B.; Lemanowicz, J. Soil properties after eight years of the use of strip-till one-pass technology. *Agronomy* **2020**, *10*, 1596. [CrossRef]
9. Jaskulska, I.; Jaskulski, D. Strip-till one-pass technology in central and Eastern Europe: A MZURI Pro-Till hybrid machine case study. *Agronomy* **2020**, *10*, 925. [CrossRef]
10. Jaskulska, I.; Lemanowicz, J.; Dębska, B.; Jaskulski, D.; Breza-Boruta, B. Changes in soil organic matter and biological parameters as a result of long-term strip-till cultivation. *Agriculture* **2023**, *13*, 2188. [CrossRef]
11. Balesdent, J.; Chenu, C.; Balabane, M. Relationship of soil organic matter dynamics to physical protection and tillage. *Soil Tillage Res.* **2000**, *53*, 215–230. [CrossRef]
12. Jaskulska, I.; Jaskulski, D. Influence of many years' fertilization on the dynamics of soil properties. *Adv. Agric. Sci.* **2003**, *4*, 21–35.
13. Blecharczyk, A.; Małecka, I.; Sierpowski, J. Long-term effect of tillage systems on physico-chemical soil properties. *Fragm. Agron.* **2007**, *1*, 7–13.
14. Aye, N.S.; Sale, P.W.G.; Tan, C. The impact of long-term liming on soil organic carbon and aggregate stability in low-input acid soils. *Biol. Fertil. Soils.* **2016**, *52*, 697–709. [CrossRef]
15. Debska, B.; Kotwica, K.; Banach-Szott, M.; Spychaj-Fabisiak, E.; Tobiašová, E. Soil fertility improvement and carbon sequestration through exogenous organic matter and biostimulant application. *Agriculture* **2022**, *12*, 1478. [CrossRef]
16. Menšík, L.; Hlisnikovský, L.; Pospíšilová, L.; Kunzová, E. The effect of application of organic manures and mineral fertilizers on the state of soil organic matter and nutrients in the long-term field experiment. *J. Soils Sediments* **2018**, *18*, 2813–2822. [CrossRef]
17. Friedrich, T.; Derpsch, R.; Kassam, A.H. Global overview of the spread of conservation agriculture. *J. Agric. Sci. Technol.* **2015**, *6*, 1–7.
18. Francaviglia, R.; Almagro, M.; Vicente-Vicente, J.L. Conservation agriculture and soil organic carbon: Principles, processes, practices and policy options. *Soil Syst.* **2023**, *7*, 17. [CrossRef]
19. Stavi, I.; Bel, G.; Zaady, E. Soil functions and ecosystem services in conventional, conservation, and integrated agricultural systems. A review. *Agron. Sustain. Dev.* **2016**, *36*, 32.
20. Busari, A.M.; Kukal, S.S.; Kaur, A.; Bhatt, R.; Dulazi, A.A. Conservation tillage impacts on soil, crop and the environment. *Int. Soil Water Conserv. Res.* **2015**, *3*, 119–129. [CrossRef]
21. Horowitz, J.; Ebel, R.; Ueda, K. *Economic Information Bulletin No. 70: "No-Till" Farming Is a Growing Practice*; Economic Research Service, U.S. Department of Agriculture: Washington, DC, USA, 2010.
22. Allam, M.; Radicetti, E.; Petroselli, V.; Mancinelli, R. Meta-analysis approach to assess the effects of soil tillage and fertilization source under different cropping systems. *Agriculture* **2021**, *11*, 823. [CrossRef]
23. Debska, B.; Jaskulska, I.; Jaskulski, D. Method of tillage with the factor determining the quality of organic matter. *Agronomy* **2020**, *10*, 1250. [CrossRef]
24. Virto, I.; Barre, P.; Burlot, A.; Chenu, C. Carbon input differences as the main factor explaining the variability in soil organic C storage in notilled compared to inversion tilled agrosystems. *Biogeochemistry* **2012**, *108*, 17–26. [CrossRef]
25. Dimassi, B.; Mary, B.; Wylleman, R.; Labreuche, J.; Couture, D.; Piraux, F.; Cohan, J.P. Long-term effect of contrasted tillage and crop management on soil carbon dynamics during 41 years. *Agric. Ecosyst. Environ.* **2014**, *188*, 134–146. [CrossRef]
26. Han, P.; Zhang, W.; Wang, G.; Sun, W.; Huang, Y. Changes in soil organic carbon in croplands subjected to fertilizer management: A global metaanalysis. *Nat. Sci. Rep.* **2016**, *6*, 27199. [CrossRef]
27. Yue, K.; Peng, Y.; Peng, C.; Yang, W.; Peng, X.; Wu, F. Stimulation of terrestrial ecosystem carbon storage by nitrogen addition: A metaanalysis. *Sci. Rep.* **2016**, *6*, 19895. [CrossRef]
28. Armas-Herrera, C.M.; Dignac, M.-F.; Rumpel, C.; Arbelo, C.D.; Chabbi, A. Management effects on composition and dynamics of cutin and suberin in topsoil under agricultural use. *Eur. J. Soil Sci.* **2016**, *67*, 360–373. [CrossRef]
29. Szostek, M.; Szpunar-Krok, E.; Pawlak, R.; Stanek-Tarkowska, J.; Ilek, A. Effect of different tillage systems on soil organic carbon and enzymatic activity. *Agronomy* **2022**, *12*, 208. [CrossRef]
30. Blanco-Canqui, H.; Ruis, S.J. No-tillage and soil physical environment. *Geoderma* **2018**, *326*, 164–200. [CrossRef]
31. Chenu, C.; Angers, D.A.; Barré, P.; Derrien, D.; Arrouays, D.; Balesdent, J. Increasing organic stocks in agricultural soils: Knowledge gaps and potential innovations. *Soil Tillage Res.* **2019**, *188*, 41–52. [CrossRef]
32. Derpsch, R.; Friedrich, T.; Kassam, A.; Li, H. Current status of adoption of no-till farming in the world and some of its main benefits. *Int. J. Agric. Biol. Eng.* **2010**, *3*, 1–25. [CrossRef]
33. Cenini, V.L.; Fornara, D.A.; McMullan, G.; Ternan, N.; Carolan, R.; Crawley, M.J.; Clément, J.-C.; Lavorel, S. Linkages between extracellular enzyme activities and the carbon and nitrogen content of grassland soils. *Soil Biol. Biochem.* **2016**, *96*, 198–206. [CrossRef]

34. Rahmati, M.; Eskandari, I.; Kouselou, M.; Feiziasl, V.; Mahdavinia, G.R.; Aliasgharzad, N.; McKenzie, B.M. Changes in soil organic carbon fractions and residence time five years after implementing conventional and conservation tillage practices. *Soil Tillage Res.* **2020**, *200*, 104632. [CrossRef]
35. Tobiašová, E.; Lemanowicz, J.; Dębska, B.; Kunkelová, M.; Sakáč, J. The Effect of reduced and conventional tillage systems on soil aggregates and organic carbon parameters of different soil types. *Agriculture* **2023**, *13*, 818. [CrossRef]
36. IUSS Working Group WRB. World Reference Base for Soil Resources. In *International Soil Classification System for Naming Soils and Creating Legends for Soil Maps*, 4th ed.; International Union of Soil Sciences (IUSS): Vienna, Austria, 2022.
37. Haddad, S.A.; Lemanowicz, J.; Abd El-Azeim, M.M. Cellulose decomposition in clay and sandy soils contaminated with heavy metals. *Int. J. Environ. Sci. Technol.* **2019**, *16*, 3275–3290. [CrossRef]
38. Tobiašová, E.; Lemanowicz, J.; Dębska, B.; Kunkelová, M.; Sakáč, J. Suitability of various parameters for the determination of the condition of soil structure with dependence to the quantity and quality of soil organic matter. *Sustainability* **2023**, *15*, 11047. [CrossRef]
39. Korec, P.; Lauko, V.; Tolmáči, L.; Zubrický, G.; Mičietová, E. *Region and Districts of Slovakia. A New Administrative Structure*; Q111: Bratislava, Slovakia, 1997; 391p.
40. Jurčová, O.; Bielek, P. *Method of Soil Organic Method Balance and the Determination of Necessity of Organic Fertilization*; VÚPÚ: Bratislava, Slovakia, 1997; 154p.
41. van Reeuwijk, L.P. *Procedures for Soil Analysis*; International Soil Reference and Information Centre: Wageningen, The Netherlands, 2002; 120p.
42. *PN-ISO 10390*; Chemical and Agricultural Analysis: Determining Soil pH. Polish Standards Committee: Warszawa, Poland, 1997.
43. Skjemstad, J.O.; Gillman, G.P.; Massis, A.; Spouncer, L.R.; Leonie, R. Measurement of the exchange capacity of the cationic fraction of organic matter from soils using the modified compulsive exchange method. *Soil Sci. Soc. Am. J.* **2008**, *39*, 926–937. [CrossRef]
44. Thalmann, A. Zur methodic derestimung der dehydrogenaseaktivität und boden mittels triphenyltetrazoliumchlorid (TTC). *Landwirtsch. Forsch.* **1968**, *21*, 249–258.
45. Johnson, J.I.; Temple, K.L. Some variables affecting the measurements of catalase activity in soil. *Soil Sci. Soc. Am.* **1964**, *28*, 207–209. [CrossRef]
46. Bartha, R.; Bordeleau, L. Cell-free peroxidases in soil. *Soil Biol. Biochem.* **1969**, *1*, 139–143. [CrossRef]
47. Tabatabai, M.A.; Bremner, J.M. Use of p–nitrophenol phosphate for assay of soil phosphatase activity. *Soil Biol. Biochem.* **1969**, *1*, 301–307. [CrossRef]
48. Eivazi, F.; Tabatabai, M.A. Glucosidases and galactosidases in soils. *Soil Biol. Biochem.* **1988**, *20*, 601–606. [CrossRef]
49. Ladd, J.N.; Jackson, R.B. Biochemistry of ammonification. In *Nitrogen in Agricultural Soil*; American Society of Agronomy, Inc. Crop Science Society of America, Inc. Soil Science Society of America Inc.: Madison, WI, USA, 1982. [CrossRef]
50. Hinojosa, M.B.; Garcia-Ruiz, R.; Viñegla, B.; Carreira, J.A. Microbiological rates and enzyme activities as indicators of functionality in soils affected by the Aznalcóllar toxic spill. *Soil Biol. Biochem.* **2004**, *36*, 1637–1644. [CrossRef]
51. Tan, X.; Xie, B.; Wang, J.; He, W.; Wang, X.; Wei, G. County-scale spatial distribution of soil enzyme activities and enzyme activity indices in agricultural land: Implications for soil quality assessment. *Sci. World J.* **2014**, *2014*, 535768. [CrossRef]
52. Shapiro, S.S.; Wilk, M.B. An analysis of variance test for normality (complete samples). *Biometrika* **1965**, *52*, 591–611. [CrossRef]
53. Hammer, Ø.; Harper, D.A.; Ryan, P.D. Past: Paleontological statistics software package for education and data anlysis. *Palaeontol. Electron.* **2001**, *4*, 9.
54. Ward, J.H. Hierarchical grouping to optimize an objective function. *J. Am. Stat. Ass.* **1963**, *58*, 236–244. [CrossRef]
55. Ahmad, N.; Virk, A.L.; Hafeez, M.B.; Ercisli, S.; Golokhvast, K.S.; Qi, Y.; Guo, X.; Zhang, Y.; Wang, R.; Wang, X.; et al. Effects of different tillage and residue management systems on soil organic carbon stock and grain yield of rice–wheat double cropping system. *Ecol. Indic.* **2024**, *158*, 111452. [CrossRef]
56. Yuan, J.; Yan, l.; Li, G.; Sadiq, M.; Rahim, N.; Wu, J.; Ma, W.; Xu, G.; Du, M. Effects of conservation tillage strategies on soil physicochemical indicators and N2O emission under spring wheat monocropping system conditions. *Sci. Rep.* **2022**, *12*, 7066. [CrossRef]
57. Mosley, L.M.; Rengasamy, P.; Fitzpatrick, R. Soil pH: Techniques, challenges and insights from a global dataset. *Eur. J. Soil Sci.* **2024**, *75*, 70021. [CrossRef]
58. He, G.; Zhang, Z.; Wu, X.; Cui, M.; Zhang, J.; Huang, X. Adsorption of heavy metals on soil collected from Lixisol of typical karst areas in the presence of $CaCO_3$ and soil clay and their competition behavior. *Sustainability* **2020**, *12*, 7315. [CrossRef]
59. Stawn, D. Sorption Mechanisms of chemicals in soils. *Soil Syst.* **2021**, *5*, 13. [CrossRef]
60. Bi, X.; Chu, H.; Fu, M.; Xu, D.; Zhao, W.; Zhong, Y.; Wang, M.; Li, K.; Zhang, Y. Distribution characteristics of organic carbon (nitrogen) content, cation exchange capacity, and specific surface area in different soil particle sizes. *Sci. Rep.* **2023**, *13*, 12242. [CrossRef]
61. Vasudevan, D.; Bruland, G.L.; Torrance, B.S.; Upchurch, W.G.; MacKay, A.A. pH-dependent ciprofloxacin sorption to soils: Interaction mechanisms and soil factors influencing sorption. *Geoderma* **2009**, *151*, 68–76. [CrossRef]

62. Lorandi, R. Evaluation of cation exchange capacity (CEC) in tropical soils using four different analytical methods. *J. Agric. Sci.* **2012**, *4*, 278–289. [CrossRef]
63. Šimansky, V.; Pollakova, N. Soil organic matter and sorption capacity under different soil management practices in a productive vineyard. *Arch. Agron. Soil Sci.* **2014**, *60*, 1145–1154. [CrossRef]
64. Tian, K.; Zhao, Y.; Xu, X.; Hai, N.; Huang, B.; Deng, W. Effects of long-term fertilization and residue management on soil organic carbon changes in paddy soils of China: A meta-analysis. *Agric. Ecosyst. Environ.* **2015**, *204*, 40–50. [CrossRef]
65. Laufer, D.; Loibl, B.; Marlander, B.; Koch, H.J. Soil erosion and surface run-of under strip tillage for sugar beet (*Beta vulgaris* L.) in Centraf Europe. *Soil Tillage Res.* **2016**, *162*, 1–7. [CrossRef]
66. Powlson, D.S.; Bhogal, A.; Chambers, B.J.; Coleman, K.; Macdonald, A.J.; Goulding, K.W.T.; Whitmore, A.P. The potential to increase soil carbon stocks through reduced tillage or organic material additions in England and Wales: A case study. *Agric. Ecosyst. Environ.* **2012**, *146*, 23–33. [CrossRef]
67. Kumar, S.; Kadono, A.; Lal, R.; Dick, W. Long-term no-till impacts on organic carbon and properties of two contrasting soils and corn yields in Ohio. *Soil Sci. Soc. Am. J.* **2012**, *76*, 1798–1809. [CrossRef]
68. Bolan, N.S.; Adriano, D.C.; Kunhikrishnan, A.; James, T.; McDowell, R.; Senesi, N. Dissolved organic matter: Biogeochemistry, dynamics, and environmental significance in soils. *Adv. Agron.* **2011**, *110*, 1–75. [CrossRef]
69. Wright, A.; Dou, F.; Hons, M.F. Soil organic C and N distribution for wheat cropping systems after 20 years of conservation tillage in central Texas. *Agric. Ecosyst. Environ.* **2007**, *121*, 376–382. [CrossRef]
70. Liu, E.; Teclemariam, S.G.; Yan, C.; Yu, J.; Gu, R.; Liu, S.; He, W.; Liu, Q. Long-term efects no-tillage management practice on soil organic carbon and its fractions in the northern China. *Geodema* **2014**, *213*, 379–384. [CrossRef]
71. Leinweber, P.; Schulten, H.R.; Körschens, M. Hot water extracted organic matter: Chemical composition and temporal variations in a long-term field experiment. *Biol. Fertil. Soils* **1995**, *20*, 17–23. [CrossRef]
72. Shaheen, S.M.; Rinklebe, J. Geochemical fractions of chromium, copper, and zinc and their vertical distribution in floodplain soil profiles along the Central Elbe River, Germany. *Geoderma* **2014**, *228*, 142–159. [CrossRef]
73. Furtak, K.; Grządziel, J.; Gałązka, A.; Niedźwiecki, J. Analysis of soil properties, bacterial community composition, and metabolic diversity in fluvisols of a floodplain area. *Sustainability* **2019**, *11*, 3929. [CrossRef]
74. Spiegel, H.; Sandén, T.; Sandén, H.; Götzinger, S.; Miloczki, J.; Kandeler, E. Changes in biological and chemical soil properties in an Austrian long-term tillage experiment. *Eur. J. Soil Sci.* **2025**, *76*, e70037. [CrossRef]
75. Nannipieri, P.; Giagnoni, L.; Landi, L.; Renella, G. Role of phosphatase enzymes in soil. In *Phosphorus in Action*; Bünemann, E., Oberson, A., Frossard, E., Eds.; Soil Biology Series; Springer: Berlin/Heidelberg, Germany, 2011; Volume 26. [CrossRef]
76. Panettieri, M.; de Sosa, L.L.; Domínguez, M.T.; Madejón, E. Long-term impacts of conservation tillage on Mediterranean agricultural soils: Shifts in microbial communities despite limited effects on chemical properties. *Agric. Ecosyst. Environ.* **2020**, *304*, 107144. [CrossRef]
77. Fernández-Ortega, J.; Álvaro-Fuentes, J.; Delgado, A.; García-López, A.M.; Cantero-Martínez, C. Assessing management strategies for carbon storage in Mediterranean soils: Double-cropping, no-tillage, and nitrogen fertilization reduction. *Soil Tillage Res.* **2025**, *249*, 106496. [CrossRef]
78. Mirzavand, J.; Asadi-Rahmani, H.; Moradi-Talebbeigi, R. Biological indicators of soil quality under conventional, reduced, and no-tillage systems. *Arch. Agron. Soil Sci.* **2022**, *68*, 311–324. [CrossRef]
79. Dick, W.A.; Cheng, L.; Wang, P. Soil acid and alkaline phosphatase activity as pH adjustment indicators. *Soil Biol. Biochem.* **2000**, *32*, 1915–1919. [CrossRef]
80. Piazza, G.; Pellegrino, E.; Moscatelli, M.C.; Ercoli, L. Long-term conservation tillage and nitrogen fertilization effects on soil aggregate distribution, nutrient stocks and enzymatic activities in bulk soil and occluded microaggregates. *Soil Tillage Res.* **2020**, *196*, 104482. [CrossRef]
81. Wen, L.; Peng, Y.; Zhou, Y.; Cai, G.; Lin, Y.; Li, B. Effects of conservation tillage on soil enzyme activities of global cultivated land: A meta-analysis. *J. Environ. Manag.* **2023**, *345*, 118904. [CrossRef]
82. Zhu, Y.; Zhang, H.; Wang, Q.; Zhu, W.; Kang, Y. Soil extracellular enzyme activity linkage with soil organic carbon under conservation tillage: A global meta-analysis. *Eur. J. Agron.* **2024**, *155*, 127135. [CrossRef]
83. Li, T.; Li, G.; Lu, Z.; Zhao, D.; Li, Y.; Wang, Z.; Wen, X.; Liao, Y. Crop diversification increases soil extracellular enzyme activities under no tillage: A global meta-analysis. *Soil Tillage Res.* **2024**, *235*, 105870. [CrossRef]
84. Woźniak, A. Chemical properties and enzyme activity of soil as affected by tillage system and previous crop. *Agriculture* **2019**, *9*, 262. [CrossRef]
85. Lasota, J.; Błońska, E.; Piaszczyk, W. State of soil enzymatic activity in relationship to some chemical properties of Brunic Arenosols. *Soil Sci. Ann.* **2021**, *72*, 140641. [CrossRef]
86. Shao, X.; Yang, W.; Wu, M. Seasonal dynamics of soil labile organic carbon and enzyme activities in relation to vegetation types in Hangzhou Bay Tidal Flat Wetland. *PLoS ONE* **2015**, *11*, e0142677. [CrossRef]

87. Daunoras, J.; Kačergius, A.; Gudiukaitė, R. Role of soil microbiota enzymes in soil health and activity changes depending on climate change and the type of soil ecosystem. *Biology* **2024**, *13*, 85. [CrossRef]
88. Geisseler, D.; Horwath, W.R. Relationship between carbon and nitrogen availability and extracellular enzyme activities in soil. *Pedobiologia* **2009**, *53*, 87–98. [CrossRef]
89. Chaves, V.B.S.; Guimarães, T.M.; Bezerra, A.C.T.P.; da Costa, C.H.M.; Cruz, S.C.S. Enzymatic activity in different crop succession systems in the Cerrado Region. *Agronomy* **2024**, *14*, 810. [CrossRef]
90. de Almeida, R.F.; Naves, E.R.; da Mota, R.P. Soil quality: Enzymatic activity of soil β-glucosidase. *Glob. J. Agric. Res. Rev.* **2015**, *3*, 146–450.
91. Fernández-Ortega, J.; Álvaro-Fuentes, J.; Talukder, R.; Lampurlanés, J.; Cantero-Martínez, C. The use of double-cropping in combination with no-tillage and optimized nitrogen fertilization improve crop yield and water use efficiency under irrigated conditions. *Field Crops Res.* **2023**, *301*, 109017. [CrossRef]
92. Tian, L.; Dell, E.; Shi, W. Chemical composition of dissolved organic matter in agroecosystems: Correlations with soil enzyme activity and carbon and nitrogen mineralization. *Appl. Soil Ecol.* **2010**, *46*, 426–435. [CrossRef]
93. Song, Y.; Song, C.; Yang, G.; Miao, Y.; Wang, J.; Guo, Y. Changes in labile organic carbon fractions and soil enzyme activities after marshland reclamation and restoration in the Sanjiang Plain in Northeast China. *Environ. Manag.* **2012**, *50*, 418–426. [CrossRef]
94. Uwituze, Y.; Nyiraneza, J.; Fraser, T.D.; Dessureault-Rompré, J.; Ziadi, N.; Lafond, J. Carbon, nitrogen, phosphorus, and extracellular soil enzyme responses to different land use. *Front. Soil Sci.* **2022**, *2*, 814554. [CrossRef]
95. Chen, H.; Li, D.; Zhao, J.; Xiao, K.; Wang, K. Effects of nitrogen addition on activities of soil nitrogen acquisition enzymes:A meta-analysis. *Agric. Ecosyst. Environ.* **2018**, *252*, 126–131. [CrossRef]
96. Chen, T.; Cheng, R.; Xiao, W.; Shen, Y.; Wang, L.; Sun, P.; Zhang, M.; Li, J. Nitrogen addition enhances soil nitrogen mineralization through an increase in mineralizable organic nitrogen and the abundance of functional genes. *J. Soil Sci. Plant Nutr.* **2024**, *24*, 975–987. [CrossRef]
97. Landi, L.; Renella, G.; Giagnoni, L.; Nannipieri, P. Activities of proteolytic enzymes. *Methods Soil Enzymol.* **2011**, *9*, 247–260. [CrossRef]
98. Mierzwa-Hersztek, M.; Gondek, K.; Klimkowicz-Pawlas, A.; Chmiel, M.J.; Dziedzic, K.; Taras, H. Assessment of soil quality after biochar application based on enzymatic activity and microbial composition. *Int. Agrophys.* **2019**, *33*, 331–336. [CrossRef]
99. Banach-Szott, M.; Debska, B.; Tobiasova, E. Properties of humic acids depending on the land use in different parts of Slovakia. *Environ. Sci. Pollut. Res.* **2021**, *28*, 58068–58080. [CrossRef]

sustainability

MDPI

Article

The Effect of Waste Organic Matter on the Soil Chemical Composition After Three Years of *Miscanthus* × *giganteus* Cultivation in East-Central Poland

Elżbieta Malinowska * and Paweł Kania

Faculty of Agricultural Sciences, University of Siedlce, Konarskiego 2 Str., 08-110 Siedlce, Poland; pawelkania933@gmail.com
* Correspondence: elzbieta.malinowska@uws.edu.pl; Tel.: +48-25-6431319

Abstract: The circular economy practice of using waste to fertilize plants should be more widespread. It is a means to manage natural resources sustainably in agriculture. This approach is in line with organic and sustainable farming strategies, reducing the cultivation costs. Organic waste dumped into a landfill decomposes and emits greenhouse gases. This can be reduced through its application to energy crops, which not only has a positive impact on the environment but also improves the soil quality and increases yields. However, organic waste with increased content of heavy metals, when applied to the soil, can also pose a threat. Using *Miscanthus* × *giganteus* M 19 as a test plant, an experiment with a randomized block design was established in four replications in Central–Eastern Poland in 2018. Various combinations of organic waste (municipal sewage sludge and spent mushroom substrate) were applied, with each dose containing 170 kg N ha^{-1}. After three years (in 2020), the soil content of total nitrogen (N_t) and carbon (C_t) was determined by elemental analysis, with the total content of P, K, Ca, Mg, S, Na, Fe, Mn, Mo, Zn, Ni, Pb, Cr, Cd, and Cu determined by optical emission spectrometry, after wet mineralization with aqua regia. For the available forms of P and K, the Egner–Riehm method was used, and the Schachtschabel method was used for the available forms of Mg. The total content of bacteria, actinomycetes, and fungi was also measured. The application of municipal sewage sludge (SS) alone and together with spent mushroom substrate (SMS) improved the microbiological composition of the soil and increased the content of N_t and C_t and the available forms of P_2O_5 and Mg more than the application of SMS alone. SMS did not contaminate the soil with heavy metals. In the third year, their content was higher after SS than after SMS application, namely for Cd by 12.2%, Pb by 18.7%, Cr by 25.3%, Zn by 16.9%, and Ni by 14.7%.

Keywords: soil chemical composition; municipal sewage sludge; spent mushroom substrate; *Miscanthus* × *giganteus*; sustainable farming

1. Introduction

In order to obtain the appropriate quantity and quality of perennial lignocellulosic plants, alternative production technologies are applied, using various post-production waste and agricultural residues, such as sewage sludge, mushroom substrate, digestate from biogas plants, or alcohol distillery waste. One of the main problems with the use of organic waste as a fertilizer is its content of pathogenic microorganisms and heavy metals, with the latter accumulating in the soil [1–5]. The long-term presence of heavy metals in the soil can lead to their bioaccumulation in plants. Heavy metals reduce plants' ability to absorb nutrients and water and, consequently, inhibit their growth. Heavy metals can also disrupt the functions of soil microorganisms and their ecosystem [6–8]. The action of organic materials in the soil, especially of SS, is dynamic and changes over time. Initially, the soil physicochemical properties improve and plant yields increase [9–12], but, over time, depending on their chemical composition, problems related to the accumulation of pollutants (e.g., heavy metals) and reduced soil biodiversity may arise [13]. Depending on the habitat conditions and the types of materials introduced into the soil, as much as 40% to 70% of organic matter undergoes mineralization in the first year. In the following years, the rate of mineralization decreases in favor of humification [14]. Climate change has a strong impact on these processes [15]. In Poland, a large amount of waste ranging from 1 to 1.3 million tons per year is generated during mushroom production [16]. This waste is much safer in terms of heavy metal content than SS, but it can contain fungi and many other microorganisms that may negatively affect plant growth and development. Therefore, it should always be heat-treated before being introduced into the soil [17]. The substrate after mushroom cultivation is produced on the basis of organic materials (chicken or horse manure, straw, and peat) supplemented with a mineral substance.

The use of organic waste as a fertilizer always requires careful monitoring and compliance with environmental regulations to avoid long-term negative effects on human health and the soil ecosystem. In Poland, the use of SS is determined by the Regulation of the Ministry of the Environment [18] and the Act of 27 April 2001 on waste [19]. In the cultivation of energy crops, the use of SS is much less restrictive than in the cultivation of forage crops and those produced for human consumption, but this does not mean that cyclical soil and plant tests should not be performed.

Taking the above into account, this paper examines changes in the chemical composition of the soil treated with different combinations of SS and SMS, each with the same dose of N, after three years of *Miscanthus* × *giganteus* cultivation.

The aim of this research was to determine the long-term effects of SS and SMS on the soil content of carbon, nitrogen, hydrogen, macro- and micronutrients, and selected heavy metals, as well as bacteria, actinomycetes, and fungi, after three years of *Miscanthus* × *giganteus* cultivation in Central–Eastern Poland.

2. Materials and Methods

2.1. Description of the Experiment

The field experiment was established in the experimental field of the University of Siedlce (52°17′ N, 22°28′ E) in 2018, on anthropogenic soil of the culture-earth type and hortisol subtype. It was clay loam soil, with light loam as the subsoil [20]. The experiment was set up in a system of random blocks with four replications. The test plant was *Miscanthus* × *giganteus* (giant miscanthus) M 19, a perennial grass of the C4 photosynthetic type.

The experimental site consisted of the following plots and treatments (Table 1).

Both SS and SMS were applied only once before planting the grass rhizomes, namely in the spring. Sludge was obtained from the sewage treatment plant in Siedlce, with a capacity of about 24,000 m^3 of SS per day. It produced 1897 Mg of SS per year and 5.2 Mg per day. The substrate obtained after mushroom cultivation from a farm located in the Siedlce district was subjected to thermal treatment. The cultivation of the white mushroom on the farm lasted six weeks. The producer of the substrate for mushroom cultivation was Unikost, while the peat moss for the casing layer was produced by Wokas.

Table 1. The layout of the field experiment.

Experimental Object	Treatment
Control	no fertilizer treatment
(SS)	SS—170 kg N ha^{-1} (0.907 kg of sludge per plot, 4.54 Mg ha^{-1})
(SS_{75} + SMS_{25})	SS + SMS—170 kg N ha^{-1}, i.e., 75% sludge + 25% substrate (0.683 kg + 1.40 kg per plot, 3.38 Mg ha^{-1} + 7 Mg ha^{-1})
(SS_{50} + SMS_{50})	SS + SMS—170 kg N ha^{-1}, i.e., 50% sludge + 50% substrate (0.454 kg + 2.80 kg per plot, 2.25 Mg ha^{-1} + 14 Mg ha^{-1})
($SS2_5$ + SMS_{25})	SS + SMS—170 kg N ha^{-1}, i.e., 25% sludge + 75% substrate (0.227 kg + 4.20 kg per plot, 1.13 Mg ha^{-1} + 21 Mg ha^{-1})
(SMS)	SMS—170 kg N ha^{-1} (5.60 kg per plot, 28 Mg ha^{-1})

2.2. Weather Conditions During the Experiment

The meteorological conditions from 2018 to 2020 were assessed on the basis of data made available by the Institute of Meteorology and Water Management, National Research Institute (PIB) in Warsaw. In order to determine the temporal variability and the impact of precipitation and the air temperature on the growth and development of plants, Sielianinov's hydrothermal coefficient (*K*) was determined according to the following formula:

$$K = \frac{P}{0.1\Sigma t} \qquad (1)$$

where

P—monthly rainfall;

Σt—the sum of the daily air temperature values in a given month [21].

Then, nine classes of hydrothermal conditions were established using Sielianinov's coefficient (K):

$K \leq 0.4$ extreme drought (ed);
$0.4 < K \leq 0.7$ severe drought (sd);
$0.7 < K \leq 1.0$ drought (d);
$1.0 < K \leq 1.3$ moderate drought (md);
$1.3 < K \leq 1.6$ optimal (o);
$1.6 < K \leq 2.0$ moderately wet (mw);
$2.0 < K \leq 2.5$ wet (w);
$2.5 < K \leq 3.0$ severely wet (sw);
$K > 3.0$ extremely wet (ew) [22].

Optimal thermal and moisture conditions occurred only in June, July, and October 2018, i.e., in the first year of grass growth (Table 2). In the remaining growing seasons, they varied to a large degree. In June 2019 and July 2020, extremely dry conditions were observed. The most difficult growing conditions occurred in 2019, when, with the exception of May, they ranged from extremely dry to quite dry.

Table 2. Values of Sielianinov's hydrothermal coefficient (K) in individual months in 2018–2020.

Year	Month						
	April	May	June	July	August	September	October
2018	1.07 (md)	0.50 (sd)	1.38 (o)	1.5 (o)	0.44 (sd)	0.92 (d)	1.52 (o)
2019	0.32 (ed)	2.83 (sw)	0.44 (sd)	1.7 (d)	1.21 (md)	1.01 (md)	0.62 (sd)
2020	0.29 (ed)	3.24 (ew)	3.02 (ew)	0.69 (sd)	1.09 (md)	1.06 (md)	2.73 (sw)

ed—extreme drought, sd—severe drought, d—drought, md—moderate drought, o—optimal, sw—severely wet, ew—extremely wet.

2.3. Soil and Organic Material Analysis

Representative soil samples were collected from three layers, 0–20 cm, 20–40 cm, and 40–60 cm, at the beginning of the experiment and at the end of the third year (2020) only from the arable layer (0–20 cm). They were air dried, and the following were determined:

- pH_{H2O} and in 1 mol KCl L^{-1} by the potentiometric method;
- total nitrogen (N_t) and carbon (Ct) content by elemental analysis using the Perkin Elmer CHNS/O Series II 2400 autoanalyzer with a thermal conductivity detector;
- total content of P, K, Ca, Mg, S, Na, Fe, Mn, Mo, Zn, Ni, Pb, Cr, Cd, and Cu by optical emission spectrometry, after the wet mineralization of the soil samples with aqua regia, at Eurofins OBiKŚ Polska Ltd, Katowice, Poland. in Katowice, formerly the Centre for Environmental Research and Control;
- available forms of P and K by the Enger–Riehm method, at the District Chemical-Agricultural Station in Lublin, according to the Polish standards, respectively [23,24];
- available forms of Mg by the Schachtschabel method, at the District Chemical and Agricultural Station in Lublin, according to the Polish standard [25].

In addition, after harvesting *Miscanthus × giganteus*, the total soil content of bacteria and actinomycetes was determined, expressed as the number of colony-forming units (10^7 $CFUg^{-1}$ DM of soil), via the plate method on LB medium after incubation at 28 °C for 5 days. Petri dishes were seeded with bacterial colonies using the flood plate method. The total number of fungi (10^4 $CFUg^{-1}$ DM of soil) was determined on Martin's substrate.

The experiment was conducted on soil with high total C and N content and a neutral pH (Table 3, (a)). The content of Cr, Cd, Cu, and Ni at the beginning of the experiment was several times lower than the limits specified by the Ministry of the Environment's Regulation [18] for light soils when using SS, while the amounts of Zn and Pb were within the standards (Table 3, (b)). The content of available forms of P, K, and Mg in the soil top layer was as follows: P_2O_5—117; K_2O—47.5; Mg—10.04 mg kg^{-1}. This indicated their large amounts (Table 4).

Table 3. Soil chemical properties before the experiment.

(a)

Layer of Soil Profile	pH_{H2O}	pH_{KCl}	C_t	N_t	C:N	P	K	Ca	Mg	S	Na
					(g kg^{-1})						
0–20 (A1)	6.93	6.60	40.50	2.85	14.21	1.19	0.736	9.42	0.973	0.377	0.066
20–40 (A2)	6.25	5.85	20.30	1.65	12.30	0.784	0.680	6.72	0.738	0.301	0.071
40–60 (A3)	6.10	5.50	19.15	1.60	11.97	0.581	0.502	2.99	0.427	0.109	0.065

(b)

Layer of Soil Profile	Fe	Mn	Mo	Pb	Cd	Cr	Cu	Zn	Ni
					(mg kg^{-1})				
0–20 (A1)	5186.5	145.8	0.231	48.98	0.959	8.95	18.85	149.7	5.53
20–40 (A2)	3918.4	160.0	0.070	41.71	0.511	5.35	13.24	136.1	4.29
40–60 (A3)	2274.3	121.4	0.108	24.18	0.318	3.28	6.09	46.59	2.27

Table 4. Content of available macronutrients in the soil (mg 100 g^{-1} of soil) before the experiment.

Level of the Soil Profile	P_2O_5	K_2O	Mg
		(mg 100 g^{-1} Soil)	
0–20 (A1)	117.0	47.5	10.04
20–40 (A2)	109.3	44.8	8.20
40–60 (A3)	90.5	39.3	7.41

In a representative sample of waste organic material, the following were determined:

- dry matter, after drying the sample at 105 °C until a constant mass was reached;
- pH value in 1 mol KCl L^{-1} by the potentiometric method;
- total nitrogen (N_t) content by the modified Kjeldahl method, after mineralization in concentrated sulfuric acid (VI), in the presence of a selenium mixture [26];
- organic C (C_{org}) content by the redox titration method [27];
- total content of P, K, Ca, Mg, S, Na, Fe, Mn, Mo, Zn, Ni, Pb, Cr, Cd, and Cu by optical emission spectrometry after the wet mineralization of the samples using aqua regia.

The municipal sludge used in the experiment was characterized by high content of dry matter (93%) and macroelements (Table 5), with low content of heavy metals. Their permissible amounts were provided by the Regulation of the Ministry of the Environment [18], allowing the use of sewage sludge in the cultivation of *Miscanthus* × *giganteus* (in accordance with Directive 86/278/EEC). Of all heavy metals in SS, the Zn content was the highest (987.2 mg kg^{-1} DM) and Cd (1.81 mg kg^{-1} DM) the lowest. The content of heavy metals in the organic material was ranked in a sequence of decreasing values (mg kg^{-1}): Zn (987.2) > Cu (88.01) > Ni (44.23) > Pb (36.12) > Cr (15.44) > Co (3.58) > Cd (1.81).

Table 5. Chemical composition of SS.

pH_{KCl}	DM (%)	C_{org} (g kg^{-1} DM)	C:N	N	P	K	Ca	Mg	S	Na
							(g kg^{-1} DM)			
6.4	93.0	348	7.8	40.50	19.81	2.56	34.25	6.25	5.36	0.589
Fe	Mn	B	Mo	Co	Pb	Cd	Cr	Cu	Zn	Ni
				(mg kg^{-1} DM)						
8950	602	6.15	2.53	3.58	36.12	1.81	15.44	88.01	987.2	44.23

The spent substrate left after the production of white mushrooms was characterized by relatively low content of heavy metals (Table 6). Among them, Zn (156.9 mg kg^{-1}) was the most abundant, and the cobalt (0.415 mg kg^{-1}) and cadmium content (0.287 mg kg^{-1}) was the lowest. The content of all heavy metals in SMS was several times lower than in SS. The chemical composition of the SMS was determined by the materials used for its production.

Table 6. Chemical composition of SMS.

pH_{KCl}	DM (%)	C_{org} (g kg^{-1} DM)	C:N	N	P	K	Ca	Mg	S	Na
							(g kg^{-1} DM)			
6.41	30.00	284	13.59	20.9	8.86	11.21	78.83	4.21	18.77	1.16
Fe	Mn	B	Mo	Co	Pb	Cd	Cr	Cu	Zn	Ni
				(mg kg^{-1} DM)						
2383	334.3	12.60	1.44	0.415	3.98	0.287	3.08	12.59	156.9	4.84

2.4. Statistical Processing

The results were developed statistically using the analysis of variance for a univariate experiment. The significance of the fertilizer treatment's effect on the values of the characteristics was checked on the basis of the Fisher–Snedecor F test. The $LSD_{0.05}$ value (for a detailed comparison of the means) was calculated using Tukey's test. The Statistica StatSoft 13.1 [28] program was used for calculation

3. Results and Discussion

After the third year of *Miscanthus* × *giganteus* cultivation, an increase in the soil content of total C and N was noted only in two plots (Table 7, Table 3 (a)), one with SS used on its own and the other treated with a mixture of SS at the highest dose and SMS (SS_{75} + SMS_{25}). Applied on its own, SS increased the N_t content in the soil to the largest extent, by 12.3%, while, in the plot with SS_{75} + SMS_{25}, it rose only by 6.0%. The content of C_t in both plots increased much less than that of N_t, only by 1.5%. In the remaining fertilized plots, a decrease in the content of N_t and C_t was noted, and, in the case of the latter, it was the largest in the control plot and in response to SMS used on its own.

Table 7. The content of N_t, C_t, and H (g kg^{-1} DM of soil) after the third year of *Miscanthus* × *giganteus* cultivation and the N_t and C_t content changes.

Experimental Object	N_t	N_t Content Change	C_t	C_t Content Change	H
	(g kg^{-1} DM Soil)				
Control plot	2.65	−0.20	33.6	−6.90	6.10
SS	3.20	0.35	41.1	0.60	5.80
SS_{75} + SMS_{25}	3.02	0.17	41.2	0.70	5.70
SS_{50} + SMS_{50}	2.80	−0.05	38.9	−1.60	5.20
SS_{25} + SMS_{75}	2.71	−0.14	37.8	−2.70	5.20
SMS	2.65	−0.20	35.6	−4.90	4.90
mean	2.84		38.0		5.48
$LSD_{0.05}$	0.461		3.59		NS

NS—not significant; SS—sewage sludge dose introducing 170 kg N ha^{-1}; SMS—spent mushroom substrate dose introducing 170 kg N ha^{-1}; SS used together with SMS in various proportions: SS_{75} + SMS_{25}, SS_{50} + SMS_{50}, and SS_{25} + SMS_{75}, with each dose introducing 170 kg N ha^{-1}.

A significant difference was noted between the SS plot and the control concerning the content of N_t. The total N content in the former was the highest, with 3.20 g kg^{-1} DM of soil, while, in the control plot, it was 2.65 g kg^{-1} DM. The content of N_t in the control plot was similar to that in the SMS plot. This may have been the result of the initial immobilization or inhibition of the N-nitrification process in the soil [29,30]. The nitrogen content in the remaining fertilized plots was higher than in the soil from the control plot, but these differences were not statistically significant. Bik-Małodzińska et al. [31] indicated a significant increase in soil nitrogen content after the application of SS, while Grzywnowicz and Strutyński [32] recorded an increase that was more than twice as high in response to manure application. Kuziemska et al. [33] stated that SS, compared to other organic wastes, had the greatest impact on soil nitrogen enrichment, which was confirmed by the results of the present research.

The lowest content of C_t was found in the control plot (33.6 g kg^{-1} DM of soil) and after applying SMS (35.6 g kg^{-1} DM of soil). The average content of C_t in the soil from all fertilized plots after three years of research was 38.0 g kg^{-1} DM of soil. On all fertilized plots, except for that with SMS, a significant increase in C_t was found in relation to the control plot.

Many authors stress the beneficial effect of SS in increasing the content of organic C in the soil [34,35]. Additionally, its content can be increased by organic matter development, limited agrotechnical treatments, and leaf shedding before harvest due to wind, the temperature, and heavy precipitation [36]. The accumulation of C in the soil is largely determined by the air temperature and the amount of precipitation. Any increase in temperature leads to an increase in evaporation and to a water deficit, which may be closely related to a decrease in soil C [14]. During the experiment, difficult growing conditions prevailed, especially in 2019 (Table 2), which probably contributed to the reduction in the soil C content in the control plot.

The content of H in the soil after three years of *Miscanthus* × *giganteus* cultivation was, on average, 5.48 g kg^{-1} (Table 7). No significant differences were noted in response to organic waste application. The highest H content was recorded in the control plot (6.10 g kg^{-1} DM of soil) and the lowest after the application of SMS. Grzywnowicz and Strutyński [32] found an increase in the active and potential acidity of soil treated with SS. The exchangeable acidity increased in response to increasing amounts of exchangeable hydrogen.

The average content of P in the soil after three years of *Miscanthus* × *giganteus* cultivation was 1.14 g kg^{-1} (Table 3 (a), Table 8), slightly lower than before the experiment (1.19 g kg^{-1}). After three years of research, the lowest content of this macronutrient was found in the control plot (1.08 g kg^{-1}) and in the plot with SS and SMS at the dose of SS_{25} + SMS_{75} (1.09 g kg^{-1}). In relation to the other fertilized plots, the soil content of P was most favorably affected by two combinations of SS and SMS, namely SS_{75} + SMS_{25} and SS_{50} + SMS_{50}, with 1.24 and 1.20 g kg^{-1} DM of soil, respectively. The differences in the soil P content in response to organic waste were not statistically significant. Xu et al. [37] reported that about 90% of the P found in sewage sludge was strongly bound to iron or aluminum. On the other hand, Grzywnowicz, Strutyński [32] and Kuziemska et al. [33] indicated that the P content in the soil increased with the use of SS.

Table 8. The content of selected macronutrients (g kg^{-1} DM of soil) after the third year of *Miscanthus* × *giganteus* cultivation.

Experimental Object	P	K	Ca	Mg	S
	(g kg^{-1} DM of Soil)				
Control plot	1.08	0.739	9.40	0.968	0.402
SS	1.11	0.825	10.58	1.23	0.451
SS_{75} + SMS_{25}	1.24	0.788	10.94	1.08	0.431
SS_{50} + SMS_{50}	1.20	0.769	11.23	1.35	0.474
SS_{25} + SMS_{75}	1.09	0.897	10.78	1.21	0.459
SMS	1.14	0.891	10.08	1.65	0.412
Mean	1.14	0.818	10.05	1.25	0.438
$LSD_{0.05}$	NS	0.061	0.392	0.544	NS

NS—not significant; SS—sewage sludge dose introducing 170 kg N ha^{-1}; SMS—spent mushroom substrate dose introducing 170 kg N ha^{-1}; SS used together with SMS in various proportions: SS_{75} + SMS_{25}, SS_{50} + SMS_{50}, and SS_{25} + SMS_{75}, with each dose introducing 170 kg N ha^{-1}.

The soil content of K was, on average, 0.818 g kg^{-1} and was higher than before the experiment was established (Table 3 (a), Table 8). This increase was due to the introduced organic materials, but also to the shedding of large quantities of *Miscanthus* × *giganteus* leaves [38]. The lowest content of total K in the soil was recorded in the control plot (0.739 g kg^{-1}). The highest content (0.897 g kg^{-1}) was reported in the plot where a combination of the lowest dose of SS and the highest dose of SMS (SS_{25} + SMS_{75}) was applied, but also in that with SMS (0.891 g kg^{-1}). Generally, in its composition, SS contains a small amount of K; therefore, after its application, a small increase in soil K is observed—or, in the case of small doses of sludge, no changes are noticed [37].

The average content of total Ca in the soil after three years increased compared to the start of the experiment and was 10.05 g kg^{-1} DM (Table 3 (a), Table 8). However, its amount in the control plot was the lowest; it was nearly the same as its content in the humus layer before the experiment was established. The highest content of total Ca in the soil after the third year (11.23 g kg^{-1} DM) was found in the plot treated with equal doses of SS and SMS (SS_{50} + SMS_{50}). Martyn et al. [39] reported that the use of SS in the production of energy crops increased the Ca content in the soil. SMS contains fairly large amounts of Ca [40], which is the result of its addition to the casing and to the substrate itself.

The average content of total Mg in the soil increased in relation to the start of the experiment and was 1.25 g kg^{-1} DM (Table 3 (a), Table 8). The highest value (1.65 g kg^{-1} DM of soil) was noted in the plot with SMS. The effect of other organic materials resulted in a slight, statistically insignificant increase compared to the control plot. Martyn et al. [39] found that an increase in the amount of Mg was dependent on the SS dose.

The total content of S in the soil after three years of *Miscanthus* × *giganteus* cultivation was, on average, 0.438 g kg^{-1} and increased by 0.061 g kg^{-1} in relation to the start of the experiment (Table 3 (a), Table 8). The lowest value was recorded in the control plot (0.402 g kg^{-1} DM of soil). On the other hand, the highest content (0.474 g kg^{-1} DM of soil) was noted in the soil treated with SS and SMS together (SS_{50} + SMS_{50}), both containing the same amounts of nitrogen. However, no significant effect of waste organic materials on the soil S content was found. Similarly, Czekała [41] stated that SS did not significantly affect its amounts in the soil.

The soil pH in H_2O after three years of *Miscanthus* × *giganteus* cultivation was neutral, ranging from 6.8 to 7.0 (Table 9). Organic materials did not significantly change its value, with SMS having deacidifying properties due to its composition. On the other hand, SS might have reduced the soil pH slightly [13].

Table 9. The content of available macronutrients (mg 100 g^{-1} DM of soil) and the soil pH after the third year of *Miscanthus* × *giganteus* cultivation.

Experimental Object	P_2O_5	K_2O	Mg	pH_{H2O}
	(mg 100 g^{-1} DM of Soil)			
Control plot	115.0	44.2	9.90	6.9
SS	128.0	41.7	11.8	6.8
SS_{75} + SMS_{25}	117.0	33.6	9.70	7.0
SS_{50} + SMS_{50}	116.0	35.8	10.3	7.0
SS_{25} + SMS_{75}	116.0	37.1	9.20	7.0
SMS	127.0	27.2	10.2	7.0
Mean	120.4	36.6	10.2	
$LSD_{0.05}$	NS	2.79	2.32	

NS—not significant; SS—sewage sludge dose introducing 170 kg N ha^{-1}; SMS—spent mushroom substrate dose introducing 170 kg N ha^{-1}; SS used together with SMS in various proportions: SS_{75} + SMS_{25}, SS_{50} + SMS_{50}, and SS_{25} + SMS_{75}, with each dose introducing 170 kg N ha^{-1}.

The content of available P in the soil after three years of research was the lowest in the control plot, with 115 mg 100 g^{-1} of soil (Table 9). The soil with SS and SMS applied on their own contained 128.0 and 127.0 mg P_2O_5, which were greater than in the control plot. On the other hand, the differences between the fertilized plots and the control were not significant. However, some authors have reported an increase in the content of available P forms in the soil in the first and subsequent years after the application of SS [13]. Other studies have indicated slight changes despite the high content of P in this waste. According to many authors, it results from P precipitation by Fe and Al ions [34].

The highest content of available K in the soil after the end of the three-year research was found in the control plot, with 44.2 mg of K_2O in 100 g^{-1} of soil (Table 9). However, its average content decreased significantly throughout the experiment to 36.6 mg 100 g^{-1} of soil. The lowest amount of available K was recorded in the plot with SMS (27.2 mg 100 g^{-1} of soil). According to Krzywy-Gawrońska [42], compared to other composted organic materials, SS composts contain less available K, which is released during the decomposition of this waste. Hajduk [43] found that, after the introduction of SS into the soil, the content of available K increased to a lesser extent due to its low content in this organic waste. The decreasing content of chemical elements in the soil treated with organic materials can be explained by the removal of nutrients with the biomass yield.

The highest content of available Mg (11.8 mg 100 g^{-1} of soil) was recorded on plots with SS (Table 9). The authors also noted a positive effect of SS on the content of available forms of P and K, which was observed by other researchers [44]. The process of the mineralization of organic materials introduced into the soil depends on many factors, including the soil pH, temperature, moisture, microbiological activity, and nutrient content [45].

After the third year of *Miscanthus* × *giganteus* cultivation, the increased content of some heavy metals (Pb, Cr, Zn, and Ni) was observed in the soil treated with waste organic materials (Table 3 (b), Table 10). In the control plot soil, the content of these elements was the lowest. The largest amounts of Pb, Cr, and Zn were recorded in the soil treated with SS. The content of Ni increased the most in response to SS and SMS applied together, both containing equal doses of N (SS_{50} + SMS_{50}). A twofold increase in soil Zn content after SS application was confirmed by Malinowska [46]. The average content of heavy metals in the soil after the third year could be listed in a series of decreasing values (mg kg^{-1}): Zn (273.4) > Pb (58.95) > Cr (11.68) > Ni (8.50) > Cd (0.432). An increase in the content of most heavy metals after the application of SS was also reported by Kalembasa and Malinowska [47]. The authors found decreased Cd content three years after SS application. Similar results were observed in the present research. The soil content of heavy metals after the third year was higher in response to SS than to SMS: for Cd, by 12.2%; for Pb, by 18.7%; for Cr, by 25.3%; for Zn, by 16.9%; and for Ni, by 14.7%. This resulted from the differences in the chemical composition of the organic waste materials.

Table 10. The content of selected soil elements and heavy metals (mg kg^{-1} DM of soil) after the third year of *Miscanthus* × *giganteus* cultivation.

Experimental Object	Cl	Fe	Mn	Cd	Pb	Cr	Zn	Ni
	(mg kg^{-1} DM of Soil)							
Control plot	0.078	5025	143.2	0.417	48.98	8.01	142.7	5.09
SS	0.084	7710	223.2	0.466	68.70	14.73	326.1	9.46
SS_{75} + SMS_{25}	0.092	10,500	201.3	0.460	63.50	12.82	304.0	8.36
SS_{50} + SMS_{50}	0.099	8960	214.0	0.402	58.81	12.71	312.4	11.70
SS_{25} + SMS_{75}	0.083	9910	222.8	0.438	57.91	10.80	287.3	8.33
SMS	0.089	9730	199.4	0.409	55.82	11.00	271.1	8.07
Mean	0.088	8639	200.7	0.432	58.95	11.68	273.4	8.50
$LSD_{0.05}$	NS	555.2	28.04	0.040	2.17	2.23	21.70	0.671

NS—not significant; SS—sewage sludge dose introducing 170 kg N ha^{-1}; SMS—spent mushroom substrate dose introducing 170 kg N ha^{-1}; SS used together with SMS in various proportions: SS_{75} + SMS_{25}, SS_{50} + SMS_{50}, and SS_{25} + SMS_{75}, with each dose introducing 170 kg N ha^{-1}.

After the third year, the content of heavy metals in the soil of the control plot in relation to the start of the experiment decreased for some elements by the following percentages:

Cr, 10.5%; Zn, 4.68%; and Ni, 7.96%. These changes were affected by their bioaccumulation by plants and their removal from the soil together with the biomass yield. However, no difference was found in the Pb content. On the other hand, an almost twofold decrease in the Cd content in the soil was recorded on all experimental plots after three years of research.

All fertilizer combinations increased the content of Fe and Mn in the soil at the end of the third year of *Miscanthus* × *giganteus* cultivation (Table 10). The highest Fe concentration (10,500 mg kg^{-1} DM of soil) was recorded after the combined application of SS and SMS (SS_{75} + SMS_{25}), as well as Mn (223.2 mg kg^{-1} DM of soil) after applying SS on its own. Data in the literature indicate increased content of Mn and Fe in soil treated with SS [48]. The content of Mn and Fe was the lowest on the control plot (Table 10).

The average Cl content in the soil after three years was 0.088 mg kg^{-1} DM (Table 10), with the highest value (0.099 mg kg^{-1} DM) in the plot with SS and SMS applied together (SS_{50} + SMS_{50}), both with the same nitrogen dose, and the lowest in the soil from the control plot (0.078 mg kg^{-1} DM). The content of Cl did not significantly vary across the experimental plots. According to Kabata-Pendias and Pendias [49], most Polish soils do not contain increased amounts of Cl, but regions with dry climates, coastal areas, and those close to communication routes are exposed to this element. Fertilizers with KCl, sometimes containing as much as 50% of Cl, increase its content in the soil. According to Burzała [50], Cl constitutes only 0.101% of the dry mass of SS, with a similar amount (0.1%) in SMS.

Soil microorganisms contribute to harvest residue decomposition and mineralization, which determines the availability of nutrients to plants [51,52]. Bacteria and fungi affect the formation of humus, its sorption properties, and the amounts of soil organic components. They also participate in the formation of a lumpy soil structure. The content of bacteria and fungi depends on the soil physicochemical properties, organic matter content, and plant species and on the tillage of the soil and its treatment with mineral and organic fertilizers [53]. The number of soil microorganisms also depends on the temperature, as well as on the climatic zone. Their largest number is seen at the humus level.

The data presented in Table 11 indicate that, after three years, the number of microorganisms varied in response to the different combinations of organic materials. The treatment had a positive effect on the development of soil bacteria, actinomycetes, and fungi, compared to the control.

Table 11. Total number of soil bacteria and actinomycetes (10^7 CFU g^{-1} DM of soil) and fungi (10^4 CFU g^{-1} DM of soil) after the third year of *Miscanthus* × *giganteus* cultivation.

Experimental Object	Total Number of Bacteria and Actinomycetes (10^7 CFU g^{-1} DM of Soil)	Total Number of Fungi (10^4 CFU g^{-1} DM of Soil)	Ratio of Bacteria and Actinomycetes to Fungi
Control plot	60.08	15.64	3841
SS	118.9	27.16	4345
SS_{75} + SMS_{25}	150.9	22.36	6749
SS_{50} + SMS_{50}	135.1	51.16	2641
SS_{25} + SMS_{75}	270.5	27.14	9969
SMS	96.67	21.81	4432
Mean	138.7	27.55	5329
$LSD_{0.05}$	13.62	4.17	102.3

SS—sewage sludge dose introducing 170 kg N ha^{-1}; SMS—spent mushroom substrate dose introducing 170 kg N ha^{-1}; SS used together with SMS in various proportions: SS_{75} + SMS_{25}, SS_{50} + SMS_{50}, and SS_{25} + SMS_{75}, with each dose introducing 170 kg N ha^{-1}.

The highest density of bacteria and actinomycetes (270.55×10^7 CFU g^{-1} DM of soil) was found in the soil treated with both SS and SMS in the combination of SS_{25} + SMS_{75}, and the lowest (more than four times lower) was found in the soil from the control plot (60.08×10^7 CFU g^{-1} DM of soil). The strongest effect on the total number of fungi was observed in the plot with SS and SMS applied together (51.16×10^4 CFU g^{-1} DM of soil), both with equal amounts of N (SS_{50} + SMS_{50}). The lower content of soil microorganisms after the application of SMS on its own compared to other fertilizer combinations (Table 11) might have been due to the fact that the SMS had a narrow C:N ratio.

In a pot experiment, Fijałkowski and Kacprzak [54] found that the number of soil fungi and actinomycetes increased after SS application. In a similar way, according to Wydro et al. [55], SS in the form of granules contributed to an increase in the number of bacteria and actinomycetes, while the soil fungal content depended on the date of sampling.

One of the problems encountered when using SMS in agriculture is the possibility of the presence of disease-causing fungi, so its chemical composition and the content of unfavorable microorganisms should be tested. This is why it is subjected to thermal treatment [17,56]. According to Wlazło et al. [56], thermal treatment reduces the total number of microorganisms from 1.7×10^7 CFU g^{-1} before the disinfection process to 2.7×10^6 CFU g^{-1} after treatment. Becher [57] stated that properly stored and disinfected SMS maintained its sanitary safety, without the development of pathogens or fungi. In the present research, after three years of plant cultivation, the number of soil fungi in the plot with SMS applied on its own was the lowest among all combinations (excluding the control plot).

The soil quality also depends on the ratio of bacteria to fungi [58]. The latter produce toxins and phytopathogenic substances, reducing the soil quality. On the other hand, they decompose organic matter, releasing mineral nutrients.

After three years of *Miscanthus* × *giganteus* cultivation, the soil ratio of bacteria to fungi ranged from 2641 to 9969 (Table 11), with high values indicating a limited number of fungi [55]. If this ratio is high, the soil has good microbiological properties. A decrease in the ratio of bacterial to fungal content is not desirable [58,59].

4. Conclusions

Organic waste applied in Central–Eastern Poland affected the soil chemical composition. After three years of *Miscanthus* × *giganteus* cultivation, the content of N_t and C_t changed. Their largest amounts were recorded in response to SS used on its own and together with SMS (SS_{75} + SMS_{25}). Compared to the N_t and C_t amounts before organic waste treatment, SS increased the soil N_t content by 12.3% and the C_t content by approximately 1.5%. In the remaining plots treated with SMS on its own and with SMS + SS, the latter one applied in lower doses, the N_t content decreased; it declined the most in the control plot and in the one with SMS (by 8%). Compared to the control, in all fertilized plots, except the one with SMS, a significant increase in the content of C_t in the soil was noted. Organic matter content is an integral indicator of soil quality and should be maintained by applying appropriate organic materials. The content of heavy metals in the soil remained at a low level, but it was higher after SS treatment than in response to SMS. Generally, the long-term effect of SS and of its combination with SMS (SS + SMS) improved the soil chemical and microbiological composition more than SMS used on its own. Compared to others, in plots with SMS only, an increase in the total Mg and K soil content was noted. Replacing traditional mineral fertilizers with post-production waste contributes to increasing the sustainability of lignocellulosic crop production. The use of recyclable waste in agriculture supports the circular economy and reduces the amounts of materials that need to be disposed of. This approach is in line with organic and sustainable farming strategies, reducing the cultivation costs. Using organic waste as a fertilizer can support sustainable farming by improving the soil quality, increasing its fertility, and reducing the negative impacts of agriculture on the environment. In the case of SMS applied on its own, additional fertilizer treatment should be applied to *Miscanthus × giganteus*. In the long run, the application of post-production organic waste can lead to soil degradation and reduced yields. It is also necessary to control the doses and types of materials applied to the soil to avoid adverse changes in the nutrient balance.

Author Contributions: Conceptualization, E.M.; data curation, E.M.; formal analysis, E.M.; writing—original draft, E.M.; writing—review and editing, E.M. and P.K.; investigation, P.K.; resources, P.K.; software, E.M. All authors have read and agreed to the published version of the manuscript.

Funding: The research was carried out under research theme No. 161/23/B, financed by a science grant provided by the Ministry of Education.

Institutional Review Board Statement: Not applicable.

Informed Consent Statement: Not applicable.

Data Availability Statement: The datasets generated and analysed during the current study are not publicly available as they are the authors' own data, but they are available from the corresponding author on reasonable request.

Conflicts of Interest: The authors declare no conflicts of interest.

References

1. Emmerich, W.E.; Lund, L.J.; Page, A.L.; Chang, A.C. Movement of heavy metals in sewage sludge-treated soils. American Society of Agronomy, Crop Science Society of America, and Soil Science Society of America. *J. Environ. Qual.* **1982**, *11*, 174–178. [CrossRef]
2. Alloway, B.J.; Jackson, A.P. The behaviour of heavy metals in sewage sludge-amended soils. *Sci. Total Environ.* **1991**, *100*, 151–176. [CrossRef] [PubMed]
3. Nunes, N.; Ragonezi, C.; Gouveia, C.S.; Pinheiro de Carvalho, M.Â. Review of sewage sludge as a soil amendment in relation to current international guidelines: A heavy metal perspective. *Sustainability* **2021**, *13*, 2317. [CrossRef]
4. Zia ur Rehman, Z.; Junaid, M.F.; Ijaz, N.; Khalid, U.; Ijaz, Z. Remediation methods of heavy metal contaminated soils from environmental and geotechnical standpoints. *Sci. Total Environ.* **2023**, *867*, 161468. [CrossRef]
5. Zeeshan Manzoor, M.; Sarwar, G.; Ibrahim, M.; Rehan, S.S.; Hasnain, Z.; Rais, A.; Gul, S.; Alfagham, A.T.; Manono, B.O.; Mehmood, K.; et al. Remediation quantum of organic amendments to immobilize potentially toxic heavy metals in wastewater-contaminated soils through maize cultivation. *Front. Environ. Sci.* **2024**, *12*, 1420705. [CrossRef]
6. Sharma, R.K.; Agrawal, M. Biological effects of heavy metals: An overview. *J. Environ. Biol.* **2005**, *26*, 301–313.
7. Giller, K.E.; Witter, E.; McGrath, S.P. Heavy metals and soil microorganisms. *Soil Biol. Biochem.* **2009**, *41*, 2041–2052. [CrossRef]
8. Han, Z.; Wenjie, L.; Ying, X.; Guowen, L.; Jianglong, C.; Zhao, C.; Zhang, L. Effects of dissolved organic matter on distribution characteristics of heavy metals and their interactions with microorganisms in soil under long-term exogenous effects. *Sci. Total Environ.* **2024**, *947*, 174565.
9. Harlt, W.; Putz, B.; Erhart, E. Influence of rates and timing of biowaste compost application on rye yield and soil nitrate levels. *Eur. J. Soil Biol.* **2003**, *39*, 129–139.
10. Bolan, N.; Kunhikrishnan, A.; Thangarajan, R.; Kumpiene, J.; Park, J.; Makino, T.; Kirkham, M.B.; Scheckel, K. Remediation of heavy metal(loid)s contaminated soils—To mobilize or to immobilize? *J. Hazard Mater.* **2014**, *266*, 141–166. [CrossRef]
11. Tadayyon, A.; Naeimi, M.M.; Pessarakli, M. Effects of vermicompost and vermiwash biofertilizers on fenugreek (*Trigonella foenum*) plant. *Commun. Soil Sci. Plant Anal.* **2018**, *49*, 2396–2405. [CrossRef]
12. Malinowska, E.; Wiśniewska-Kadżajan, B. The effects of different doses of organic waste on prairie cordgrass (*Spartina pectinata* L.) yield and selected energy parameters. *Energies* **2023**, *16*, 5599. [CrossRef]
13. Vaca, R.; Lugo, J.; Martínez, R.; Esteller, M.V.; Zavaleta, H. Effects of sewage sludge and sewage sludge compost amendment on soil properties and *Zea mays* L. plants (heavy metals, quality and productivity). *Rev. Int. Contam. Amb.* **2011**, *27*, 303–311.
14. Pikuła, D. Practices to prevent losses of soil organic carbon. *Stud. I Rap. IUNG-PIB Zesz.* **2019**, *59*, 77–91. (In Polish)
15. Berg, B. Decomposition of root litter and some factors regulating the process: Long term root litter decomposition in a Scots pine forest. *Soil Biol. Biochem.* **1984**, *16*, 609–618. [CrossRef]
16. Agricultural Report. Institute of Food Economics and Management, PIB 2023. Available online: https://ierigz.waw.pl (accessed on 20 January 2025).

17. Söchtig, H.; Grabbe, K. The production and utilization of organic-mineral fertilizer from spent mushroom compost. *Mushroom Sci* **1995**, *14*, 907–915.
18. Regulations of the Ministry of Climate and the Environment of December 31 on the Use of Municipal Sewage Sludge. Dz. U. 2022 poz. 89. Available online: https://isap.sejm.gov.pl/isap.nsf/DocDetails.xsp?id=WDU20220000089 (accessed on 3 January 2025)
19. Regulation of the Minister of the Environment on the Waste Catalogue of 27 September 2001 (Dz. U. nr 112 poz. 1206). Available online: https://isap.sejm.gov.pl/isap.nsf/DocDetails.xsp?id=wdu20011121206 (accessed on 3 January 2025).
20. Polish Soil Science Society. Grading classification of soils and mineral formations. *Rocz. Glebozn.* **2009**, *60*, 5–16.
21. Ziernicka-Wojtaszek, A. Comparison of selected indices for the assessment of atmospheric drought in the Podkarpackie Province in the years 1901–2000. *Woda ŚrodowiskoObszaryWiejskie* **2012**, *12*, 365–376. (In Polish)
22. Skowera, B.; Puła, J. Pluviometric extreme conditions in spring season in Poland in the years 1971–2000. *Acta Agroph.* **2004**, *1* 171–177. (In Polish)
23. *PN-R-04023:1996*; Agricultural Chemistry—Determination of Available Phosphorus in Mineral Soils. Polish Committee for Standardization: Warsaw, Poland, 1996.
24. *PN-R-04022:1996+Az1:2004*; Agricultural Chemistry—Determination of Available Potassium in Mineral Soils. Polish Committee for Standardization: Warsaw, Poland, 2004.
25. *PN-R-04020:1994+Az1:2004*; Agricultural Chemistry—Determination of Available Magnesium in Mineral Soils. Polish Committee for Standardization: Warsaw, Poland, 2004.
26. Kalembasa, S.; Carlson, R.W.; Kalembasa, D. A new method for the reduction in nitrates in total nitrogen determination according to the Kjeldahl method. *Pol. J. Soil Sci.* **1989**, *22*, 21–26.
27. Kalembasa, S.; Kalembasa, D. A quick method for determination of C/N ratio in mineral soils. *Pol. J. Soil Sci.* **1992**, *25*, 41–46.
28. StatSoft, Inc. Statistica (Data Analysis Software System), Version 13.1. 2021. Available online: https://www.statsoft.com (accessed on 9 January 2025).
29. Mahler, R.L.; Hamid, A. Evaluation of water potential, fertilizer placement and incubation time on volatilization losses of urea in two northern Idaho soils. *Commun. Soil Sci. Plant Anal.* **1994**, *25*, 1991–2004. [CrossRef]
30. Stewart, D.P.C.; Cameron, K.C.; Cornforth, I.S. Effects of spent mushroom substrate on soil chemical conditions and plant growth in an intensive horticultural system: A comparison with inorganic fertiliser. *Soil Res.* **1998**, *36*, 185–198. [CrossRef]
31. Bik-Małodzińska, M.; Myszura, M.; Żukowska, G.; Wójcikowska-Kapusta, A.; Wesołowska, S. Changes in the content of organic matter and mineral nitrogen in soil reclaimed by sewage sludge and mineral wool. *Przem. Chem.* **2021**, *100*, 99–102. (In Polish)
32. Grzywnowicz, I.; Strutyński, J. Changes in some chemical properties of the soil after sludge application for fertilization purposes. *Zesz. Probl. Post. Nauk Rol.* **1999**, *467*, 299–306. (In Polish)
33. Kuziemska, B.; Kalembasa, S.; Jaremko, D.; Popek, M. Influence of liming and organic waste materials on nitrogen, phosphorus and potassium content in the biomass of cocksfoot grown on soil contaminated with nickel. *Fragm. Agron.* **2013**, *30*, 65–73. (In Polish)
34. Debosz, K.; Petersen, S.O.; Kure, L.K.; Ambus, P. Evaluating effects of sewage sludge and household compost on soil physical, chemical and microbiological properties. *Appl. Soil Ecol.* **2002**, *19*, 237–248. [CrossRef]
35. Kirchmann, H.; Börjesson, G.; Kätterer, T.; Cohen, Y. From agricultural use of sewage sludge to nutrient extraction: A soil science outlook. *Ambio* **2017**, *46*, 143–154. [CrossRef]
36. Blanco-Canqui, H.; Shapiro, C.A.; Wortmann, C.S.; Drijber, R.A.; Mamo, M.; Shaver, T.M.; Ferguson, R.B. Soil organic carbon: The value to soil properties. *J. Soil Water Conserv.* **2013**, *68*, 129A–134A. [CrossRef]
37. Xu, H.; Zhang, H.; Shao, L.; He, P. Fraction distributions of phosphorus in sewage sludge and sludge ash. *Waste Biomass Valori.* **2012**, *3*, 355–361. [CrossRef]
38. Ociepa, A.; Bień, J.; Kacprzak, M. The influence of diversified fertilization of soil on the amount of crops of: *Miscanthus giganteus* and *Virginia mallow*. *Inż. Ochr. Środ.* **2008**, *11*, 255–260. (In Polish)
39. Martyn, W.; Wyłupek, T.; Czerwiński, A. Contents of some macroelements in soil and energy plants fertilized with sewage. *Łąkarstwo Polsce* **2007**, *10*, 149–158. (In Polish)
40. Medina, E.; Paredes, C.; Pérez-Murcia, M.D.; Bustamante, M.A.; Moral, R. Spent mushroom substrates as component of growing media for germination and growth of horticultural plants. *Bioresour. Technol.* **2009**, *100*, 4227–4232. [CrossRef] [PubMed]
41. Czekała, J. The influence of sewage sludge on some chemical properties of the soil. *Zesz. Probl. Post. Nauk. Roln.* **2004**, *499*, 39–46. (In Polish)
42. Krzywy-Gawrońska, E. Changes of pH and macronutrient contents in composts prepared from potato pulp and municipal sewage sludge during the time of their decomposition. *Zesz. Probl. Post. Nauk. Roln.* **2006**, *512*, 369–379. (In Polish)
43. Hajduk, E. The influence of sewage sludge on the bioaccumulation of trace elements in the biomass of energy willow (*Salix viminalis* L.) and the properties of light soil. Wyd. Uniwersytetu Rzeszowskiego. *Rozpr. Hab.* **2019**, 1–173. (In Polish)
44. Kulhánek, M.; Balík, J.; Černý, J.; Vašák, F.; Shejbalová, Š. Influence of long-term fertilizer application on changes of the content of Mehlich-3 estimated soil macronutrients. *Plant Soil Environ.* **2014**, *60*, 151–157. [CrossRef]
45. Gasco, G.; Martinez-Inigo, M.; Lobo, M. Soil organic matter transformation after a sewage sludge application. *EJEAFChe* **2004**, *3*, 3.

6. Malinowska, E. Zinc speciation in soil under various rates of sewage sludge and liming. *Environ. Prot. Eng.* **2016**, *42*, 5–15. [CrossRef]
7. Kalembasa, D.; Malinowska, E. Influence of sewage sludge fertilization on heavy metal content in biomass of silver grass during field experiment. *Environ. Prot. Engin.* **2009**, *35*, 149–155.
8. Rajmund, A.; Bożym, M. The iron and manganese content changes in light soil fertilized with sewage sludge or composts during the 6-years lysimeter experiment. *Woda-Środowisko-Obszary Wiejskie* **2017**, *17*, 101–113. (In Polish)
9. Kabata-Pendias, A.; Pendias, H. *Biogechemistry of Trace Elements*; Polish Scientific Publishing Company: Warsaw, Poland, 1999; p. 400. (In Polish)
0. Burzała, B. Thermal treatment of sewage sludge as one of the methods of its disposal. *Nowa Energ.* **2014**, *1*, 29–32. (In Polish)
1. Compant, S.; Duffy, B.; Nowak, J.; Clement, C.; Barka, E.A. Use of plant growth-promoting bacteria for biocontrol of plant diseases: Principles, mechanisms of action, and future prospects. *Appl. Environ. Microbiol.* **2005**, *71*, 4951–4959. [CrossRef] [PubMed]
2. Masto, R.E.; Chhonkar, P.K.; Singh, D.; Patra, A.K. Changes in soil biological and biochemical characteristics in long-term field trial on a sub-tropical inceptisol. *Soil Biol. Biochem.* **2006**, *38*, 1577–1584. [CrossRef]
3. Acosta−Martinez, V.; Mikha, M.M.; Vigil, M.F. Microbial communities and enzyme activities in soils under alternative crop rotations compared to wheat−fallow for the Central Great Plains. *Appl. Soil Ecol.* **2007**, *37*, 41–52. [CrossRef]
4. Fijałkowski, K.; Kacprzak, M. The influence of sewage sludge addition on selected physico-chemical parameters of degraded soils. *Inż. Ochr. Środ.* **2009**, *12*, 133–141. (In Polish)
5. Wydro, U.; Wołejko, E.; Jabłońska-Trypuć, A.; Butarewicz, A.; Łoboda, T. The impact of sewage sludge on a biological activity of urban soil. *Civil Environ. Eng.* **2017**, *8*, 103–108. (In Polish)
6. Wlazło, Ł.; Nowakowicz-Dębek, B.; Czech, A.; Ossowski, M.; Bis-Wencel, H.; Sasáková, N. Analysis of the microbiological quality of the substrate after mushroom production. *J. Anim. Sci. Biol. Bioecon.* **2019**, *37*, 25–30. [CrossRef]
7. Becher, M. Chemical composition of spent mushroom substrate after cultivation of agaricus bisporus as a waste organic material. *Ekon. Środow.* **2013**, *4*, 207–213. (In Polish)
8. Skwaryło-Bednarz, B. Ocena właściwości biologicznych gleby pod uprawą szarłatu (*Amaranthus cruentus* L.). *Acta Agroph.* **2008**, *12*, 527–534. (In Polish)
9. Wielgosz, E.; Szember, A. The effect of selected plants on the number and activity of soil microorganisms. *Ann. UMCS Sec. E.* **2006**, *61*, 107–119. (In Polish)

sustainability

MDPI

Article

Response of Soil Chemical and Biological Properties to Cement Dust Emissions: Insights for Sustainable Soil Management

Serdar Bilen [1,*], Murat Bilen [2], Mudahir Ozgul [1], Ekrem Ozlu [3] and Ugur Simsek [4]

[1] Department of Soil Science and Plant Nutrition, Faculty of Agriculture, Ataturk University, Erzurum 25030, Türkiye; mozgul@atauni.edu.tr
[2] Department of Chemical Engineering, Faculty of Engineering, Gazi University, Ankara 06560, Türkiye; muratbilen@gazi.edu.tr
[3] Vernon G. James Research and Extension Center, Department of Crop and Soil Sciences, North Carolina State University, Plymouth, NC 27692, USA; eozlu@ncsu.edu
[4] Department of Soil Science and Plant Nutrition, Faculty of Agriculture, Igdir University, Igdir 76000, Türkiye; ugur.simsek@igdir.edu.tr
* Correspondence: sbilen@atauni.edu.tr

Abstract: Land use change is associated with both higher fossil fuel usage and global cement production, significantly impacting environmental sustainability. Cement dust emission is the third-largest source of anthropogenic CO_2 emissions, right behind fossil fuel usage due to intense agricultural practices like aggressive tillage management. This study's aim is to determine cement dust emissions impacts on various tillage management methods and the formation of cement dust-affected CO_2 emissions, soil pH, soil organic matter content, total nitrogen content, available phosphorus, $CaCO_3$ content, bacteria and fungi populations, and enzyme activities. The target of this study is to evaluate how cement dust emissions impact the soil properties and sustainability of different tillage practices. Composite soils from wheat–sugar beet (potato)–fallow cropping sequences under conventional tillage (CT) and no-till (NT) management were collected (0–30 cm depth) with three replications at varying distances from a cement factory (1, 2, 4, 6, 8, and 10 km). To find differences among individual treatments and distances, a two-way ANOVA was employed along with Duncan's LSD test comparing the various effects of tillage techniques. The associations between soil chemical and biological properties and CO_2 fluxes under the impact of cement dust were examined using Pearson's correlation analysis. There were notable relationships between soil microbial population, enzyme activities, pH, $CaCO_3$, and CO_2 fluxes. The sampling distance from the cement plant had a substantial correlation with soil organic carbon, urease activity, pH, $CaCO_3$, and bacterial populations. According to the study, different tillage methods (CT and NT) affected the diversity and abundance of microorganisms within the soil ecosystem. CT was more beneficial for the microbial population and for sustainable management.

Keywords: CO_2 emissions; microbial population; soil enzymes; soil health indicators; cement dust; sustainable management practices

1. Introduction

Environmental pollutants such as dust may cause dust-laden air, degrade soil and water quality, and hence lead to unprofitable farming practices [1]. Fine solid particles in dust can be elevated and transported for long distances by wind [2]. As a result, the atmosphere's total suspended particle content is greater [3]. Regularly, industrial activities emit fine particulates and their depositions, which in turn may potentially lead to soil, vegetation, human, and animal health issues [4]. These issues explained above are considerably reported under the impacts of cement dust emissions, which gather significant attention in most developing countries [3].

Industrial cement dust production is result of thermal processes that increase CO_2 fluxes and detriment soil and environmental quality [5]. Alkalization and differentiation in soil chemical composition are the primary results of cement dust pollution [5,6]. The soil's physicochemical properties may be weakened, and crop growth might be negatively impacted by the pollution particles, which may penetrate as humid or dry deposits [6,7]. By affecting the concentration of activators or inhibitors in the soil solution and the effective concentration of the substrate, soil pH can have an impact on enzyme activity [8].

Thus, cement dust pollution harms the soil. Likewise, the buildup of cement dust in soils may affect the microbial activity, biomass, and compositions of microbial communities (bacteria population, fungi population, bacteria/fungi ratio) due to pH, temperature, and moisture in the soil [9–11]. The biogeochemical cycle of nutrients depends on soil microbial activity, which is adversely impacted by cement dust pollution. According to Uzma et al. [12], soil respiration, soil enzyme activity, and microbial biomass are the most often utilized microbial activity markers for tracking soil health. The ideal conditions for moisture, pH, temperature, and substrate concentration all affect soil enzyme activity. Enzyme activities (alkaline phosphatase (PAlk), acid phosphatase (PAcd), urease (UAc), and dehydrogenase activity (DAc)) are essential to soil biological health indicators [13] that are important to determine the influence of agricultural management activities [14,15].

How soil properties respond to cement dust accumulation under various tillage practices is still not well understood. Moreover, tillage practices can disturb soil and mix upper layer materials with the lower-level soil body differently. Differences in tillage practices may lead to translocation of cement dust to different depths of the soil profile. This study aims to determine cement dust emission impacts on soil biochemical properties and CO_2 fluxes at different distances from a cement plant and under different tillage managements. The hypotheses of this study include: (i) cement dust emissions significantly alter soil biochemical properties and microbial activity, with the extent of these changes varying according to tillage management practices (conventional tillage vs. no-till) and proximity to cement production facilities, and (ii) the differential incorporation and redistribution of cement dust within soil profiles under conventional tillage (CT) and no-till (NT) systems lead to measurable variations in CO_2 fluxes, soil pH, organic matter, microbial populations, and enzyme activities, impacting soil health and sustainability.

2. Materials and Methods

2.1. Study Area and Sampling Sites

The study area was in the vicinity of Turkey's Gümüşhane Cement Factory. The study soils show that since 2000 (14 years before sample collection), the well-drained and flat (slope < 2%) clay loam soils have been planted with wheat (*Triticum vulgare*), sugar beet (*Beta vulgaris*), potatoes (*Solanum tuberosum*), and fallow. The province of Gümüşhane serves as a transition zone between the Eastern Anatolia and Black Sea regions in terms of climate characteristics. Summers are cool, while winters are cold, with an average annual temperature of 10 °C. In Gümüşhane, which exhibits typical features of continental climates, there are only two months where the average temperature exceeds 20 °C. Although summer temperatures are high in Gümüşhane, rainfall begins from the first month onward. The average annual precipitation is approximately 409.2 mm. In Gümüşhane, the months where temperatures do not drop below zero extend from March to December. Based on long-term averages, the highest temperature is recorded in August (28.6 °C), while January sees the lowest temperature (2.5 °C). In terms of rainy days per month, May has the most rainy days (15.8), whereas August has the least (4.2). Additionally, regarding total monthly rainfall amounts, May receives the most rain (67.6 kg m^{-2}), while July receives the least (12.5 kg m^{-2}).

Three soil samples from depths ranging from 0 to 30 cm were taken in 2014, during the growing season of wheat and sugar beet. Under each tillage management system (conventional tillage, CT; no-till, NT), soil samples were gathered at each distance (1, 2, 4, 6, 8, and 10 km). Each treatment with three replicates totaled 36 soil samples that were collected in the vicinity of the cement mill. When choosing each sampling location, the wind speed (2.1 m sn^{-1}) and direction (from northwest to northeast) were considered.

2.2. Laboratory Analysis

The soil dilution plate method [16] was used to analyze soil microbial (bacteria, B_{pop}; fungi, F_{pop}) populations. For B_{pop}, colony-forming units (CFU) g^{-1} of oven-dried equivalent field-moist soil were measured using an automatic colony counter. Under a microscope, fungal colonies developing on agar were spotted at 10–30×. Viable fungal spores per milligram of oven-dried soil were used to quantify the total amount of fungus [17].

Basal Respiration (BR) was measured as the CO_2 evolution from unamended field-moist soil adjusted to 60% WHC for an incubation period of 10 days at 25 ± 1 °C in the dark. The NaOH static incubation technique was used to calculate basal respiration [18]. Using the formula basal respiration (mg CO_2 kg $soil^{-1}$ day^{-1}) = (CO_2 soil-CO_2 air)/20-day, the basal respiration rate (mg CO_2 kg soil day^{-1}) of the soil was determined.

The enzyme activity of alkaline phosphatase (P_{Alk}) and acid phosphatase (P_{Acd}) were measured using p-nitrophenyl phosphate substrate (μg pNP g $soil^{-1}$ h^{-1}). A urea solution (mg NH_4-N kg $soil^{-1}$ 2 h^{-1}) was used to measure the urease activity (UAc). According to Tabatabai [19], triphenyl tetrazolium chloride was reduced to triphenyl formazan (mg TPF kg $soil^{-1}$ 24 h^{-1}) in order to measure the soil's dehydrogenase activity (DAc).

The loss-on-ignition approach [20] was utilized to measure the content of soil organic matter (SOM). SOM was converted into total carbon (TC) content using a factor of 1.724 using a pressure calcimeter to determine the soil's $CaCO_3$ concentration [21] The available P (P_2O_5) (P_{Av}) in the soil was measured using the ammonium molybdate–ascorbic acid method [22], the soil pH was measured using a pH meter with a glass electrode meter in a 1:2.5 soil: water ratio [23], and the soil particle size distribution was measured using a hydrometer method [24].

2.3. Statistical Analysis

Two-way ANOVA was utilized to ascertain differences in particular treatments and distances while Duncan's LSD test was used to evaluate differences in overall effects at $\alpha < 0.05$ [25]. Using R Studio software (Version 2024.04.2+764), Pearson's correlation analysis was performed to look at links between soil health indicators and CO_2 fluxes under conditions of cement dust buildup and tillage techniques.

3. Results

Changes in soil chemical and physical properties under cement dust accumulation and tillage practices effects were significant (Table 1). However, soil microbial properties and enzymatic activities under cement dust accumulation and different tillage management practices were also significant (Table 2). Pearson's correlation analysis showed significant correlations between soil chemical and biological properties and CO_2 fluxes under the impact of cement dust and tillage management (Figure 1).

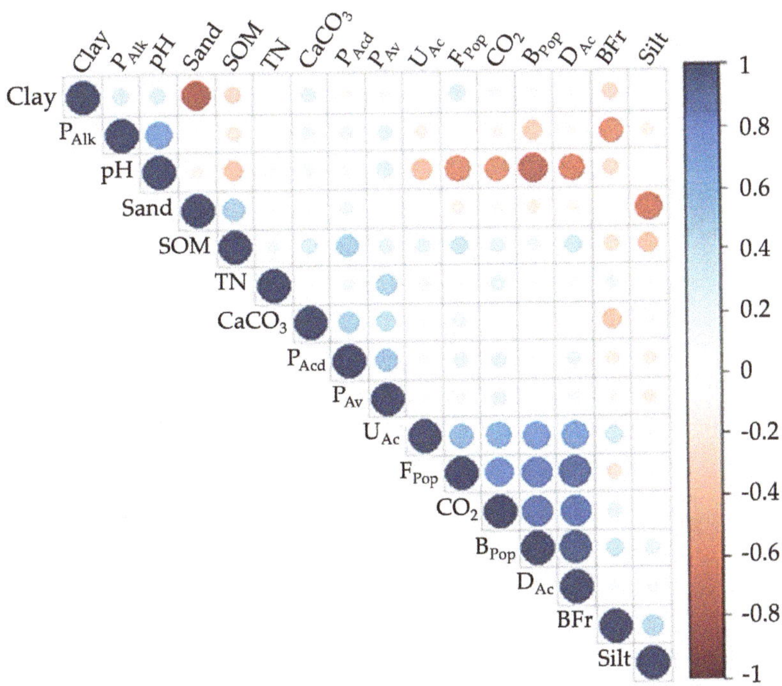

Figure 1. Pearson's correlation analysis of soil health indicators and CO_2 fluxes under cement dust accumulation at 0–30 cm depth in Gümüşhane district. The color and size of the circle denotes the magnitude and direction of the relationship. (SOM, soil organic matter; TN, total nitrogen content; PAv, available phosphorus content; Bpop, bacteria population; Fpop, fungi population; BFr, bacteria/fungi ratio; PAcd, acid phosphatase activity; PAlk, alkaline phosphatase activity; UAc, urease activity; DAc, dehydrogenase activity; CO_2, CO_2 fluxes).

Table 1. Some soil chemical and physical properties by cement dust accumulation and tillage practice in Gümüşhane district.

Treatments		Clay, %	Silt, %	Sand, %	pH	SOM, %	TN, %	P_{Av}, kg da^{-1}	$CaCO_3$, %
Distance, km	1	39.5 $^{a+}$	31.8 ^{ab}	28.7 ^{ba}	7.53 ^{b}	4.03 ^{ba}	0.18 ^{NS}	17.7 ^{NS}	4.03 ^{ab}
	2	39.8 ^{a}	31.2 ^{ab}	29.0 ^{ab}	7.67 ^{a}	3.74 ^{b}	0.18	19.8	3.74 ^{b}
	4	31.6 ^{b}	32.9 ^{ab}	35.5 ^{a}	7.38 ^{c}	4.21 ^{a}	0.19	20.2	4.21 ^{a}
	6	36.3 ^{ab}	33.0 ^{ab}	30.7 ^{ab}	7.22 ^{d}	4.18 ^{a}	0.19	19.0	4.18 ^{a}
	8	39.9 ^{a}	36.4 ^{a}	23.8 ^{b}	7.18 ^{d}	3.77 ^{b}	0.19	19.3	3.77 ^{b}
	10	34.5 ^{ab}	31.1 ^{ab}	34.5 ^{a}	6.90 ^{e}	3.75 ^{b}	0.18	18.3	3.75 ^{b}
Tillage	CT	39.0 ^{A}	32.1 ^{B}	29.0 ^{B}	7.37 ^{A}	4.11 ^{NS}	0.19 ^{NS}	19.8 ^{A}	4.11 ^{A}
	NT	34.9 ^{B}	33.4 ^{A}	31.8 ^{A}	7.26 ^{A}	3.78	0.18	18.3 ^{B}	3.78 ^{B}
Analysis of Variance									
Distance		<0.0001	<0.0001	<0.0001	<0.0001	<0.0001	0.06	0.09	<0.0001
Tillage		<0.0001	<0.0001	<0.0001	0.01	0.18	0.23	0.05	0.0001

$^{+}$ Different letters indicate significant differences between different distances at significant level of 0.05. Clay, clay content; Silt, silt content; Sand, sand content; pH, soil pH; SOM, soil organic matter; TN, total nitrogen content; P_{Av}, available phosphorus content; $CaCO_3$, calcium carbonate content; CT, conventional tillage; NT, no-till; NS, not significant.

Table 2. The significant impacts of cement dust on microbial populations and enzyme activities by cement dust accumulation and tillage practice in Gümüşhane district.

Treatments		B_{pop},	F_{pop},	BFr	P_{Acd},	P_{Alk},	UAc,	DAc,	CO_2,
		(×10^7 CFU)	(×10^4 Spore g^{-1})		µg pNP g $Soil^{-1}$ h^{-1}		mg NH_4-N kg $Soil^{-1}$ 2 h^{-1}	mg TPF kg $Soil^{-1}$ 24 h^{-1}	mg CO_2 kg $Soil^{-1}$ Day^{-1}
Distance, km	1	105 $^{e+}$	79 ^{c}	1.33 ^{b}	73 ^{c}	74 ^{a}	9.9 ^{c}	31 ^{d}	30.1 ^{c}
	2	119 ^{de}	80 ^{c}	1.49 ^{ab}	80 ^{cb}	75 ^{a}	11.7 ^{cb}	49 ^{cd}	41.3 ^{cb}
	4	145 ^{d}	94 ^{bc}	1.54 ^{ab}	84 ^{b}	67 ^{ab}	12.2 ^{b}	55 ^{cd}	44.5 ^{b}
	6	187 ^{c}	132 ^{ab}	1.42 ^{b}	94 ^{a}	58 ^{bc}	12.6 ^{b}	71 ^{cb}	51.1 ^{b}
	8	242 ^{b}	139 ^{ab}	1.74 ^{a}	76 ^{cb}	61 ^{bc}	13.5 ^{b}	83 ^{b}	64.9 ^{a}
	10	290 ^{a}	168 ^{a}	1.73 ^{a}	77 ^{cb}	54 ^{c}	15.9 ^{a}	116 ^{a}	69.1 ^{a}
Tillage	CT	206 ^{A}	144 ^{A}	1.44 ^{NS}	86 ^{A}	72 ^{A}	13.3 ^{NS}	84 ^{A}	57.5 ^{NS}
	NT	157 ^{B}	87 ^{B}	1.80	76 ^{B}	57 ^{B}	11.9	51 ^{B}	44.2
					Analysis of Variance				
Distance		<0.0001	<0.0001	0.0005	0.0004	<0.0001	0.002	<0.0001	<0.0001
Tillage		0.002	0.01	0.08	0.04	0.01	0.17	<0.0001	0.06

$^{+}$ Different letters indicate significant differences between different distances at significant level of 0.05. B_{pop}, bacteria population; F_{pop}, fungi population; BFr, bacteria/fungi ratio; P_{Acd}, acid phosphatase activity; P_{Alk}, alkaline phosphatase activity; UAc, urease activity; DAc, dehydrogenase activity; CO_2, CO_2 fluxes; CT, conventional tillage; NT, no-till; NS, not significant.

3.1. Soil Chemical Properties Impacted by Cement Dust Accumulation and Tillage Management

The distance from the cement factory significantly influenced soil pH ($p < 0.0001$), which ranged from 6.90 to 7.67 at 10 km and 2 km, respectively. Tillage practices significantly influenced the soil pH ($p < 0.01$) whereas soil pH under CT (7.37) was higher in than NT (7.26) (Table 1).

The SOM content statistically changed with distance; however, there was not a clear trend showing whether SOM increases or decreases with distance. The highest SOM content was obtained at 4 km (4.21%) followed by 6 km (4.18%), 2 km (4.03%), 8 km (3.77%), 10 km (3.75%), and 1 km (3.74%). However, SOM did not show any significant difference across different tillage practices (Table 1).

Although differences in soil particle size distribution were significant due to the impacts of both distance and tillage practice, soil texture classes did not show a clear trend associated with distance.

Total nitrogen (TN) and available phosphorus (P_{AV}) did not differ significantly with distance and tillage practices. However, both distance ($p < 0.0001$) and tillage managements ($p < 0.0001$) significantly influenced soil $CaCO_3$ content. Soil $CaCO_3$ ranged from 3.74% to 4.21% at 2 km and 4 km, respectively. Soil $CaCO_3$ content in CT (4.11%) was higher than NT (3.78%) (Table 1).

3.2. Soil Microbial Population Impacted by Cement Dust Accumulation and Tillage Management

Soil B_{pop}, F_{pop}, and their ratio (BFr) were significantly sensitive to both tillage practice and distance. Soils further away from the cement factory had higher microbial populations (B_{pop}, $p < 0.0001$; and F_{pop}, $p < 0.0001$) and BFr ($p < 0.0005$). B_{pop} at 10 km (290 cfu g^{-1}) was significantly higher than at 8 km (242 cfu g^{-1}) followed by those at 6, 4, 2, and 1 km by 19.83%, 55.08%, 2-, and 2.44-times lower values, respectively. Similar to distance, tillage management also impacted microbial populations where B_{pop} under CT (206 cfu g^{-1}) was significantly higher ($p < 0.002$) than NT (157 cfu g^{-1}) by 31.21%. Soil F_{pop} showed similar trends to B_{pop} under different tillage practices and at different distances (Table 2). The aforementioned findings show that CT systems dilute the effects of cement accumulation within plowing depth while NT led greater accumulation of cement dust on the soil surface.

The highest soil BFr was significantly ($p < 0.001$) higher at 10 km (1.73), followed by 8 km (1.74), 6 km (1.42), 4 km (1.54), 2 km (1.49), and 1 km (1.33). Duncan's test showed that differences by distance were not significant for the comparison between those at 10 km and 8 km, and also 6 km and 4 km. This result showed that any distance shorter than 4 km was not able to show significant differences. Moreover, tillage management practices also did not show significant differences for BFr ($p < 0.08$), Table 2.

3.3. Soil Enzyme Activity Impacted by Cement Dust Accumulation and Tillage Management

The enzyme activities statistically varied due to distance and tillage practices. Enzyme activities tended to rise with distance from the factory until 8 km. The soil P_{Acd} at 6 km (94 µg pNP g soil^{-1} h^{-1}) was significantly higher than those at 4, 2, 10, 8, and 1 km by 11.90%, 17.50%, 22.08%, 23.68%, and 28.77%, respectively. Also, soil P_{Acd} under CT (86 µg pNP g soil^{-1} h^{-1}) was significantly enhanced ($p < 0.04$) compared to that of NT (76 µg pNP g soil^{-1} h^{-1}). In addition, soil P_{Alk} ranged from 54 µg pNP g soil^{-1} h^{-1} to 75 µg pNP g soil^{-1} h^{-1} at 10 km and 2 km. This documents that the impacts of distance on P_{Alk} is statistically significant ($p < 0.0001$). The soil P_{Alk} under CT (72 µg pNP g soil^{-1} h^{-1}) was significantly greater ($p < 0.01$) compared to those of NT (57 µg pNP g soil^{-1} h^{-1}), Table 2.

Furthermore, soil UAc was significantly enhanced with increases in distance ($p < 0.002$), but tillage management did not influence soil UAc (Table 2). The soil UAc at 10 km (15.9 mg NH_4-N kg soil^{-1} 2h^{-1}) was significantly higher than those at 8, 6, 4, 2, and 1 km by 17.78%, 26.19%, 30.33%, 35.90%, and 59.64%, respectively. A similar trend was also detected for soil DAc. However, DAc in CT (84 mg TPF kg soil^{-1} 2 h^{-1}) was significantly higher ($p < 0.0001$) than in NT (51 mg TPF kg soil^{-1} 2 h^{-1}).

3.4. Soil CO_2 Fluxes Impacted by Cement Dust Accumulation and Tillage Management

Distance significantly influenced the soil surface CO_2 emissions ($p < 0.0001$), but not tillage practice ($p < 0.06$). Soil surface CO_2 emissions increased with distance (Table 2). CO_2 emissions at 1 km (30.1 mg CO_2 kg soil^{-1} day^{-1}) were significantly lower than those at 2 (41.3 mg CO_2 kg soil^{-1} day^{-1}), 4 (44.5 mg CO_2 kg soil^{-1} day^{-1}), 6 (51.1 mg CO_2 kg soil^{-1} day^{-1}), 8 (64.9 mg CO_2 kg soil^{-1} day^{-1}), and 10 km (69.1 mg CO_2 kg soil^{-1} day^{-1}). CO_2 emissions in CT (57.5 mg CO_2 kg soil^{-1} day^{-1}) were significantly higher ($p < 0.01$) than in NT (44.2 mg CO_2 kg soil^{-1} day^{-1}) (Table 2).

4. Discussion

4.1. Soil Biochemical Properties Impacted by Cement Dust Accumulation and Tillage Management

While addressing the impacts of cement dust emissions, one should consider soil microbial community compositions and activities that may influence CO_2 emissions associated with soil health indicators. Data from this study documented that soil particle size distribution was significantly impacted by distance; however, trends in soil sand, clay, and silt contents did not have a clear trend by distance. In general, the clay content in CT was higher than in NT, while sand content in NT was higher than in CT (Table 1). In areas where cement dust has accumulated heavily, implementing CT might help restore some structural integrity back into soils by promoting diverse microflora that contribute positively towards degradation processes. Conversely, frequent mechanical disturbances from intensive tilling might further expose soils to additional sedimentation from airborne particulates like cement dust. Soil texture plays a critical role in investigating microbial biomass due to its impacts on changing SOC and soil microclimate [26]. Soil clay content was associated with SOC level and microbial biomass [27]. Kara and Bolat [28] also documented similar observations that soil clay content was not associated with soil microbial biomass carbon.

The raw materials in cement dust contain about 80% limestone and 20% clay [29]. This may change some of the relationships of clay with other soil properties and reduce the significance of the correlations.

There was not any correlation between clay content and soil microbial biomass or SOM observed in this study; however, silt content had a significantly positive correlation with BFr, but a negative correlation with SOM. In addition, SOM had a significantly positive correlation with proportion of sand content, P_{Acd}, F_{pop}, and DAc, but a negative correlation with the proportion of silt and soil pH (Figure 1).

Increases in distance from the cement factory also increased SOM. This means fields where dust accumulation seemed to be lower had more SOM. This might be due to changes in microbial activities, the intensity of crop growth, and the impacts of tillage practices. No-till systems tend to preserve organic matter better than conventional tillage systems because they reduce erosion and maintain a layer of crop residue on the surface that supports various microorganisms. Microbial biomass is a critical part of SOM that can be used to determine the production capacity of fields under dust impact [28]. Soils under the impacts of alkaline cement dust can significantly reduce their microbial biomass carbon [7,9]. Soil $CaCO_3$ may play a critical role in protecting soil by binding soil particles and creating physical barriers between carbon and microbes. SOM under tillage practices was not statistically different. It is documented that such differences under different tillage practices may not be significant [30]. However, previous studies also showed that the NT tillage practice attempted to increase SOM in comparison to CT [31].

The pH of the soil and the distance from the cement mill were similarly inversely correlated. As the pH is more neutral at a higher distance, this may be the result of increased $CaCO_3$ buildup from cement dust deposition [28,32–34]. Additionally, the pH of the soil was greater under CT than it was under NT. Our results do not overlap existing research, which found that soil pH was considerably higher under NT than under CT [30,35]. Busari et al. [36] found that tillage techniques typically have an impact on the pH, TN, and chemical characteristics of the soil. However, tillage management and distance had a considerable impact on the $CaCO_3$ composition of the soil. It was stated that 80% of the basic ingredients in cement dust were limestone and 20% was clay [29]. At least 50% $CaCO_3$ is present in limestone [37]. This suggests that $CaCO_3$ accumulation is a sign of an attempt to gather more cement dust in the vicinity of the cement mill. Furthermore, under CT, soil $CaCO_3$ level was significantly greater than under NT. This higher $CaCO_3$ accumulation may result from the larger $CaCO_3$ accumulation closer to the cement plant, due to the interaction between tillage practice and distance from the factory [38,39]. P_{Av} content and TN had a favorable correlation. Similar positive correlations were also observed between P_{Av} and P_{Alk} activities, $CaCO_3$, and TN content. Under the impacts of distance and tillage management, differences in TN and P_{Av} activity were not statistically significant (Figure 1).

4.2. Soil Microbial Communities Impacted by Cement Dust Accumulation and Tillage Management

The balance between soil health and GHG (greenhouse gas) emissions determines how soil structure and carbon are stabilized, which contributes to biological diversity and enzyme properties that are investigated for soil fertility and management practices [40]. Therefore, microbial community compositions and microbial activities are important for determining the structure, capacity, and functionality of the soil. Increases in distance from the cement factory also increase B_{pop} and F_{pop} as well as BFr. Cement dust contains a high lime concentration, which changes soil microbial biomass [28,41]. This can be a reason for how cement dust changes soil microbial biomass [7,33]. It was documented by Bilen, Bilen and Turan [38] that increases in distance from the cement factory also increase B_{pop} in both CT and NT. Soil B_{pop} was positively correlated with UAc, CO_2, DAc, BFr, and F_{pop} and negatively correlated with P_{Alk} and pH (Figure 1). The unfavorable impacts on microbial community compositions closer to cement dust might be due to increases in soil pH from $CaCO_3$-containing dust [38,39]. Soil F_{pop} was positively correlated with SOM, UAc, CO_2, DAc, and B_{pop} but negatively correlated with pH. Moreover, B_{pop} and F_{pop} biomass under CT were significantly higher than those under NT. However, there was no significant contrast between CT and NT in terms of BFr. BFr was positively correlated with B_{pop} and silt content and negatively correlated with P_{Alk} and $CaCO_3$ content. Changes in soil fertility and management may quickly influence soil microbial community compositions and activities due to alterations in dominance or species of microbes [42]. The interplay between cement dust accumulation and different tillage practices could either exacerbate or mitigate negative effects on microbial communities. According to our study, different tillage methods (CT and NT) affected the diversity and abundance of microorganisms within the soil ecosystem. CT's effects were more beneficial for the microbial population and sustainable management.

4.3. Soil Enzyme Activity Impacted by Cement Dust Accumulation and Tillage Management

Microbial community compositions may impact soil enzyme activities [43]. Soil enzyme activities were increased with increases in distance from the cement factory except for P_{Alk}, whereas CT increased enzyme activities, except for UAc. CT may increase soil enzyme activities compared to NT because of better adaptation of microbes to disturbance stress [44]. Soil P_{Acd} activity was increased with CT and distance from the cement factory. Bilen, Bilen and Turan [38] reported that CT increases P_{Acd} and UAc by 13% and 12% compared to NT. In the present study's soil, P_{Acd} activity was positively correlated with SOM, $CaCO_3$, and P_{Av}. Similarly, soil P_{Alk} was decreased with increasing distance from the cement factory but increased with CT. P_{Alk} was reported to have a significant negative linear relationship with distance [38]. Soil P_{Alk} was positively correlated with soil pH but negatively correlated with B_{pop} and BFr (Figure 1). The correlation between soil pH and P_{Acd} and P_{Alk} activities was reported by Kalembasa and Symanowicz [10]. In addition, P_{Alk} activity was found to be more resistant to greater soil pH [45].

Soil UAc was increased by distance from the cement factory but not influenced by tillage practice. These results were also supported by Bilen, Bilen and Turan [38]. Soil UAc and DAc were positively correlated with pH whereas they were negatively correlated with F_{pop}, B_{pop}, CO_2, and DAc (Figure 1). The negative correlation of UAc with total bacteria and specifically G− bacteria was also reported by Ozlu, Sandhu, Kumar and Arriaga [40]. The optimum UAc was reported at a soil pH of 7.0 whereas the maximum DAc was documented at pH between 6.6 and 7.2 [46]. The lower substrate availability and labile pools of SOC and N significantly decrease both UAc and DAc due to higher pH conditions at a lower distance from the cement factory [43,47].

4.4. Soil CO_2 Fluxes Impacted by Cement Dust Accumulation and Tillage Management

Soil CO_2 fluxes were significantly enhanced by increases in distance from the cement factory ($p < 0.0001$). However, tillage practice did not influence CO_2 fluxes ($p < 0.06$). The combination of lower pH, and higher UAc, F_{pop}, B_{pop}, and DAc in the present study might lead to increased CO_2 fluxes. Intensive tillage practices and degradation of soil structure due to higher accumulation of dust may increase CO_2 fluxes. This might be owing to significant increases in B_{pop} and BFr as a result of greater disorder in the soil physical structure and lower carbon and nutrient contents [38]. Intensive tillage practices and higher crop residues enhance porosity and soil temperature, damage soil aggregation, and deprotect carbon from microbial attacks and hence enhance CO_2 fluxes due to the higher decomposition rate and chemical oxidation of carbon [18]. Also, there is a significant association between SOM content and CO_2 emissions, which could be further enhanced by improving soil structure and its functionality [48]. SOC-SOM may indirectly influence CO_2 fluxes related to other soil properties [49].

In conclusion, data showed that the distance from the cement factory significantly impacts SOC, TC, UAc, and B_{pop}. Soil surface CO_2 emissions were increased by increases in distance. Soil B_{pop}, SOM, and DAc seem to play a critical role in differentiating the impacts of cement dust accumulation at different distances. Soil pH was significantly influenced by tillage management; CT increased pH, clay content, $CaCO_3$, and B_{pop} content compared to NT. Soil TN and P_{Av} activity were not influenced by tillage practices. Soil F_{pop}, P_{Alk}, P_{Acd}, and DAc showed similar trends to B_{pop} under different tillage practices. However, CO_2 emissions and BFr were not significantly impacted by different tillage management practices. Moreover, significant correlations were seen between soil microbial community compositions, enzyme activities, soil pH, $CaCO_3$, and P_{Alk}. This study indicates that cement dust accumulation and tillage management negatively impact soil chemical and biological properties and CO_2 fluxes.

5. Conclusions

This study demonstrates the significant impact of cement dust accumulation and tillage management practices on soil chemical and biological properties, as well as CO_2 fluxes. The findings underline the importance of distance from pollution sources and tillage methods in maintaining soil health and supporting sustainable agricultural practices. Conventional tillage showed higher pH, clay content, and bacterial populations, while no-till practices reduced CO_2 emissions. These insights contribute to informed decisions on managing soil in areas affected by industrial emissions.

Author Contributions: S.B. and E.O. conceived and designed the experiments. S.B., M.B., M.O., E.O. and U.S. wrote the manuscript. S.B. and E.O. performed the experiments. E.O. analyzed the data; S.B., M.B., M.O., E.O. and U.S. reviewed the manuscript. All authors have read and agreed to the published version of the manuscript.

Funding: This research received no external funding.

Data Availability Statement: The data for this study are unavailable due to privacy restrictions.

Conflicts of Interest: The authors declare no conflicts of interest.

References

1. FAO. *Water Pollution from Agriculture. A Global Review. Executive Summary*; Food and Agriculture Organization of the United Nations: Rome, Italy, 2017.
2. Shao, J.; Mao, J. Dust Particle Size Distributions during Spring in Yinchuan, China. *Adv. Meteorol.* **2016**, *2016*, 6940502. [CrossRef]
3. Wufem, B.M.; Ibrahim, A.; Maina, H.M.; Gungsat, N.J.; Barnabas, N.J. The impacts of cement dust deposits on soil available micronutrients. *Int. J. Sci. Engi. Appl.* **2014**, *3*, 53–62. [CrossRef]
4. Vhahangwele, M.; Khathutshelo, L.M. Environmental Contamination by Heavy Metals. In *Heavy Metals*; Hosam El-Din, M.S., Refaat, F.A., Eds.; IntechOpen: Rijeka, Croatia, 2018; Chapter 7.
5. Mlitan, A.B.; Alajtal, A.I.; Alsadawy, A.M. Toxicity of heavy metals and microbial analysis of soil samples collected from the area around Zliten cement factory. *Open J. Air Pollut.* **2013**, *2*, 25–28. [CrossRef]
6. Lamare, R.E.; Singh, O.P. Effect of cement dust on soil physico-chemical properties around cement plants in Jaintia Hills, Meghalaya. *Environ. Eng. Res.* **2020**, *25*, 409–417. [CrossRef]
7. Arul, A.; Nelson, R. Effect of cement dust pollution on morphology and photosynthetic pigments of some legume plants grown in Ariyalur district, Tamil Nadu. *Int. J. Adv. Multidiscip. Res.* **2015**, *2*, 12–59.
8. Dick, W.A.; Cheng, L.; Wang, P. Soil acid and alkaline phosphatase activity as pH adjustment indicators. *Soil Bio. Biochem.* **2000**, *32*, 1915–1919. [CrossRef]
9. Alavi, M. Evaluation of cement dust effects on soil microbial biomass and chlorophyll content of *Triticum aestivum* L. and *Hordeum vulgare* L. *Int. J. Hum. Cap. Manag.* **2017**, *2*, 113–124.

10. Kalembasa, S.; Symanowicz, B. Enzymatic activity of soil after applying various waste organic materials, ash, and mineral fertilizers. *Pol. J. Environ. Stud.* **2012**, *21*, 1635–1641.
11. Dhal, B.; Thatoi, H.N.; Das, N.N.; Pandey, B.D. Chemical and microbial remediation of hexavalent chromium from contaminated soil and mining/metallurgical solid waste: A review. *J. Hazard. Mater.* **2013**, *250–251*, 272–291. [CrossRef] [PubMed]
12. Uzma, R.; Bhat, Z.; Ali, T.; Sofi, K.; Tajamul, R. A review on impact of cement dust on soil health with special reference to Kashmir, India. *Int. J. Chem. Sci.* **2018**, *2*, 24–30.
13. Utobo, E.; Tewari, L. Soil enzymes as bioindicators of soil ecosystem status. *Appl. Ecol. Environ. Res.* **2015**, *13*, 147–169.
14. Adetunji, A.T.; Lewu, F.B.; Mulidzi, R.; Ncube, B. The biological activities of β-glucosidase, phosphatase and urease as soil quality indicators: A review. *J. Soil Sci. Plant Nutr.* **2017**, *17*, 794–807. [CrossRef]
15. Ozlu, E. *Long-Term Impacts of Annual Cattle Manure and Fertilizer on Soil Quality Under Corn-Soybean Rotation in Eastern South Dakota*; South Dakota State University: Brookings, SD, USA, 2016.
16. Nandhini, B.; Josephine, R.M. A study on bacterial and fungal diversity in potted soil. *Int. J. Curr. Microbiol. Appl. Sci.* **2013**, *2*, 1–5.
17. Maiti, S.K. Soil Microbiological Parameters. In *Ecorestoration of the Coalmine Degraded Lands*; Maiti, S.K., Ed.; Springer: New Delhi, India, 2013; pp. 319–338.
18. Islam, K.R.; Weil, R.R. Land use effects on soil quality in a tropical forest ecosystem of Bangladesh. *Agric. Ecosyst. Environ.* **2000**, *79*, 9–16. [CrossRef]
19. Tabatabai, M.A. Soil Enzymes. In *Methods of Soil Analysis*; SSSA Book Series; Soil Science Society of America, Inc.: Madison, WI, USA, 1994; pp. 775–833.
20. Nelson, D.W.; Sommers, L.E. Total Carbon, Organic Carbon, and Organic Matter. In *Methods of Soil Analysis*; SSSA Book Series; Soil Science Society of America, Inc.: Madison, WI, USA, 1996; pp. 961–1010.
21. Loeppert, R.H.; Suarez, D.L. Carbonate and gypsum. In *Methods of Soil Analysis Part 3 Chemical Methods*; Soil Science Society of America, Inc.: Madison, WI, USA, 1996; Volume 5, pp. 437–474.
22. Combs, S.; Nathan, M. Recommended chemical soil test procedures for the North Central Region. *North Cent. Reg. Publ.* **1998**, *221*, 55–57.
23. Hendershot, W.; Lalande, H.; Duquette, M. Soil reaction and exchangeable acidity. In *Soil Sampling and Methods of Analysis*, 2nd ed.; CRC Press: Boca Raton, FL, USA, 2007.
24. Gee, G. Particle size analysis. In *Methods of Soil Analysis/ASA and SSSA*; Klute, A., Weaver, R.W., Sparks, D.L., Dane, J.H., Topp, G.C., Eds.; Soil Science Society of America, Inc.: Madison, WI, USA, 1986; pp. 383–409.
25. SAS Institute. *SAS 9.4 Output Delivery System: User's Guide*; SAS Institute: Hong Kong, China, 2014.
26. Hedo, J.; Lucas-Borja, M.E.; Wic, C.; Andrés-Abellán, M.; de Las Heras, J. Soil microbiological properties and enzymatic activities of long-term post-fire recovery in dry and semiarid Aleppo pine (*Pinus halepensis* M.) forest stands. *Solid Earth.* **2015**, *6*, 243–252. [CrossRef]
27. Wei, H.; Guenet, B.; Vicca, S.; Nunan, N.; Asard, H.; AbdElgawad, H.; Shen, W.; Janssens, I.A. High clay content accelerates the decomposition of fresh organic matter in artificial soils. *Soil Biol. Biochem.* **2014**, *77*, 100–108. [CrossRef]
28. Kara, Ö.; Bolat, I. Impact of alkaline dust pollution on soil microbial biomass carbon. *Turk. J. Agric. For.* **2007**, *31*, 181–187.
29. Dąbkowska-Naskręt, H.; Jaworska, H.; Długosz, J. Assessment of the total nickel content and its available forms in the soils around cement plant Lafarge Poland. *Int. J. Environ. Res.* **2014**, *8*, 231–236.
30. Ghimire, R.; Machado, S.; Bista, P. Soil pH, Soil Organic Matter, and Crop Yields in Winter Wheat–Summer Fallow Systems. *Agron. J.* **2017**, *109*, 706–717. [CrossRef]
31. Sapkota, T.B.; Mazzoncini, M.; Bàrberi, P.; Antichi, D.; Silvestri, N. Fifteen years of no till increase soil organic matter, microbial biomass and arthropod diversity in cover crop-based arable cropping systems. *Agron. Sustain. Dev.* **2012**, *32*, 853–863. [CrossRef]
32. Zerrouqi, Z.; Sbaa, M.; Oujidi, M.; Elkharmouz, M.; Bengamra, S.; Zerrouqi, A. Assessment of cement's dust impact on the soil using principal component analysis and GIS. *Int. J. Environ. Sci. Technol.* **2008**, *5*, 125–134. [CrossRef]
33. Bilen, S. Effect of cement dust pollution on microbial properties and enzyme activities in cultivated and no-till soils. *Afr. J. Microbiol. Res.* **2010**, *4*, 2418–2425.
34. Khamparia, A.; Chattergee, S.; Sharma, G. Assessment on effect of cement dust pollution on soil health. *Environ. Sci.* **2012**, *7*, 368–374.
35. Busari, M.; Salako, F. Effect of tillage, poultry manure and NPK fertilizer on soil chemical properties and maize yield on an Alfisol at Abeokuta, south-western Nigeria. *Niger. J. Soil Sci.* **2013**, *23*, 206–218.
36. Busari, M.A.; Kukal, S.S.; Kaur, A.; Bhatt, R.; Dulazi, A.A. Conservation tillage impacts on soil, crop and the environment. *Int. Soil Water Conserv. Res.* **2015**, *3*, 119–129. [CrossRef]
37. Sen Gupta, A.; McNeil, B. Chapter 6—Variability and Change in the Ocean. In *The Future of the World's Climate*, 2nd ed.; Henderson-Sellers, A., McGuffie, K., Eds.; Elsevier: Boston, MA, USA, 2012; pp. 141–165.

38. Bilen, S.; Bilen, M.; Turan, V. Relationships between Cement Dust Emissions and Soil Properties. *Pol. J. Environ. Stud.* **2019**, *28*, 3089–3098. [CrossRef]
39. Bilen, S.; Özgül, M.; Özlü, E.; Bilen, M. Impacts of Gümüşhane cement dust emissions on soil elemental compositions. *Turk. J. Agric. For.* **2021**, *45*, 766–774. [CrossRef]
40. Ozlu, E.; Sandhu, S.S.; Kumar, S.; Arriaga, F.J. Soil health indicators impacted by long-term cattle manure and inorganic fertilizer application in a corn-soybean rotation of South Dakota. *Sci. Rep.* **2019**, *9*, 11776. [CrossRef] [PubMed]
41. Lorenz, K.; Feger, K.-H.; Kandeler, E. The response of soil microbial biomass and activity of a Norway spruce forest to liming and drought. *J. Plant Nutr. Soil Sci.* **2001**, *164*, 9–19. [CrossRef]
42. Heidari, G.; Mohammadi, K.; Sohrabi, Y. Responses of Soil Microbial Biomass and Enzyme Activities to Tillage and Fertilization Systems in Soybean (*Glycine max* L.) Production. *Front. Plant Sci.* **2016**, *7*, 1730. [CrossRef] [PubMed]
43. Frederick, O.O.; Iroha, A.E.; Oswald, E.C. Evaluation of the Concentration of selected heavy metals and the effects on soil enzymatic activities in an Abandoned Cement Factory Nigercem Nkalagu and its environs. *Int. J. Biochem. Res. Rev.* **2013**, *4*, 16–27. [CrossRef] [PubMed]
44. Acosta-Martínez, V.; Zobeck, T.M.; Allen, V. Soil Microbial, Chemical and Physical Properties in Continuous Cotton and Integrated Crop–Livestock Systems. *Soil Sci. Soc. Am. J.* **2004**, *68*, 1875–1884. [CrossRef]
45. Turner Benjamin, L. Variation in pH Optima of Hydrolytic Enzyme Activities in Tropical Rain Forest Soils. *Appl. Environ. Microbiol.* **2010**, *76*, 6485–6493. [CrossRef] [PubMed]
46. Natywa, M.; Selwet, M. Respiratory and dehydrogenase activities in the soil under maize growth in the conditions of irrigated and nonirrigated fields. *Acta Sci. Polonorum. Agric.* **2011**, *10*, 3.
47. Liu, X.; Guo, K.; Huang, L.; Ji, Z.; Jiang, H.; Li, H.; Zhang, J. Responses of absolute and specific enzyme activity to consecutive application of composted sewage sludge in a Fluventic Ustochrept. *PLoS ONE* **2017**, *12*, e0177796. [CrossRef] [PubMed]
48. Ozlu, E.; Kumar, S. Response of surface GHG fluxes to long-term manure and inorganic fertilizer application in corn and soybean rotation. *Sci. Total Environ.* **2018**, *626*, 817–825. [CrossRef] [PubMed]
49. Mbonimpa, E.G.; Hong, C.O.; Owens, V.N.; Lehman, R.M.; Osborne, S.L.; Schumacher, T.E.; Clay, D.E.; Kumar, S. Nitrogen fertilizer and landscape position impacts on CO_2 and CH_4 fluxes from a landscape seeded to switchgrass. *GCB Bioenergy* **2015**, *7*, 836–849. [CrossRef]

sustainability

MDPI

Article

Effects of Solid Dairy Manure Application on Greenhouse Gas Emissions and Corn Yield in the Upper Midwest, USA

Eric Young * **and Jessica Sherman**

USDA-ARS Dairy Forage Research Center, Institute for Environmentally Integrated Dairy Management, 2615 Yellowstone Drive, Marshfield, WI 54449, USA; jessica.sherman@usda.gov
* Correspondence: eric.young@usda.gov; Tel.: +1-715-384-9673

Abstract: Dairy manure is an important nitrogen (N) source for crops, but its role in greenhouse gas (GHG) emissions and farm sustainability is not fully understood. We evaluated the effects of application of two dairy manure sources (bedded pack heifer, BP, and separated dairy solids, SDS) on corn silage yield and GHG emissions (carbon dioxide, CO_2; methane, CH_4; nitrous oxide, N_2O) compared to a urea-fertilizer-only control (80 kg N ha^{-1} yr^{-1}). The BP and SDS were applied at 18.4 and 19.4 Mg dry matter ha^{-1} in fall 2020 in the final year of ryegrass production. No-till corn was planted from 2021 to 2023, and GHG emissions were measured each season (from May to November). The results showed significantly greater CO_2-C emissions for BP in 2021 and no differences in 2022 or 2023. A small N_2O-N emission increase for BP occurred in the spring after application; however, seasonal fluxes were low or negative. Mean CH_4-C emissions ranged from 2 to 7 kg ha^{-1} yr^{-1} with no treatment differences. Lack of soil aeration appeared to be an important factor affecting seasonal N_2O-N and CH_4-C emissions. The results suggest that GHG models should account for field-level nutrient management factors in addition to soil aeration status.

Keywords: dairy manure; greenhouse gas emissions; nutrient management; soil aeration; no till

1. Introduction

Climate change poses major sustainability challenges to agriculture. Greater frequency of weather extremes (excess heat and moisture or prolonged drought) associated with global climate change increases crop production risk and loss of nutrients in general. Agriculture itself contributes substantially to greenhouse gas (GHG) emissions in the US and globally, accounting for 9.6% of total US GHG emissions in 2019 [1–4]. Between 21 and 37% of GHG emissions are linked to food production systems globally [2]. The main agricultural GHGs are carbon dioxide (CO_2), methane (CH_4), and nitrous oxide (N_2O). Approximately 60% of anthropogenic N_2O emissions are attributed to agricultural soils [5].

Anthropogenic CH_4 sources contributed approximately 61% of total global CH_4 emissions in 2017, and 44% of the observed CH_4 increases between 2006 and 2017 were attributed to enteric fermentation and manure associated with livestock production [6]. Per mole, N_2O has 273 times the global warming potential (GWP) of CO_2 (GWP = 1) for a 100-year timescale, whereas CH_4 has a GWP of 27 [7].

Cropland GHG emissions are affected by climate, cover type, and land management, which interact with soil properties (i.e., drainage/texture, organic carbon content, and N fertility) [8–13]. Gross CO_2 emissions reflect soil and root microbiological respiration and are broadly a function of hydroclimatic and soil properties; however, agronomic management, including tillage, manure application, and cropping system, also affects CO_2 emissions [9,14,15]. While increased tillage intensity can increase CO_2 emissions, lack of aeration from reduced and/or no-tillage systems can increase CH_4 emissions [16].

In contrast to CO_2, N_2O emissions are more episodic in nature, with weaker correlations with temperature and weather-related variables [9,17,18]. Emission of N_2O is associated with the denitrification process, whereby nitrate (NO_3^-) is reduced to NO, N_2O, or N_2 [19,20]. While agroecosystem N_2O emissions are generally small (0.1–3.1%), they are an important N loss mechanism and a potent GHG. Several studies note episodic N_2O pulses during the growing season when rainfall follows N applications [13,20,21].

A water-filled pore space between 70 and 80% is considered ideal for N_2O emission; however, emission rates also depend on the supply of nitrate (electron acceptor), gas diffusion rates, N availability, and soil pH [17,18,21–24]. Saha et al. [19] developed a machine-learning model that accounted for 65–89% of daily N_2O flux variance based on a relatively small set of variables that included soil moisture, days after fertilization, soil texture, air temperature, soil carbon, precipitation, and N fertilizer rate. Manure type (solid or liquid) and nutrient characteristics are also important [23]. For example, Gregorich et al. [24] reported nearly 3-fold greater N_2O-N losses after liquid manure was applied compared to semi-solid manure.

Emission of CH_4 generally requires longer term water saturation and lower soil redox potentials [25,26] to stimulate methanogenesis, the microbial energy transduction pathway that reduces CO_2 to CH_4 in anaerobic soil zones [6]. The difference between CH_4 production and oxidation to CO_2 governs the CH_4 available for diffusion to the atmosphere [26]. While wetlands are known CH_4 sources, poorly drained agricultural soils can also be CH_4 sources [12]. The amounts of CH_4 lost directly from manure application processes are generally quite low compared to soil methanogenesis [8]. Fields in arid climatic regions tend to have lower CH_4 emissions compared to humid areas [27,28]. Tillage intensity also affects CH_4 emissions [24,29]. Omonode et al. [16] showed that chisel or moldboard plowed fields functioned as CH_4 sinks, whereas no-till fields generated an average of 7.7 kg CH_4-C ha^{-1} yr^{-1} from lack of tillage.

With respect to farm sustainability, animal manure is an important nutrient source for cropland; however, there is an urgent need to better quantify manure application impacts on GHG emissions and field nutrient use efficiency. Previous research in Wisconsin indicated that GHG emissions for conventionally tilled fields were greater in a high N growing environment; however, emissions were unaffected by tillage at a site with poor drainage and lower N availability [30]. We hypothesized that a large, one-time application of solid manure (bedded heifer pack manure, BP, or separated solids from liquid dairy manure, SDS) would increase CO_2 and N_2O emissions compared to a fertilizer N only control. Here, we quantified seasonal GHG emission differences for corn (*Zea mays*) plots receiving a large one-time fall application of BP or SDS compared to a fertilizer N only control.

2. Materials and Methods

The experiment was performed at the Marshfield Agricultural Research Station (MARS) in Stratford, Wisconsin. The station manages 350 ha of mixed forage crop and pastureland for research and is owned by the University of Wisconsin and the USDA-ARS Dairy Forage Research Center, Environmentally Integrated Dairy Management Research Unit in Marshfield, Wisconsin. The 30-year average (1986–2015) annual temperature and precipitation are 6.9 °C and 875 mm, respectively. Soils at the site and in the region are largely derived from loess and silty alluvium mapped as Alfisols [31]. Our experiment was conducted on a Withee silt loam (fine-loamy, mixed, superactive, frigid Aquic Glossudalfs) with a dense B horizon restricting internal drainage at approximately 30 cm.

Gleization is typical in the top 25 cm, with glossic horizons and stark redoximorphic features/mottling indicating a shallow, seasonally high water table [31]. Twelve plots (9.2 × 4.6 m) arranged in four blocks were established in spring 2020 when the whole field (10 ha) was managed in ryegrass (*Lolium multiflorum*). Treatments (control, BP, and SS) were randomly assigned to plots, and a 1.5 m buffer that received no manure or fertilizer separated blocks and plots within blocks.

2.1. Manure Treatments and Agronomic Considerations

The MARS facility includes a 150-cow dairy with heifer raising facilities. Lactating cow manure is pumped through a screw-press separation system (Fan/Bauer, Upper Bavaria, Germany), and liquid manure is stored in a clay-lined earthen lagoon. Separated dairy solids (SDS) are used for bedding or land applied. Manure from the heifer facility (BP) was a mix of manure and sawdust bedding, scraped daily from pens and pushed out of the barn. Samples of each manure type were collected in 2020 and analyzed for dry matter/solids content (dried for 48 h at 55 °C) and total N and total carbon (C) via high-temperature combustion using a carbon/nitrogen analyzer (Elementar, Ronkonkoma, NY, USA) to determine field application rates based on providing a total N application rate of 300 kg N ha^{-1}. Manure samples were also sent to the University of Wisconsin's Soil and Forage Analysis Laboratory for determination of potassium (K), total P, and sulfur (S) analysis [32]. The application rates for BP and SDS, respectively, were 18.4 and 19.4 Mg ha^{-1} of dry manure solids.

University of Wisconsin's soil fertility system [33] was used to estimate manure N availability, where 25% of total manure N mass applied (75 kg N ha^{-1}) is assumed to mineralize to plant-available N (i.e., nitrate-N and ammonium-N) during the growing season. Residual manure N credits for second (10% of manure total N applied) and third year (5% of manure total N applied) contributions (2022 and 2023). Since manure was applied during fall, when temperatures were cooling down, the 25% manure N availability was assumed for the next spring's rotation to corn in 2021. Manure was hand-applied on 14 September 2020.

Manure was transferred from two premixed piles to experimental plots using 18.9 L buckets. After all manure was applied, garden rakes were used to evenly distribute manure across plots. Plots remained undisturbed until the following spring when corn was no-till planted, except for GHG sampling. It is important to note that manure was applied in the fall because it is a common practice for dairy farms in the region, as fields are generally too wet in the spring to apply manure without compaction damage.

In spring 2021, plots were sprayed with 0.39 L ha^{-1} of glyphosate, 0.91 kg ha^{-1} of ammonium sulfate, and 1.2 L ha^{-1} of haloxyfop on 13 May 2021. The same herbicide mixture and application rates were used in 2022 and 2023, applied about one week before planting. Corn was planted (for silage harvest) with a six-row no-till corn planter (John Deere, Moline, IL, USA) at a 5 cm planting depth. In 2021, corn (Prairie Estates, G2922) was planted on 13 May 2021 at a seeding rate of 80,275 seeds ha^{-1}. Corn (Pioneer 9188) was planted on 6 June 2022 at 71,000 seeds ha^{-1} (5 cm depth). In 2021 and 2022, 190 kg ha^{-1} of 15-8-21 (N-P_2O_5-K_2O) dry granular fertilizer was applied via the corn planter as a starter fertilizer. In 2023, corn (Dairyland 3022AM) was planted on 2 May 2023 at 80,275 seeds ha^{-1} (5 cm depth), and 202 kg ha^{-1} of 18-10-21-6 (N-P_2O_5-K_2O-S) granular fertilizer was applied as a starter through the planter. Control plots received the same starter fertilizer and 80 kg N ha^{-1} of surface-applied urea (46-0-0) each season when corn was between 60 and 80 cm in height (typically mid-July). No additional N was applied to BP or SDS plots, except the small amount of starter fertilizer at planting.

Corn yield was estimated by hand-harvesting the middle two corn rows each fall when test samples taken from the outside rows were approaching 35% moisture content. Corn plants were cut to leave 30 cm of stalk above the ground. Corn samples were run through a chopper, collected in tubs, and weighed using a portable scale. Freshly chopped subsamples at each harvest (2021–2023) were taken back to the laboratory and dried at 55 °C to determine moisture content.

2.2. Greenhouse Gas Emission Measurements

Nitrous oxide, CO_2, and CH_4 were measured using the static vented chamber technique following the GRACEnet protocol [34,35]. One stainless-steel chamber base was installed per plot (61 × 38.1 × 10.2 cm). Details on the chambers and the method have been reported previously [36–39]. Briefly, the chamber bases were inserted in the ground, so that approximately 3 cm remained above the soil surface. Insulated stainless-steel lids were sealed with weather stripping on top of the bases during measurement by clipping the steel lids to bases with binder clips. Frames were connected to a portable closed-path Fourier transform infrared spectrophotometer (FTIR, Gasmet DX4040, Vantaa, Finland) with 0.64 cm quick-connect fittings for FTIR tubing to measure GHG concentrations. Gas samples were pumped through 4 m long 4 mm ID PTFE tubing to the FTIR and back into the chamber as part of the closed loop. For each event, GHG concentrations were determined over a 7 min period and with field measurements approximately once every 10–14 days (between 0800 h and 1300 h).

Gas fluxes were computed from the rate of change in concentration over the sampling period using linear regression with R^2 values typically >0.99. Annual cumulative flux was estimated using linear interpolation between events [35]. It is important to note that the method permits estimates of both GHG production and consumption (sometimes referred to as negative fluxes), which is important for CH_4 and N_2O fluxes [34,35]. Precipitation and air temperature were measured by a weather station at the field edge (Spectrum Technologies, Aurora, IL, USA). Plot-level soil volumetric moisture content and temperature were measured (Delta-T Devices, Cambridge, Cambridgeshire, UK) at a 5 cm depth. In 2021, day 167 was the last measurement before a 100-day gap when the FTIR had a technical issue; measurements resumed on day 267 after FTIR was repaired.

2.3. *Agronomic Soil Testing*

Soil samples were taken at multiple depth intervals from each plot in September 2020 while plots were still in grass to determine background differences in soil fertility prior to manure addition. Soil samples were taken from each plot (2 cm diameter) at depths of 0–5, 5–15, and 15–30 cm. Samples were sent to the University of Wisconsin for standard soil test analysis (organic matter content, 1:1 pH, and Bray-1 extractable P and K) following standard procedures [32,33]. In addition, samples were also analyzed at the ARS laboratory for total carbon and N by dry combustion (Elementar, Ronkonkoma, NY, USA). During the corn phase (2021–2023), 30 cm long × 2 cm wide soil sample cores were taken from control plots to determine extractable (2 M KCl) soil nitrate-N (NO_3-N) and ammonium-N (NH_4^+-N) concentrations via automated flow injection (Lachat 8500 series II, Hach, Loveland, CO) using standard techniques [40,41]. This was performed to determine fertilizer N requirements. Since no further N additions were planned for BP and SDS plots, NH_4^+-N and NO_3^--N were only measured in the control plots.

2.4. *Statistical Analysis*

Plots were arranged in a randomized complete block design with 4 replications/ treatment. All statistical analyses were conducted using the Statistical Analysis System [42]. The dependent variables included pre-manure soil fertility, event-based GHG emissions (2020–2023), and seasonal GHG emissions. The main effect of manure type on GHG emission was assessed using the generalized linear mixed modeling procedure (proc glimmix) in SAS. Manure type was treated as a fixed effect and block as a random effect. Proc glimmix accommodates normally (i.e., Gaussian) and non-normally distributed variables. The log-link function and dual quasi-Newton optimization were used with the proc glimmix procedure. When glimmix models failed to converge, proc mixed was used with restricted maximum likelihood estimation. To simplify event-based GHG emission analyses, proc glimmix was also applied on an event basis in addition to repeated measures. Least square means were separated using the smm statement. Statistical significance was declared at $p \leq 0.05$. Pearson correlation coefficients and simple linear regression were also performed on select variables.

3. Results and Discussion

3.1. Weather Conditions

Weather conditions were characterized by substantially cooler temperatures compared to 30-year means (averaged over 1981–2010; Table 1). June and October 2021 were the exceptions and were 1.5 and 2.5 °C warmer than the long-term average. In addition, 2021 was wetter than normal from May to August. In contrast, much of the spring and summer of 2022 and 2023 were drier than normal. September and October 2023 had above-normal precipitation. The daily temperature and rainfall are presented in Appendix A.

Table 1. Long-term average monthly temperature and precipitation totals (1981–2010) at the Marshfield Agricultural Research Station and numerical differences between measured and long-term average values for each study year.

Month	1981–2010 †	2020	2021	2022	2023	1981–2010	2020	2021	2022	2023
	Monthly Average Temperature (°C)					**Monthly Precipitation Totals (mm)**				
April	7.2	−2.9	0.0	−3.8	−1.5	71.1	−34.3	−6.8	23.9	21.8
May	13.7	−1.5	−1.1	0.5	0.5	93.2	−12.9	11.4	−15.5	−36.2
June	18.9	0.1	1.5	−0.7	0.4	113.0	16.5	46.8	−42.1	−44.7
July	21.2	0.3	−0.8	−0.9	−1.4	101.0	28.3	37.9	−53.2	−2.6
August	20.1	−0.5	−0.3	−0.5	−0.6	109.0	−9.7	109.9	−18.3	−75.8
September	15.4	−1.9	0.0	−0.3	1.1	100.0	−44.0	−54.8	28.3	47.3
October	8.6	−4.3	2.6	−0.5	−0.2	66.3	−6.9	−13.7	−35.3	104.7
November	0.8	1.9	−0.2	0.2	0.2	56.4	1.7	−27.7	0.6	−55.4

† Long-term mean computed from historical weather data.

3.2. Soil Properties

Soil organic matter content, pH, Bray-1 extractable P and K, total C, and total N did not differ statistically before manure application (Figure 1). Soil pH was slightly lower at the 0–5 cm depth compared to the two deeper depths (Figure 1). Organic matter content, total C and N, Bray-1 P, and Bray-1 K were greater at 0–5 cm, showing marked nutrient stratification, which is typical of hay production/no-till fields from the lack of tillage and mixing of soil layers [24].

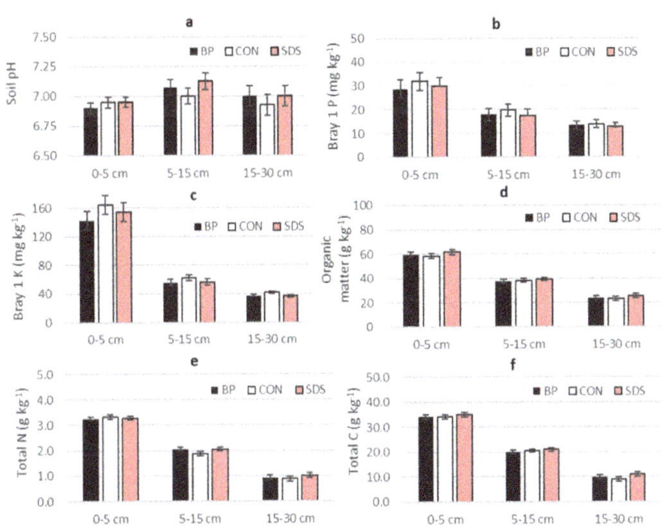

Figure 1. Select soil properties from samples taken in fall 2020 before manure application. Average soil pH (**a**), Bray-1 extractable phosphorus (**b**), Bray 1 extractable potassium (**c**), soil organic matter

content (**d**), total nitrogen (**e**), and total carbon (**f**) at three depth intervals (0–5 cm, 5–15 cm, and 15–30 cm). Error bars are one standard error of the mean of four replicated plots. Note: These were baseline soil analyses prior to manure application to assess consistency. Dairy heifer bedded pack manure designated plots = BP; separated dairy manure solids designated plots = SDS; control (fertilizer only) = CON.

3.3. Applied Manure Nutrients

Application of BP and SDS resulted in substantial inputs of dry matter solids, C, N, P, K, and S to manure-treated plots (Figure 2). The total N applied was close to the target rate of 300 kg ha^{-1}. Since manure application rates were based on total N application, slightly more total C, dry matter solids, and S were applied in the SDS treatment. The greatest difference in nutrient application was K, where BP application resulted in 142 kg ha^{-1} more total K added than SDS (Figure 2).

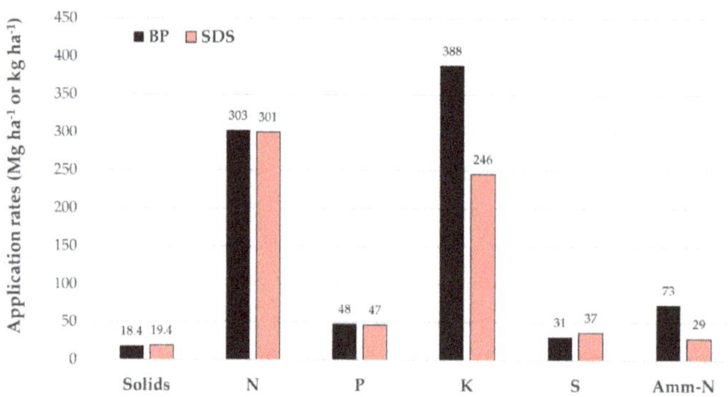

Figure 2. Dry matter solids and nutrients applied from dairy heifer bedded pack manure (BP) and separated dairy manure solids (SDS). Solids = manure solids (Mg ha^{-1}); N = total nitrogen (kg ha^{-1}); P = total phosphorus (kg ha^{-1}); K = potassium (kg ha^{-1}); S = sulfur (kg ha^{-1}); Amm-N = ammonium-nitrogen (kg ha^{-1}).

3.4. Corn Silage Yield

Corn silage yield ranged from 10 to >20 Mg ha^{-1} (dry yield) across the years (Figure 3). BP and SDS yields did not differ from the control yield in the first year after application in 2021. In 2022, the corn yields were lower and likely related to below-average precipitation for June and July (Table 1). The control plot corn silage yield was significantly greater than BP and SDS in 2022, indicating a lack of available N in manured plots compared to control, which received 80 kg N ha^{-1} yr^{-1} broadcast as urea. The lack of moisture in June and July 2022 probably reduced yields that season. Thus, it is clear that manure application contributed the available N to the corn crop but was insufficient to maximize yield under our study conditions.

Yields were greater in 2023 compared to 2021 or 2022, likely related to weather conditions; however, there were no significant differences among corn yields (Figure 3).

Soil NO_3^- and NH_4^+ concentrations in the control plots showed low available N status each year (Figure 3). Since 30 cm soil cores were taken for these samples when corn was between V3 and V6, and both NO_3^- and NH_4^+ were measured, these concentrations reflect the pre-sidedress soil N test (PSNT) for assessing additional corn N need [43]. The PSNT critical level is between 20 and 25 mg N kg^{-1} or approximately 2.5-fold greater than the average concentrations in our study, confirming the low soil-available N status of these soils (Figure 3).

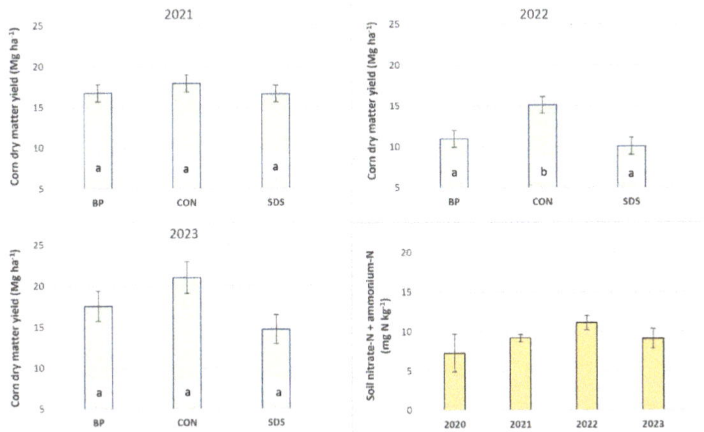

Figure 3. Corn silage dry matter yield in 2021, 2022, and 2023 along with mean soil inorganic nitrogen concentration (ammonium-N + nitrate-N) taken from control plots when corn was approximately at the V5 growth stage. Means with different lowercase letters differ at $p \leq 0.05$. Dairy heifer bedded pack manure = BP; separated dairy manure solids = SDS; control (fertilizer only) = CON.

3.5. Greenhouse Gas Emissions: Carbon Dioxide

In 2020, when plots were in ryegrass, the control tended to have numerically larger CO_2-C emissions than BP or SDS between day 150 and manure application on day 265 (Figure 4).

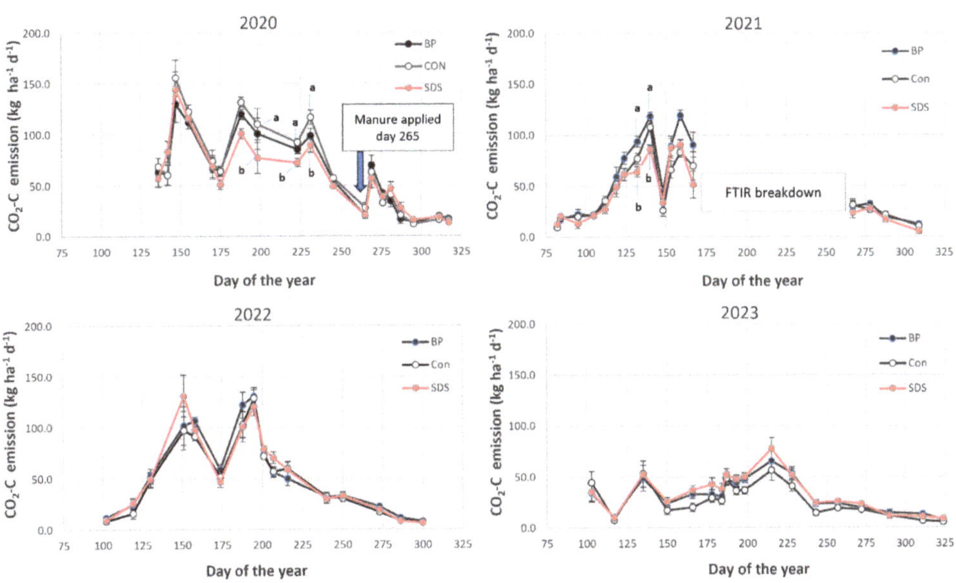

Figure 4. Mean carbon dioxide–carbon (CO_2-C) emissions for each study year. Means notated with different lowercase letters for an event differ ($p \leq 0.05$). Means without letters do not differ ($p > 0.05$). Error bars are one standard error of the mean of four replicated plots. Dairy heifer bedded pack manure = BP; separated dairy manure solids = SDS; control (fertilizer only) = CON. FTIR breakdown denotes the time during which the FTIR instrument was under repair, and field measurements were not taken.

There were no differences among treatments from day 269 to day 317 (November 19) in 2020. The larger CO_2-C emission from day 265 to day 269 was likely driven by the 2.7 °C increase in soil temperature from the previous event.

The mean CO_2-C emissions for BP from day 119 to day 264 in 2021 were numerically larger (Figure 4). The mean CO_2-C emission was numerically larger for BP on days 153, 159, and 167 in 2021 and on days 158, 273, and 286 in 2022 (Figure 4). In 2023, CO_2-C emissions for BP and SDS were consistently larger than control plots between days 150 and 272. Overall, the results indicate that CO_2-C emission increased with BP and SDS addition compared to urea only. With respect to seasonal totals, BP had significantly greater CO_2-C emission than the control or SDS in 2021, but it displayed no differences in other years, and cumulative emissions did not differ in the study (Figure 5). Greater manure C inputs likely contributed to larger CO_2-C emission for BP and SDS, whereas the control plots received no exogenous organic C, only fertilizer.

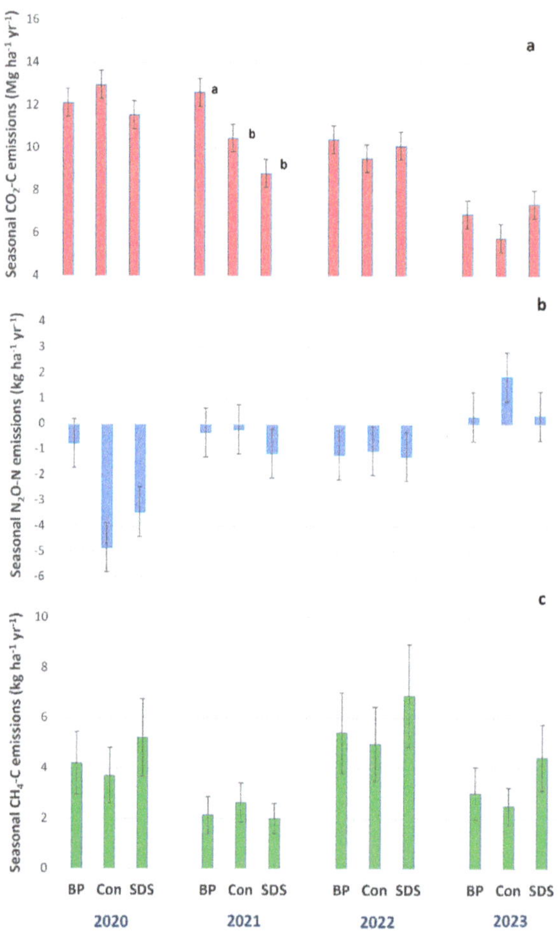

Figure 5. Cumulative carbon dioxide–carbon (CO_2-C) (**a**), nitrous oxide–nitrogen (N_2O-N) (**b**), and methane–carbon (CH_4-C) (**c**) emissions in each study year. Means notated with different lowercase letters differ ($p \leq 0.05$). Error bars are one standard error of the mean of four replicated plots. Dairy heifer bedded pack manure = BP; separated dairy manure solids = SDS; control (fertilizer only) = CON.

The correlations between soil temperature and CO_2 emissions are well documented [8,9,14,15,26]. Combining the treatments for the study duration showed that changes in CO_2 emissions were linearly related to soil temperatures ($R^2 = 0.33$, $p \leq 0.001$). Omonode et al. [16] also reported significant correlations between CO_2-C fluxes and soil temperature and moisture, although the variation accounted for by either variable was $\leq$27%. There was no correlation between soil moisture and CO_2-C emission in our study, which could be related to the poor drainage and relative abundance of soil water that could facilitate respiration.

Emissions of CO_2-C decreased linearly ($R^2 = 0.79$, $p < 0.0001$) from 2020 to 2023 across treatments at an average rate of -0.43 Mg CO_2-C ha^{-1} yr^{-1}, indicating that respiration decreased with the number of years in corn silage production. Other studies have also reported larger CO_2 emissions for perennial crops compared to corn, and they support our findings [14,27,44]. Dungan et al. [28] reported nearly 2-fold greater CO_2 emissions from alfalfa compared to corn for an irrigated semi-arid climate in Idaho. Ryegrass termination in our experiment likely contributed to elevated CO_2 emissions in 2020 and 2021 and may have confounded our ability to clearly detect additional CO_2 fluxes from BP and SDS decomposition/mineralization.

The effects of manure application on CO_2 emission are variable and depend on site-specific factors, including the hydroclimatic region, weather, soil type/properties, and crop management. For example, Omonode et al. [16] showed little impact of tillage type (chisel, moldboard, and no-till) on the annual GHG emissions from corn plots in Indiana. Hernandez-Ramirez et al. [8] evaluated GHG emissions in continuous corn and corn rotated with soybeans and either fertilized during the season with urea-ammonium nitrate or liquid swine manure and reported no treatment differences. In a semi-arid cropping system in southern Idaho, Dungan et al. [27] showed that CO_2-C emissions were greater in fields receiving fall and spring dairy manure compared to applications of urea only, and their results support our findings here.

3.6. Nitrous Oxide Emissions

The average N_2O-N fluxes before manure application in 2020 were low or negative, with plots following similar emission patterns until manure was applied on day 265 (Figure 6). There was a clear effect of BP application on N_2O-N emission, with the largest individual N_2O-N flux (0.61 kg N_2O-N ha^{-1} for BP) measured for the study occurring 4 days after application (day 269) (Figure 6). While BP induced a few positive N_2O-N fluxes in the fall after application, the mean N_2O-N emission remained close to zero.

Interestingly, both the control and SDS had their largest N_2O consumption rates (negative fluxes) in the fall after application. Several authors have stressed the importance of quantifying N_2O-N consumption and including it when estimating seasonal N_2O-N emissions [13,17,18,21].

The greatest N_2O consumption for the control and SDS was measured in the fall after application. It is possible that this was related to a lack of mineralization/nitrification with decreasing fall temperatures or that denitrification was able to effectively reduce NO_3^- efficiently to N_2 without N_2O emission. Poorly drained soils transform NO_3^- to N_2 more efficiently than well-drained, aerated soils [16–18,45]. It is important to note that saturated soils are not necessary for denitrification and reduction of NO_3^- to N_2. Studies suggest that a water-filled pore space between 70 and 80% is ideal for N_2O emission; however, this depends on other soil properties, including the pH, gas diffusion rates, and labile carbon [17,18].

In 2021, N_2O-N emission remained consistently low to negative, ranging from -0.05 to < 0.10 kg N_2O-N ha^{-1}. The patterns of N_2O-N emission in 2022 and 2023 were similar, with no significant differences between treatments on any sampling date and similar N_2O-N emission ranges over the season (Figure 7). When summed over seasons, the mean N_2O-N emissions were negative, except for 2023, when they were close to zero (Figure 5).

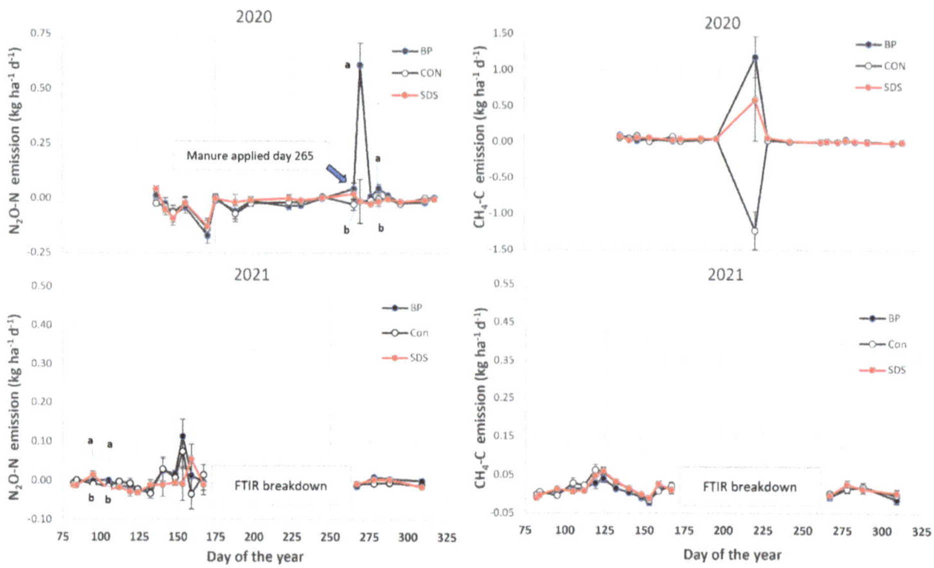

Figure 6. Mean nitrous oxide–nitrogen (N_2O-N) and methane–carbon (CH_4-C) emissions for 2020 and 2021. Means notated with different lowercase letters for an event differ ($p \leq 0.05$). Means without letters do not differ ($p > 0.05$). Error bars are one standard error of the mean of four replicated plots. Dairy heifer bedded pack manure = BP; separated dairy manure solids = SDS; control (fertilizer only) = CON. FTIR breakdown denotes the time during which the FTIR instrument was under repair, and field measurements were not taken.

Across seasons, N_2O consumption decreased significantly (R^2 = 0.53, p = 0.007), indicating that N_2O consumption rates decreased with time. We also found a moderate but significant positive relationship between corn silage yield and N_2O-N emissions (R^2 = 0.24, p = 0.006) across all study years. In a two-year study, Sadeghpour et al. [23] evaluated N_2O and mean corn grain yields in New York for separated manure solids, liquid dairy manure, and a fertilizer N only treatment and reported highly significant relationships between corn grain yield and N_2O emissions (R^2 = 0.89–0.94). While the relationship in our study was weaker, it nonetheless suggests that soil moisture variance influenced both corn yields and N_2O dynamics, as the Sadeghpour et al. [23] study demonstrated, with higher corn yields and N_2O emissions for the better drained plots on average.

There was no significant correlation between soil moisture and N_2O-N emission in our study. As discussed, poorly drained soils in general have a higher propensity to convert NO_3^- to N_2 relative to better aerated soils [17,18,20]. It is possible that the poorly drained conditions in our experiment promoted more denitrification than expected based on soil moisture alone, since denitrification can occur readily across a range of soil moisture contents, including field capacity [18,20]. Moreover, in our experiment, no tillage in combination with poor drainage and relatively high surface-soil organic matter content presumably provided ample labile C for denitrifying bacteria to convert NO_3^- to N_2, hence limiting N_2O emissions in general and possible correlations with changes in water content.

The broadcast application of solid manure may have further reduced N_2O-N emissions. Liquid dairy manure has a higher inorganic N/organic N ratio and greater N_2O-N emission potential in general than solid dairy manure, particularly when injected into the soil [22,23,36–39]. Whereas we found net N_2O consumption, Kazula and Lauer [30] reported N_2O-N emissions ranging from 1.2 to 1.7 kg N_2O-N ha^{-1} for tilled corn with fertilizer N applied, with no emission differences among three rotations (continuous corn, corn–soybean, or corn–soybean–wheat).

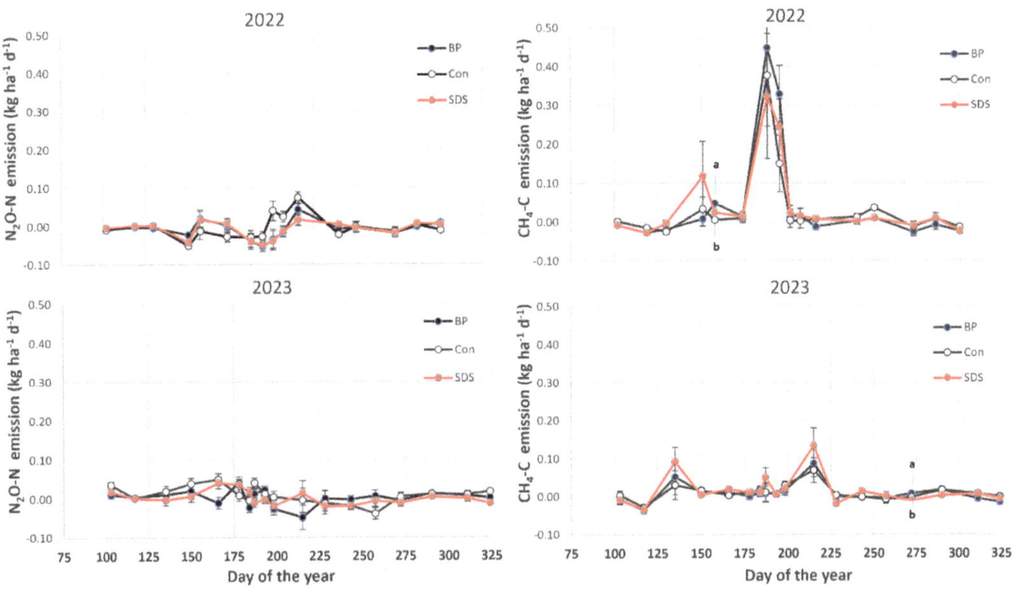

Figure 7. Mean nitrous oxide–nitrogen (N_2O-N) and methane–carbon (CH_4-C) emissions for 2022 and 2023. Means notated with different lowercase letters for an event differ ($p \leq 0.05$). Means without letters do not differ ($p > 0.05$). Error bars are one standard error of the mean of four replicated plots. Dairy heifer bedded pack manure = BP; separated dairy manure solids = SDS; control (fertilizer only) = CON.

These differences point to the importance of GHG models accounting for soil type to improve N_2O emission predictions, especially in large, variably drained fields. Better quantification of N_2O and other tradeoffs from manure application, such as runoff water quality and soil health, are critical to foster higher adoption of sustainable farming practices. Saha et al. [46] showed that manure application and termination of a legume-rich cover crop caused a microbial "priming effect" that increased respiration and oxygen consumption, inducing large N_2O-N emission spikes in a Pennsylvania organic cropping systems trial. When the cover crop was harvested before corn planting to prevent co-location of manure and biomass, N_2O-N emissions decreased by 60%. Therefore, managing manure and organic inputs for well-drained soils is more critical for N_2O-N emission mitigation compared to poorly drained conditions.

3.7. Methane Emissions

Unlike N_2O, CH_4-C emissions were consistently positive (i.e., more emission than consumption), with no differences among treatments prior to manure application in 2020 (Figure 6). There was also considerable CH_4-C emission in 2021, 2022, and 2023 (Figures 6 and 7). A large spike between days 175 and 200 for all treatments occurred in 2022, likely related to rainfall during this period; 13 out of 26 days had rainfall, with a total of 59.9 mm of precipitation (Figures 8, 9 and A1). The increase in rainfall/soil moisture under warm soil conditions likely drove higher CH_4-C emission due to a decrease in redox potential [25].

Interestingly, CH_4-C emission occurred across a range of soil moisture contents, indicating active methanogenesis even though soils were often in the range of field capacity with respect to soil moisture status. This background methanogenesis resulted in considerable seasonal fluxes (2–7 kg CH_4-C^{-1} yr^{-1}) measured in our study (Figure 5). Drainage affects CH_4 background emissions from agricultural fields and adjoining lands. Kandel et al. [47] reported high CH_4-C emission (172 kg CH_4-C ha^{-1} yr^{-1}) from a natural undrained peatland in Denmark, whereas nearby crop fields with tile drainage had close to net zero emissions (-1.5 to 1.5 CH_4-C ha^{-1} yr^{-1}). Such difference stresses the critical importance of factoring in soil type and drainage conditions in agroecosystem CH_4-C fluxes. Interactions between hydroclimatic and agronomic factors can confound independent assessments of manure and tillage effects. In more semi-arid regions, for example, a greater likelihood of CH_4-C oxidation may prevail, regardless of manure addition [27,28], whereas manure addition in more humid climates may increase CH_4-C emissions above background levels [24].

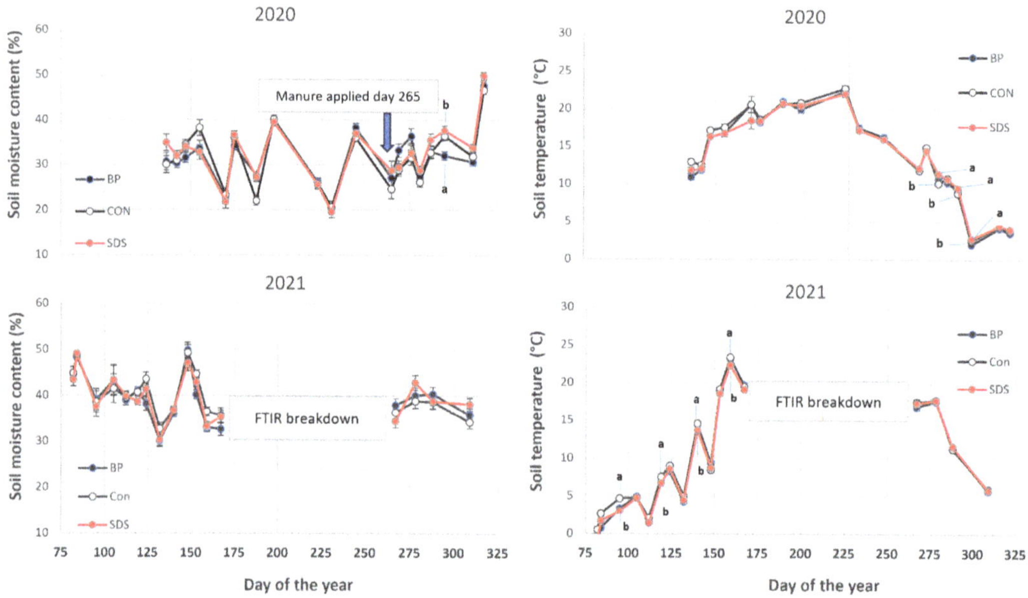

Figure 8. Changes in plot soil moisture content and temperature for 2020 and 2021. Means notated with different lowercase letters for an event differ ($p \leq 0.05$). Means without letters do not differ ($p > 0.05$). Error bars are one standard error of the mean of four replicated plots. Dairy heifer bedded pack manure = BP; separated dairy manure solids = SDS; control (fertilizer only) = CON. FTIR breakdown denotes the time during which the FTIR instrument was under repair, and field measurements were not taken.

The consistent CH_4-C emission measured in our experiment confirms limited aeration in these soils and suggests that the redox potentials were low enough to support active methanogenesis (≈ -150 to -160 mV) in portions of the profile. In a laboratory study, Wang et al. [25] demonstrated a 475-fold increase in CH_4-C emission with a change in redox potential from +10 to -230 mV. The lack of tillage in our study may have further enhanced CH_4-C emission, since tilling imperfectly drained soils enhances aeration, at least in the short term [8,16,24,26]. Alluvione et al. [48] reported nearly 20-fold greater CH_4-C emission from no-till treatments compared to conventionally tilled plots, with no effect of N fertilization or crop rotation. We hypothesize that poor drainage was the overriding factor affecting CH_4-C emissions at this site. Moreover, no-till leaves more residue on the surface, conserving moisture and C in surface-soil layers compared to tilled fields, which may have further enhanced CH_4 emission. The results stress the need for GHG models to capture changes in soil type and landscape heterogeneity to improve delineation among CH_4 sources and sinks in agricultural fields [49].

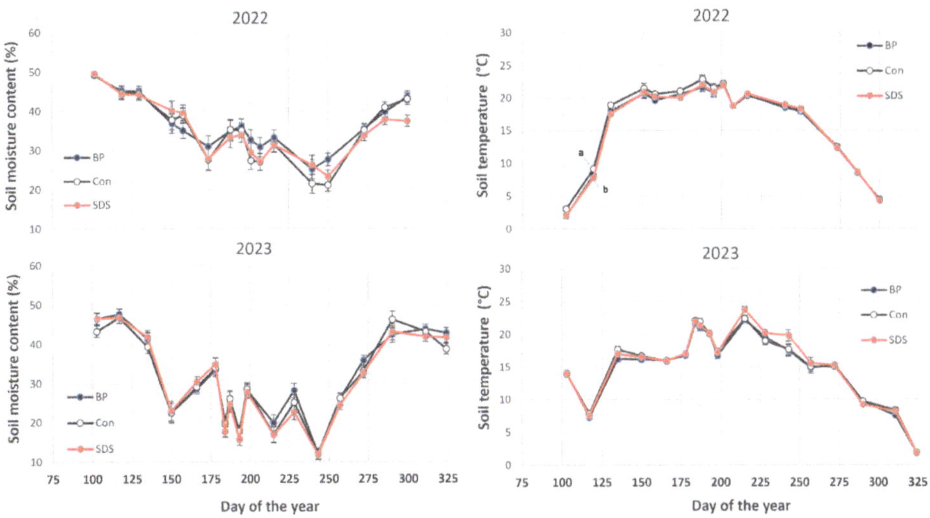

Figure 9. Changes in plot soil moisture content and temperature for 2020 and 2021. Means notated with different lowercase letters for an event differ ($p \leq 0.05$). Means without letters do not differ ($p > 0.05$). Error bars are one standard error of the mean of four replicated plots. Dairy heifer bedded pack manure = BP; separated dairy manure solids = SDS; control (fertilizer only) = CON.

3.8. Soil Moisture and Temperature

Plot-level soil volumetric moisture content and temperature were measured from 2020 to 2024 for each GHG measurement (Figures 8 and 9). There were a few significant differences noted among treatments in 2020 after manure application and again in 2021. In 2020, SDS had significantly higher soil moisture content than BP on day 295. The SDS treatment had a higher temperature than BP on days 276, 287, and 295 (Figure 8).

Control soil temperatures were consistently greater than SDS and BP between days 84 and 159 and in spring 2021. The results suggest that manure solids may have altered soil thermal properties. Combining temperature and soil moisture data over the study period revealed a significant inverse curvilinear relationship ($y = -0.0093x^2 + 0.1733x + 19.408$; $R^2 = 0.33$, $p < 0.001$), indicating that wetter plots were also significantly cooler on average. While the temperature and moisture differences among treatments were relatively minor in our experiment, the results illustrate the importance of capturing changes in soil temperature and moisture when implementing site-specific GHG modeling approaches [49].

4. Conclusions

Field experiments are critical for calibrating and improving GHG models aimed at developing more sustainable farming practices. In our study, the lack of N_2O emission and propensity for consumption were noteworthy and unlike the results observed in better drained soils, where N_2O fluxes tended to be positive. In the poorly drained field conditions of our study, N_2O emissions were negligible or negative, while CH_4 fluxes were considerable. In contrast to N_2O, CH_4 emission was consistent, likely due to a combination of poor drainage, lack of tillage, and abundant labile C. No tillage may have acted synergistically by providing a layer of undisturbed residue and C, with no mechanical aeration to oxidize CH_4. The results suggest that GHG models should account for both field nutrient management factors as well as soil aeration status.

Author Contributions: Conceptualization, E.Y. and J.S.; methodology, J.S. and E.Y.; software, E.Y.; validation, J.S. and E.Y.; formal analysis, E.Y.; investigation, E.Y.; resources, J.S. and E.Y.; data curation, J.S. and E.Y.; writing—original draft preparation, E.Y.; writing—review and editing, E.Y. and J.S.;

visualization, E.Y. and J.S.; supervision, E.Y.; project administration, E.Y.; funding acquisition, E.Y. All authors have read and agreed to the published version of the manuscript.

Funding: This research received no external funding.

Institutional Review Board Statement: Not applicable.

Informed Consent Statement: Not applicable.

Data Availability Statement: Data are contained within the article.

Conflicts of Interest: The authors declare no conflict of interest.

Abbreviations

The following abbreviations are used in this manuscript:

NH_4^+	Ammonia
BP	Bedded pack heifer manure
C	Carbon
CO_2	Carbon dioxide
N_2	Dinitrogen gas
MARS	Marshfield Agricultural Research Station
CH_4	Methane
GHG	Greenhouse gas
GWP	Global warming potential
N	Nitrogen
NO_3^-	Nitrate
N_2O	Nitrous oxide
K	Potassium
SDS	Separated dairy solids
S	Sulfur

Appendix A

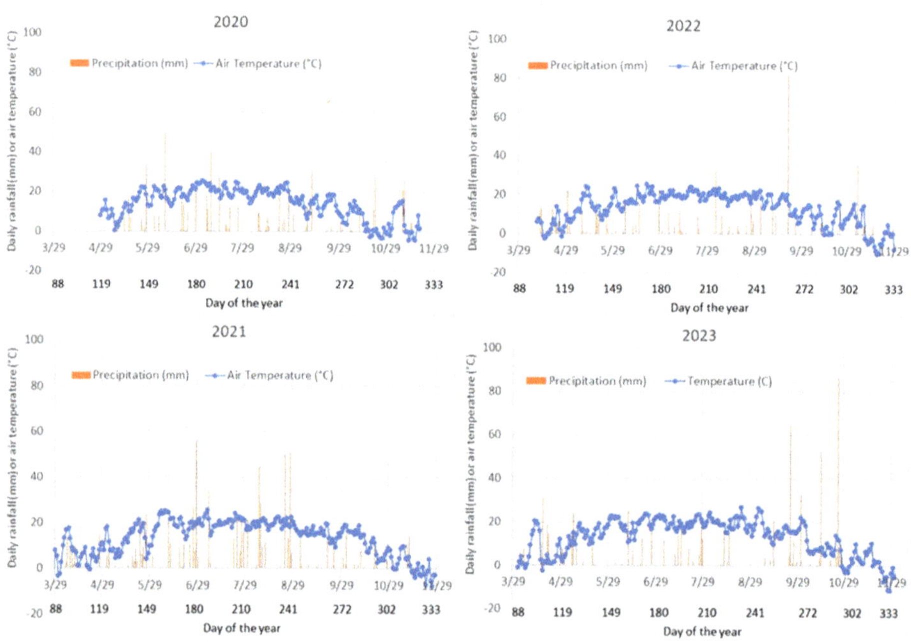

Figure A1. Daily total precipitation and temperature at the study site for 2020–2023.

References

1. IPCC. Intergovernmental Panel on Climate Change Special Report. Climate Change and Land. Available online: https://www.ipcc.ch/srccl/ (accessed on 20 March 2024).
2. USEPA. United States Environmental Protection Agency. Inventory of U.S. Greenhouse Gas Emissions and Sinks. National Emissions and Removals. 2024. Available online: https://www.epa.gov/ghgemissions/inventory-us-greenhouse-gas-emissions-and-sinks (accessed on 20 March 2024).
3. USEPA. United States Environmental Protection Agency. Sources of Greenhouse Gas Emissions. Available online: https://www.epa.gov/ghgemissions/sources-greenhouse-gas-emissions (accessed on 20 March 2024).
4. USEPA. United States Environmental Protection Agency. Causes of Climate Change. 2024. Available online: https://www.epa.gov/climatechange-science/causes-climate-change (accessed on 20 March 2024).
5. Del Grosso, S.J.; Ogle, S.M.; Nevison, C.; Gurung, R.; Parton, W.J.; Wagner-Riddle, C.; Smith, W.; Winiwarter, W.; Grant, B.; Tenuta, M.; et al. A gap nitrous oxide emission reporting complicates long- term climate mitigation. *Proc. Natl. Acad. Sci. USA* **2022**, *119*, e2200354119. [CrossRef] [PubMed]
6. Jackson, R.B.; Saunois, M.; Bousquet, P.; Canadell, J.G.; Poulter, B.; Stavert, A.R.; Bergamaschi, P.; Niwa, Y.; Segers, A.; Tsuruta, A. Increasing anthropogenic methane emissions arise equally from agricultural and fossil fuel sources. *Environ. Res. Lett.* **2020**, *15*, 071002. [CrossRef]
7. IPCC. Intergovernmental Panel on Climate Change (IPCC). *Climate Change 2021–The Physical Science Basis: Working Group I Contribution to the Sixth Assessment Report of the Intergovernmental Panel on Climate Change*; Cambridge University Press: Cambridge, UK, 2023. [CrossRef]
8. Hernandez-Ramirez, G.; Brouder, S.M.; Smith, D.R.; Van Scoyoc, G.E. Greenhouse gas fluxes in an eastern Corn Belt soil: Weather, nitrogen source, and rotation. *J. Environ. Qual.* **2009**, *38*, 841–854. [CrossRef] [PubMed]
9. Joshi, D.R.; Clay, D.E.; Clay, S.A.; Moriles-Miller, J.; Daigh, A.L.M.; Reicks, G.; Westhoff, S. Quantification and machine learning based N_2O–N and CO_2–C emissions predictions from a decomposing rye cover crop. *Agron. J.* **2022**, *116*, 1–15. [CrossRef]
10. Joshi, D.R.; Sieverding, H.L.; Xu, H.; Kwon, H.; Wang, M.; Clay, S.A.; Johnson, J.M.; Thapa, R.; Westhoff, S.; Clay, D.E. A global meta-analysis of cover crop response on soil carbon storage within a corn production system. *Agron. J.* **2023**, *115*, 1543–1556. [CrossRef]
11. Kwon, H.; Liu, X.; Xu, H.; Wang, M. Greenhouse gas mitigation strategies and opportunities for agriculture. *Agron. J.* **2021**, *113*, 4639–4647. [CrossRef]
12. McDaniel, M.D.; Saha, D.; Dumont, M.G.; Hernandez, M.; Adams, M.A. The effect of land-use change on soil CH_4 and N_2O fluxes: A global meta-analysis. *Ecosystems* **2019**, *22*, 1424–1443. [CrossRef]
13. Miller, L.T.; Griffis, T.J.; Erickson, M.D.; Turner, P.A.; Deventer, M.J.; Chen, Z.; Yu, Z.; Venterea, R.T.; Baker, J.M.; Frie, A.L. Response of nitrous oxide emissions to individual rain events and future changes in precipitation. *J. Environ. Qual.* **2022**, *51*, 312–324. [CrossRef]
14. Abdalla, M.; Kumar, S.; Jones, M.; Burke, J.; Williams, M. Testing DNDC model for simulating soil respiration and assessing the effects of climate change on the CO_2 gas flux from Irish agriculture. *Glob. Planet. Chang.* **2011**, *78*, 106–115. [CrossRef]
15. Faimon, J.; Lang, M. What actually controls the minute to hour changes in soil carbon dioxide concentrations? *Geoderma* **2018**, *323*, 52–64. [CrossRef]
16. Omonode, R.A.; Vyn, T.J.; Smith, D.R.; Hegymegi, P.; Gal, A. Soil carbon dioxide and methane fluxes from long-term tillage systems in continuous corn and corn-soybean rotations. *Soil Tillage Res.* **2007**, *95*, 182–195. [CrossRef]
17. Chapuis-Lardy, L.; Wrage, N.; Metay, A.; Chotte, J.L.; Bernoux, M. Soils, a sink for N2O? A review. *Glob. Chang. Biol.* **2007**, *13*, 1–17. [CrossRef]
18. Butterbach-Bahl, K.; Baggs, E.M.; Dannenmann, M.; Kiese, R.; Zechmeister-Boltenstern, S. Nitrous oxide emissions from soils: How well do we understand the processes and their controls? *Phil. Trans. R. Soc.* **2013**, *368*, 20130122. [CrossRef] [PubMed]
19. Saha, D.; Basso, B.; Robertson, G.P. Machine learning improves predictions of agricultural nitrous oxide (N_2O) emissions from intensively managed cropping systems. *Environ. Res. Lett.* **2021**, *16*, 024004. [CrossRef]
20. Grosz, B.; Matson, A.; Butterbach-Bahl, K.; Clough, T.; Davidson, E.A.; Dechow, R.; DelGrosso, S.; Diamantopoulos, E.; Dörsch, P.; Haas, E.; et al. Modeling denitrification: Can we report what we won't know? *AGU Adv.* **2023**, *4*, e2023AV000990. [CrossRef]
21. Xia, Y.; Wander, M. Management zone-based estimation of positive and negative nitrous oxide flux in organic corn fields. *Soil Sci. Amer. J.* **2022**, *86*, 1043–1057. [CrossRef]
22. Miller, D.J.; Chai, J.; Guo, F.; de León, M.A.P.; Ryals, R.; Dell, C.J.; Karsten, H.; Hastings, M.G. Cropland soil nitrogen oxide emissions vary with dairy manure incorporation methods. *AGE* **2024**, *7*, e20485. [CrossRef]
23. Sadeghpour, A.; Ketterings, Q.M.; Godwin, G.; Czymmek, K.J. Vermeylen, Agro-environmental consequences of shifting from nitrogen-to phosphorus-Based manure management of corn. *Soil Sci. Soc. Am. J.* **2017**, *81*, 1127–11138. [CrossRef]
24. Gregorich, P.; Rochette, A.J.; VandenBygaart, A.J.; Angers, D. Greenhouse gas contributions of agricultural soils and potential mitigation practices in Eastern Canada. *Soil Tillage Res.* **2004**, *83*, 53–72. [CrossRef]
25. Wang, Z.P.; DeLaune, D.; Masscheleyen, P.H.; Patrick, W.H. Soil redox and pH effects on methane production in a flooded rice soil. *Soil Sci. Soc. Am. J.* **1993**, *57*, 382–385. [CrossRef]
26. Smith, K.A.; Ball, T.; Conen, F.; Dobbie, K.E.; Massheder, J.; Rey, A. Exchange of greenhouse gases between soil and atmosphere. Interactions of soil physical factors and biological processes. *Eur. J. Soil Sci.* **2003**, *54*, 779–791. [CrossRef]

27. Dungan, R.S.; Leytem, A.B.; Tarkalson, D.D.; Ippolito, J.A.; Bjorneberg, D.L. Greenhouse gas emissions from an irrigated dairy forage rotation as influenced by fertilizer and manure applications. *Soil Sci. Soc. Am. J.* **2017**, *8*, 537–545. [CrossRef]
28. Dungan, R.S.; Leytem, A.B.; Tarkalson, D.D. Greenhouse gas emissions from an irrigated cropping rotation with dairy manure utilization in a semiarid climate. *Agron. J.* **2021**, *113*, 1222–1237. [CrossRef]
29. Soane, B.D.; Ball, B.C.; Arvidsson, J.; Basch, G.; Moreno, F.; Roger-Estrade, J. No-till in northern, western and south-western Europe: A review of problems and opportunities for crop production and the environment. *Soil Tillage Res.* **2012**, *118*, 66–87. [CrossRef]
30. Kazula, M.J.; Lauer, J.G. Greenhouse gas emissions from soils in corn-based cropping systems. *J. Environ. Qual.* **2023**, *52*, 1080–1091. [CrossRef] [PubMed]
31. Bockheim, J.G.; Hartemink, A.E. *The Soils of Wisconsin*; World Soils Book Series; International Union of Soil Scientists; Springer International Publishing: Cham, Switzerland, 2017.
32. Peters, J. Lab procedures and methods. University of Wisconsin Soils and Forage Analysis Laboratory. University of Wisconsin Extension. 2013. Available online: http://uwlab.soils.wisc.edu/lab-procedures/ (accessed on 12 December 2023).
33. Laboski, C.A.M.; Peters, J.B. *Nutrient Application Guidelines for Field, Vegetable, and Fruit Crops in Wisconsin*; University of Wisconsin–Extension, Ed.; Cooperative Extension: Madison, WI, USA, 2012.
34. Parkin, T.B.; Venterea, R.T. Chamber-Based Trace Gas Flux Measurements. In *Sampling Protocols*; Follett, R.F., Ed.; USDA-ARS: Beltsville, MD, USA, 2010. Available online: https://www.ars.usda.gov/ARSUserFiles/31831/2011%20Parkin%20and%20Venterea%20Trace%20Gas%20Protocol%20Revision%20Final.pdf (accessed on 7 March 2018).
35. Venterea, R.T. Simplified method for quantifying theoretical underestimation of chamber-based trace gas fluxes. *J. Environ. Qual.* **2010**, *39*, 126–135. [CrossRef] [PubMed]
36. Sherman, J.F.; Young, E.O.; Jokela, W.E.; Cavadini, J. Impacts of low-disturbance dairy manure incorporation on ammonia and greenhouse gas fluxes in a corn silage–winter rye cover crop system. *J. Environ. Qual.* **2021**, *50*, 836–846. [CrossRef] [PubMed]
37. Sherman, J.; Young, E.; Jokela, W.; Cavadini, J. Impacts of Low Disturbance Liquid Dairy Manure Incorporation on Alfalfa Yield and Fluxes of Ammonia, Nitrous Oxide, and Methane. *Agriculture* **2021**, *11*, 750. [CrossRef]
38. Sherman, J.; Young, E.; Jokela, W.; Kieke, B. Manure Application Timing and Incorporation Effects on Ammonia and Greenhouse Gas Emissions in Corn. *Agriculture* **2022**, *12*, 1952. [CrossRef]
39. Sherman, J.F.; Young, E.O. Greenhouse gas emissions with low disturbance liquid dairy manure incorporation into a live winter cereal cover crop-corn system. *Agronomy* **2022**, *12*, 2978. [CrossRef]
40. Harbridge, J. *Determination of Nitrate in 2M KCl Soil Extracts by Flow Injection Analysis. Quikchem Method 12-107-04-1-J.*; Lachat Instruments: Loveland, CO, USA, 2007.
41. Egan, L. *Determination of Ammonia by Flow Injection Analysis. QuikChem Method 90-107-06-3-A*; Lachat Instruments: Loveland, CO, USA, 2011.
42. SAS Institute Inc. *SAS 9.4 Guide to Software Updates*; SAS Institute Inc.: Cary, NC, USA, 2013.
43. Magdoff, F.R.; Ross, D.; Amadon, J. A soil test for nitrogen availability to corn. *Soil Sci. Soc. Am. J.* **1984**, *48*, 1301–1304. [CrossRef]
44. Malek, I.; Bouteldja, M.; Posta, K.; Foti, S.; Pinter, K.; Nagy, Z.; Balogh, J. Response of soil respiration to biotic and abiotic drivers in a temperate cropland. *Eurasian J. Soil Sci.* **2021**, *54*, 1038–1048. [CrossRef]
45. Fu, C.; Lee, X.; Griffis, T.J.; Baker, J.M.; Turner, P.A. A modeling study of direct and indirect N_2O emissions from a representative catchment in the U.S. Corn Belt. *Water Resour. Res.* **2018**, *54*, 3632–3653. [CrossRef]
46. Saha, D.; Kaye, J.P.; Bhowmik, A.; Bruns, M.A.; Wallace, J.M.; Kemanian, A.R. Organic fertility inputs synergistically increase denitrification-derived nitrous oxide emissions in agroecosystems. *Ecol. Appl.* **2021**, *31*, e02403. [CrossRef]
47. Kandel, T.P.; Laerke, P.E.; Elsgaard, L. Annual emissions of CO_2, CH_4 and N_2O from a temperate peat bog: Comparison of an undrained and four drained sites under permanent grass and arable crop rotations with cereals and potato. *Agric. For. Meteorol.* **2018**, *256*, 470–481. [CrossRef]
48. Alluvione, F.; Halvorson, A.D.; Del Grosso, S.J. Nitrogen, tillage, and crop rotation effects on carbon dioxide and methane fluxes from irrigated cropping systems. *J. Environ. Qual.* **2009**, *38*, 2023–2033. [CrossRef]
49. Oertel, C.; Matschullat, J.; Zurba, K.; Zimmermann, F.; Erasmi, S. Greenhouse gas emissions from soils—A review. *Geochemistry* **2016**, *76*, 327–352. [CrossRef]

sustainability MDPI

Article

Identification of Soil Quality Factors and Indicators in Mediterranean Agro-Ecosystems

Eleftherios Evangelou [1,2,*] and Christina Giourga [2]

1 Institute of industrial and Forage Crops, Hellenic Agricultural Organization "Dimitra", 41335 Larisa, Greece
2 Department of Environment, University of the Aegean, University Hill, 81100 Mytilene, Greece; hgio@aegean.gr
* Correspondence: evangelou@elgo.gr; Tel.: +30-24-1067-1287

Abstract: Soil quality offers a holistic approach for understanding the relationships between soil's biological, chemical, and physical properties, which is crucial for sustainable land use and the management of non-renewable soil resources. This study evaluates the impact of land use on a set of 23 soil quality indicators (SQIs) across 5 land uses of the Mediterranean agro-ecosystems: forest, olive groves, wheat fields, a corn/wheat crop rotation system, and pasture. Seasonal soil sampling was carried out over two consecutive years in three conventionally managed fields representing each land use type. For each sampling, physicals SQIs (soil moisture, porosity-Vp-, bulck density-BD-, water holding capacity-WHC-, clay, silt, sand), chemical SQIs (organic carbon-Corg-, total Nitrogen-TN-, C/N, PH, electrical conductivity-EC-, ammonium-NH_4-N-, nitrate-NO_3-N- and available nitrogen-Nmin-), and biological SQIs (soil microbial biomass C-Cmic- and N-Nmic-, Cmic/Nmic, Cmic/Corg, Nmic/TN, active carbon—Cact-, Cact/Corg) were evaluated. Through multivariate analysis, five key soil quality factors—*organic matter*, *microbial biomass*, *nutrients*, *C/N ratio*, and *compaction*—were identified as indicators of soil quality changes due to land use, explaining 82.9% of the total variability in the data. Discriminant analysis identified *organic matter* and the *C/N* factors as particularly sensitive indicators of soil quality changes, reflecting the quantity and quality of soil organic matter, incorporating 87.8% of the SQIs information resulting from the 23 indicators. TN, accounting for 84% of the information on the organic matter factor, emerges as a key indicator for predicting significant changes in soil quality due to land use or management practices. The TN and C/N proposed indicators offer a simplified yet effective means of assessing soil resource sustainability in the Mediterranean agroecosystems, providing practical tools for monitoring and managing soil quality.

Keywords: soil function; minimum data set; Mediterranean agroecosystems; principal component analysis; discriminant analysis

1. Introduction

The concept of soil quality was initially defined as "the capacity of soil to function within an ecosystem and under various land uses in such a way that it sustains biological productivity, maintains water and air quality, and promotes the health of animals and plants" [1]. Larson and Pierce [2] expanded on this by suggesting that a soil's physical, chemical, and biological characteristics enable it to perform three essential functions: (1) provide a medium for plant growth, (2) control and regulate water flow in the environment, and (3) serve as an environmental filter.

While the concept of soil quality may seem straightforward, its definition and quantification pose considerable challenges [3]. Some researchers contend that the term "quality" is difficult to apply to soil, given its dynamic, complex, and variable nature [4–6]. However, a growing number of studies underscore the critical role of soil quality in environmental sustainability and human well-being [7–9]. Soil quality provides a comprehensive framework for examining the interactions among the biological, chemical, and physical properties of soil, which is essential for sustainable land use and effective soil management of non-renewable soil resources [1,3–11]. For this reason, Lal [12] recently suggested that restoring soil quality in agricultural lands could help mitigate soil degradation.

Evaluating soil quality requires an analysis of both the inherent and dynamic characteristics of soil. In any region, soil quality assessment is influenced by a combination of factors, including management practices like crop rotation and manure application, as well as climate and soil type [9]. The initial step in assessing soil quality involves identifying suitable soil quality indicators (SQIs) to create a minimum dataset (MDS) for assessment [13]. Selecting indicators that encompass a wide range of physical, chemical, and biological attributes is crucial for an accurate evaluation of soil quality. Additionally, it is important to ensure that the chosen parameters effectively convey the information offered by all relevant indicators [14].

Many soil characteristics that influence soil quality are often highly correlated, interacting with other soil properties [2,11,14]. Due to these correlations, a more robust evaluation of soil quality can be achieved using statistical approaches that take these relationships into account. Multivariate statistical analyses, for example, allow for the examination of multiple correlated variables at once, revealing patterns that might be missed when variables are assessed independently [15]. Numerous studies have utilized multivariate methods to identify a smaller set of soil quality indicators, an MDS that can effectively describe changes in soil quality [16–19], and these have been applied to various land use types, including coastal areas [20], agricultural zones [21], and grasslands [19]. Zhou et al. [22] used ANOVA and factor analysis to identify a subset of 4 key soil indicators from an initial group of 26 to build an MDS for evaluating soil quality in wheat-producing regions of China. Similarly, Brejda et al. [21] employed principal component and discriminant analyses to identify sensitive soil indicators at a regional scale.

In the Mediterranean region, only a few studies have developed specific sets of soil quality indicators for specific land uses like forest [23,24] and agricultural land uses e.g., [10,11,25–27], and even fewer have incorporated biological parameters [10,11,26]. Reis and Dintaroglou [28] employed Principal Component Analysis to evaluate dynamic soil quality in a semi-arid Mediterranean watershed for different land uses. Navaro et al. [29] used multivariate methods to select the most appropriate indicators for a soil quality index in Mediterranean ecosystems.

Most studies aimed at identifying MDS do not account for the seasonal variability of soil quality indicators, as soil sampling is typically conducted during one specific season. Soil function is significantly influenced by seasonal variations in temperature and moisture, as well as by management practices in agricultural systems. This is particularly important in Mediterranean regions, which are characterized by a pronounced seasonal contrast in temperature and rainfall between winter and summer. Soil quality indicators in Mediterranean agroecosystems often exhibit significant seasonal variability that is frequently overlooked in efforts to establish MDS at a regional scale [30,31]. Furthermore, Mediterranean agroecosystems are notable for their highly variable soil cover, spatial diversity, and long history of continuous human settlement and intensive cultivation [32], which further influence soil quality.

To address these challenges, the present study introduces a statistics-based methodology for identifying an MDS for soil quality assessment, incorporating seasonal samplings of soils, across five different land use types, over two consecutive years. The study aimed to: (i) identify regional-scale soil quality "factors" from a set of 23 biological, chemical, and physical soil quality indicators, (ii) determine which soil quality factors exhibit significant variation based on land use, and (iii) pinpoint soil properties that can serve as reliable indicators for monitoring soil quality on a regional scale, taking into account the seasonal variation of soil functions in Mediterranean agroecosystems.

2. Materials and Methods

2.1. Study Area

The study was conducted in the Kaloni Gulf watershed, located on Lesvos Island (Longitude: 26°06′51 E; Latitude: 39°12′40 N), in the northern Aegean Sea (Figure 1). Lesvos, which is the third-largest Greek island and the seventh-largest in the Mediterranean, features a diverse range of lithological units, climates, and landscapes, including forests, scrubland, and agricultural land, with olive cultivation being predominant.

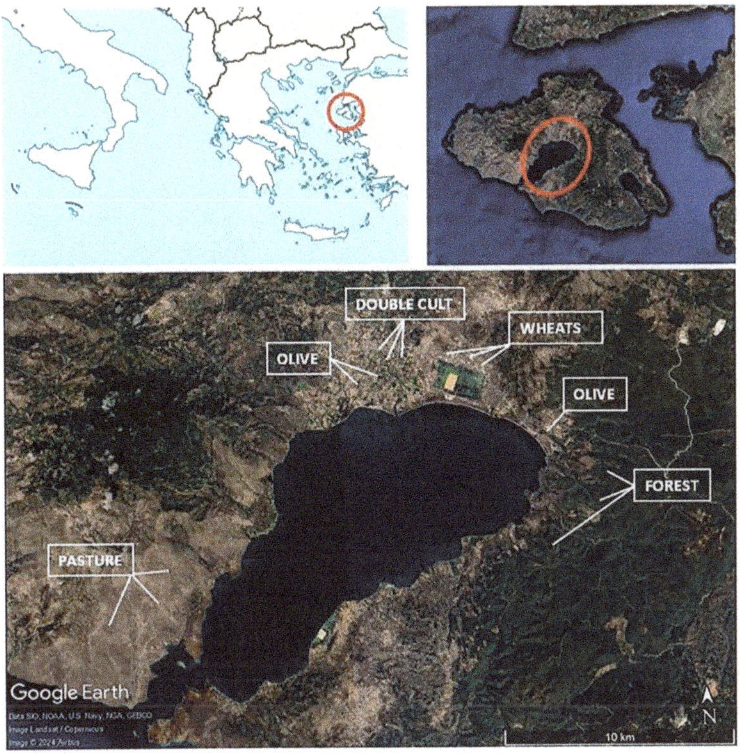

Figure 1. The study area and sampling sites in the watershed of Kaloni Gulf (Lesvos Island, Greece).

Lesvos is representative of the land-use changes that have occurred in the Mediterranean region over the past century [33]. Studies combining vegetation cover and soil geology have highlighted the complex interactions between climate, land use, and land degradation on the island [34–36]. These long-term interactions between climate, soil, and human activities have had varying impacts on land degradation and desertification. Significant land-use changes have occurred in the study area throughout the 20th century. The area dedicated to olive cultivation has expanded considerably, increasing from 26.9% of the island's total area in 1886 to 41.2% today and has been abandoned in areas of marginal productivity and moved to more fertile regions. The extent of pine forests has remained relatively constant, though their geographic distribution has shifted [33]. Pine forests have replaced some oak woodlands and grasslands due to their greater resilience and ability to regenerate after fire, as well as their capacity to thrive in shallow soils with alkaline parent material, which are less suitable for oak trees. The area under cereal cultivation has significantly declined, particularly after 1950, due to extensive migration of the local population to urban centers [37] and decreasing soil fertility [33]. Animal husbandry remains the second largest sector in primary production on the island, characterized by traditional Mediterranean practices. Extensive sheep and, to a lesser extent, goat farming is prevalent. In recent decades, grazing intensity has increased sharply due to (a) a significant reduction in grain cultivation and (b) a substantial rise in subsidies for grazing livestock. The intensity of grazing is often exacerbated by the annual setting of fires in rangelands to encourage the growth of high-quality grass, which is then subject to overgrazing. This overgrazing has led to increased soil erosion and has significantly impacted the water balance in these areas [34].

The study area is characterized by a Mediterranean climate, known for its mild, rainy winters and hot, dry summers, with a long dry season and abundant sunshine, particularly in the summer months. A notable feature of this climate is the pronounced seasonal variation in key climatic parameters: temperatures peak during the summer, while precipitation reaches its highest levels in winter. The area's climate is also influenced by the sea and the dominance of northerly winds during the summer months. The average annual temperature is 19.1 °C, and the annual precipitation averages 608.1 mm, with rainfall peaking in December and reaching its lowest levels in July or August [38]. Relative humidity follows a similar pattern, with the highest levels in December and January (72–71%) and the lowest in July (55.9%), resulting in an average annual humidity of 64.4%. The prevailing winds are predominantly from the north and northwest, occurring frequently during both winter (with the highest frequency of moderate and strong winds in February) and summer. May is typically the calmest month, experiencing the lowest wind activity [38]. Rainfall and temperature fluctuations over the two-year study period are presented in Figure 2.

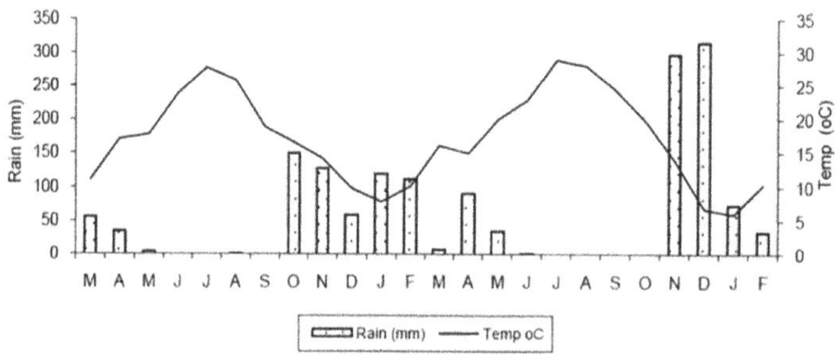

Figure 2. Mean monthly air temperature (line) and rainfall (bars), starting from March (M) 1st year to February (F) 2nd year.

The watershed covers a total area of 49,260 hectares, comprising 33.5% cropland, 39.4% pasture, 21.6% forest, and 5.5% allocated to other uses. Olive cultivation dominates the cropland, representing 70% of the area, while the remaining land is primarily used for arable crops such as wheat (Mandylas, 1998) [38]. The region's soils originate from Mio-Pliocene volcanic pyroclastic deposits, with consistent soil-forming factors observed across agricultural lands surrounding the Kalloni Gulf.

Five land covers representing forest, cropland, and pasture were selected to reflect the island's ecosystem diversity: pine forests *Pinus brutia* Ten, olive groves, wheat fields, crop rotations of corn-wheat (referred to as double cultivation), and shrubland pastures. In a preliminary screening analysis, soil samples were collected from 75 representative fields (20 pastures, 15 forests, 40 arable) covering all land use/cover types, with slopes of less than 3% to minimize erosion effects. Three fields (three hectares each) for each land cover were selected after cluster analysis to ensure similar soil texture of 23 to 30% clay and a pH 6–7. All selected soils were classified according to USDA soil taxonomy as Entisols, with Typic Xerofluvents for forest, olive trees, wheat, and wheat/maize double cultivation soils, and Lithic Xerorthents for pastures. All soil textures were sandy-clay-loam.

Soil samples were collected from conventionally farmed fields according to local practices. Forest areas have undergone minimal management, with resin collection activities discontinued approximately 30 years ago. Olive grove sites were tilled to a depth of 15 cm in

April to incorporate annual vegetation and minimize water competition. Composite 15%N—15%P_2O_5—15%K_2O fertilizer was applied every three years to each tree (45 kg N ha^{-1}). Soil samples in olive grove fields were taken the second and third year after fertilization. In wheat cultivation, seedbeds were prepared at the end of October by using a moldboard plow to turn the soil to a depth of 20–30 cm, incorporating the previous crop plant residues. No preplant fertilization was applied, and in-season ammonium nitrate (NH_4NO_3) was applied at a rate of 90–100 kg N ha^{-1} in February. Wheat is harvested in June, and fields remain bare until the next seeding in autumn. Crop rotation was implemented in a limited area for forage production, where cereals were planted in October following deep tillage using a moldboard plow to a depth of 40–50 cm. Pre-plant fertilization with 11%N—15%P_2O_5—15%K_2O fertilizer was applied (110 kg N ha^{-1}), and no in-season fertilization was performed, as harvest occurred in late April. A second deep tillage was performed between May 10th and 20th before maize seeding, with pre-plant fertilization of 110 kg ha^{-1} applied using composite 11%N—15%P_2O_5—15%K_2O fertilizer. Maize cultivation was irrigated using a drip system at intervals of 8–10 days, spanning from late June to the end of August, with a total water application ranging from 550 to 600 mm. In-season N fertilization was performed through irrigation water (fertigation) 3–4 times, totaling 165 kg N ha^{-1} as NH_4NO_3. Maize was harvested in late September. All management practices in olives, wheats, and double cultivation are consistently applied between years. Pasture sites, dominated by *Sarcopoterium spinosum* (L.) Spach, had shallow soils in a region with severe degradation through erosion. These soils did not receive any particular management except for grazing by sheep and goats.

2.2. Soil Sampling and Analysis

Over two consecutive years, soil sampling was conducted eight times, seasonally, on the same day after the last rainy day for all land uses, and when soil has been drained enough. Soil sampling was performed every year in the first days of May, August, October, and February. At each sampling site, three composite surface soil samples (0–15 cm depth) were collected randomly from the central area of the site, with each composite sample comprising 8 soil cores (2.5 cm diameter). Prior to sampling, all vegetation and plant residues were cleared from soil surface. The soil sub-samples were carefully combined to achieve homogeneity, with roots and any visible plant residues removed. The resulting composite samples were then stored at 4 °C for three days before soil microbial biomass determinations. Furthermore, soil subsamples were air-dried, ground, sieved through a 2 mm mesh, and stored separately for subsequent chemical analysis.

2.3. Soil Quality Indicators

A set of physical, chemical, and biological indicators of soil quality, originally proposed with the introduction of the soil quality concept [1,2] and still widely regarded by researchers today as critical for achieving key soil functions [39], were selected. Additionally, and particularly for the Mediterranean area where seasonal change affects soil function, indicators that have been used to monitor soil quality and are sensitive to seasonal changes, such as soil moisture, soil microbial properties, and the availability of soil nutrients, were selected.

The selected physical, chemical, and biological soil properties included in this study are well described by Brandy and Weill [40] for their importance on soil function. More specifically, the physical SQIs selected include: soil moisture, water holding capacity (WHC), soil porosity (Vp), and bulk density (BD), which influence aeration and water movement in the soil. Additionally, particle size distribution (Clay%, Silt%, Sand%) was assessed, as it is a fundamental property affecting various processes, such as moisture and nutrient retention. The chemical soil quality indicators monitored were soil organic carbon (Corg) and total nitrogen (TN), both key components of soil organic matter that influence a variety of chemical and biological processes in the soil. The C/N ratio was also included as it serves as an indicator of organic matter quality and its mineralization rate.

Other chemical indicators included the availability of essential nutrients (phosphorus (P), nitrogen (NO_3_N, NH_4_N and mineral N Nmin), and potassium (K)) as well as soil pH and electrical conductivity (EC). Biological indicators focused on soil microbial biomass carbon (Cmic) and nitrogen (Nmic), as well as active carbon (Cact), measured as permanganate oxidizable carbon (POXC), a well-established indicator of soil health in agriculture, known for its sensitivity to changes in conditions or management practices [41]. Additionally, the ratios of Cmic/Corg, Nmic/TN, Cmic/Nmic, and Cact/Corg were examined.

2.4. Laboratory Analyses

Soil moisture (gravimetric water content) was determined by drying triplicate 10 g samples at 105 °C. Bulk density was estimated using the Core method (volumetric cylinder method), after samplings of undisturbed soil samples by a specific cylinder, and estimating the soil mass in the cylinder. Maximum water holding capacity (WHC) was measured using the Gardner, 1986 method [42], where each soil sample was saturated with water in a cylinder, and WHC was calculated based on the water weight held in the sample vs. the sample dry mass (dried at 105 °C for 24 h). Soil texture was determined by physical fractionation (particle-size analysis, PSA) using the Bouyoucos method after the destruction of organic matter with hydrogen peroxide and dispersion with sodium hexametaphosphate [43]. Soil organic C was estimated by the Walkley–Black procedure [44] and total N by the semimicro-Kjeldahl method [45]. Nitrate and ammonium nitrogen were estimated chromatographically using the "cadmium reduction" and "indophenol blue method", respectively [46]. Soil-available phosphorus was extracted using the method recommended by Olsen and Sommers [47] and quantified through spectrophotometric analysis. Soil pH and electrical conductivity (EC) were measured in a 1:1 suspension with water [48]. Microbial biomass carbon (Cmic) and nitrogen (Nmic) were determined using the fumigation-extraction method developed by Vance, Brookes, and Jenkinson [49] for Cmic, and Brookes, Landman, and Jenkinson [50] for Nmic. Active carbon (Cact) was estimated using the permanganate-oxidizable carbon method [41]. All data are expressed on an oven-dry (at 105 °C) soil weight basis.

2.5. Statistical Analysis

The statistical analysis was conducted using the SPSS statistical software–(IBM SPSS Statistics 26.0). To assess differences in physical, chemical, and biological soil quality indicators across land uses, analysis of variance (ANOVA) was employed. The identification of samples with significant statistical differences was carried out using the LSD post hoc test for multiple comparisons. The Pearson correlation test was applied to assess the relationships between SQIs across all land uses. Before conducting these analyses, the data were checked for compliance with ANOVA assumptions and log-transformed where necessary. Principal component analysis (PCA) was employed to reduce the set of 23 soil quality indicators to a smaller number of factors, highlighting the most influential properties that explain the variation in the data. PCA was used to extract the factors, as it does not require prior assessment of the variance of each soil property explained by the factors [21]. PCA was performed on standardized variables using the correlation matrix to neutralize the effects of different measurement units on determining the weight of each factor loading [15,51]. Factors with eigenvalues greater than 1 were selected from the analysis using "varimax" rotation. Discriminant analysis (DA) was applied to the complete set of physical, chemical, and biological soil quality indicators, considering the impacts of land use and season as recorded by the eight seasonal samplings per land use. DA was employed to differentiate land uses based on their physical, chemical, and biological soil quality indicators, examine their spatial relationships, and identify key properties that predominantly affect this distinction.

3. Results

3.1. Land Use Effect on Soil Quality Indicators

The results of the study highlight the impact of land use on various soil quality indicators, taking into account the specific conditions associated with each land use type. Agricultural activities significantly modify soil parameters when compared to natural ecosystems like forests. Notable differences were also observed within agricultural land uses. Soil cultivation appears to have a consistent impact on the studied parameters, grouping the three crop types—olive groves, wheats, and double cultivation—together. This indicates that soil cultivation has a distinct influence on soil functions, setting it apart from other forms of agricultural use, such as pastures.

Table 1 provides all samplings average values for SQIs measured, revealing distinct patterns in soil characteristics. For example, forest shows the highest values for soil moisture content, Vp, WHC, Corg, Cmic, and Cact. Pasture soils similarly exhibit high values for Cmic and Cact but are particularly notable for the highest levels of Nmic, Nmic/Ntot ratio, and Nmin, as well as a high Cmic/Corg ratio, similar to wheat soils. Wheat soils also demonstrate high Vp and Cact/Corg ratios, while double-cropping systems show elevated values for electrical conductivity EC, NO_3-N, NH_4-N, P, and BD.

Table 1. Soil quality indicators for each land use. Average values of the eight soil samplings.

	Soil Quality Indicator	Forest	Olive Trees	Wheats	Double Cultivation	Pasture
Physical SQIs	BD, g/cm^3	1.25 c	1.39 b	1.21 c	1.49 a	1.40 b
	Vp, %	52.65 a	47.67 b	54.38 a	43.64 c	47.00 b
	Soil mois., %	19.28 a	11.07 c	13.44 bc	14.67 b	12.89 bc
	WHC, %	79.09 a	49.13 c	60.29 b	56.99 b	51.93 c
	Clay, %	30.85 a	22.96 b	26.86 a	24.42 a	29.87 a
	Silt, %	23.88 ab	24.48 ab	24.92	27.47 a	21.43 b
	Sand, %	45.27 b	52.56 a	48.22 a	48.11 a	48.71 a
Chemical SQIs	Corg, g kg^{-1}	17.49 a	9.68 d	10.55 c	8.77 e	13.44 b
	TN, g kg^{-1}	1.14 a	0.76 c	0.97 b	0.78 c	0.95 b
	C/N	15.64 a	13.08 bc	11.81 c	11.60 c	14.31 ab
	NO_3_N, mg kg^{-1}	3.75 b	6.53 b	19.52 a	20.75 a	20.59 a
	NH_4_N, mg kg^{-1}	8.75 a	2.66 c	5.68 b	5.32 b	9.49 a
	Nmin, mg kg^{-1}	12.50 c	9.19 c	25.21 b	26.07 ab	30.07 a
	P, mg kg^{-1}	2.93 d	20.76 b	24.17 a	24.74 a	6.45 c
	EC, dS m^{-1}	0.40 a	0.19 b	0.39 a	0.45 a	0.38 a
	PH	7.00 a	6.17 b	6.13 b	6.27 b	6.03 b
Biological SQIs	Cmic, mg kg^{-1}	332.26 a	164.31 c	271.57 b	159.26 c	360.13 a
	Cmic/Corg, %	1.90 b	1.70 b	2.64 a	1.85 b	2.70 a
	Nmic, mg kg^{-1}	49.95 b	24.48 c	53.28 b	23.31 c	69.62 a
	Nmic/Ntot, %	4.44 c	3.34 cd	5.68 b	3.13 d	7.29 a
	Cmic/Nmic	8.01 a	7.02 ab	6.35 ab	7.16 ab	5.78 b
	Cact, mg kg^{-1}	411.27 a	291.64 bc	331.57 b	265.55 c	374.23 a
	Cact/Corg, %	2.36 b	3.03 a	3.15 a	3.07 a	2.81 a

Values for each indicator with the same letter does not differ significantly for $p < 0.05$ according to LSD post hoc test.

Seasonal variations in SQIs were substantial, driven by the Mediterranean climate's fluctuations in temperature and rainfall, as well as agricultural practices such as irrigation, fertilization, and tillage. Of the 23 SQIs studied, significant seasonal variability in at least one land-use type was observed in the following 13 indices: soil moisture, NO_3-N, NH_4-N, Nmin, P, EC, Cmic, Nmic, and the ratios of Cmic/Nmic, Cmic/Corg, Nmic/Ntot, Cact, and Cact/Corg. Although the main objective of this study is not to present the seasonal variation of the soil quality indicators, Figure 3 presents the seasonal variation of four selected SQIs that affect soil function.

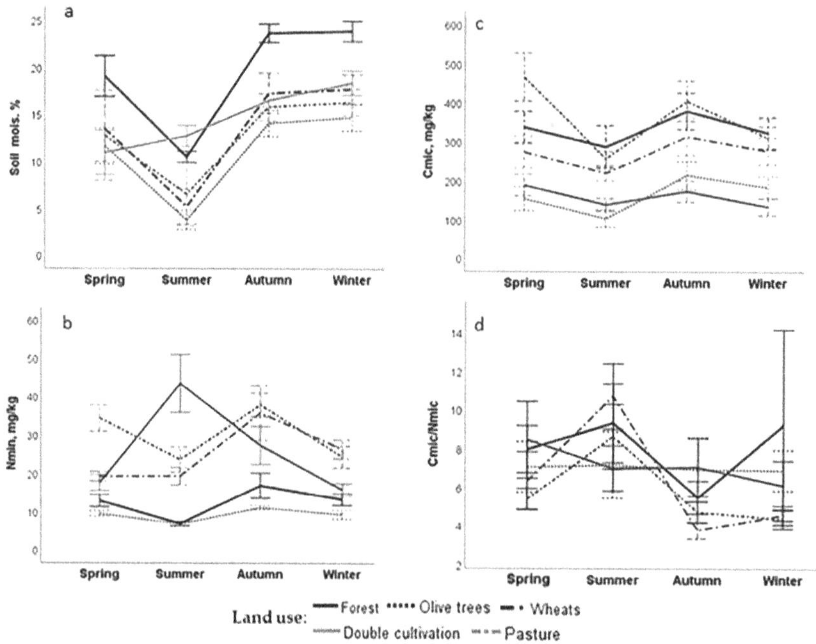

Figure 3. Seasonal variation of soil moisture Cmic (**a**), Nmic (**b**), Cmic (**c**), and the Cmic/Nim (**d**) ratio. The values represent the means of two samplings conducted during the same season across two consecutive years of the study.

As shown in Figure 3, soil moisture in all non-irrigated land uses follows a similar pattern, characterized by a notable decrease in summer and a significant increase in autumn with the onset of rainfall. In contrast, the irrigated double crop maintains more stable moisture levels during the summer. The typical hot and dry Mediterranean summer appears to impact soil microbial biomass consistently across all land uses, as the seasonal variation pattern remains similar. Available nitrogen is also affected, not only by fertilization in grain and double cropping but also by organic nitrogen mineralization, which increases significantly in autumn with the start of the rainy season. Additionally, the Cmic/Nmic ratio rises in all non-irrigated land uses during summer and declines with the onset of autumn rains.

Significant correlations between the SQIs found in 189 of 253 pairs as can be seen in Table 2, indicating that SQIs can be grouped into fewer factors, based on their correlation structure. Selected strong correlations ($r^2 > 0.6$) indicates Clay% correlations with Sand%, Corg, Cmic, P, and NH_4-N, WHC with Sand, Corg, and TN, and Corg with TN, P, Cmic and Cact.

Table 2. Correlation coefficients for soil quality indicators (n = 120).

	BD	Vp	Moist	WHC	Clay	Silt	Sand	Corg	TN	C/N	NO_3-N	NH_4-N	Nmin	P	EC	pH	Cmic	Cmc/Corg	Nmic	Nmic/Ntot	Cmic/Nmic	Cact
Vp	**−1.00 ****																					
moist	−0.13 *	0.13 *																				
WHC	−0.43 **	0.43 **	0.39 **																			
clay	−0.35 **	0.35 **	0.28 **	0.48 **																		
silt	ns	ns	ns	ns	**−0.52 ****																	
sand	0.38 **	−0.38 **	−0.32 **	**−0.62 ****	**−0.65 ****	−0.31 **																
Corg	−0.41 **	0.41 **	0.33 **	**0.69 ****	**0.77 ****	−0.37 **	**−0.53 ****															
TN	−0.32 **	0.32 **	0.40 **	**0.60 ****	**0.54 ****	−0.29 **	−0.34 **	**0.65 ****														
C/N	−0.15 *	0.15 *	ns	0.13 *	0.29 **	ns	−0.23 **	0.38 **	−0.30 **													
NO_3-N	ns	ns	ns	−0.13 *	ns	ns	ns	−0.25 **	ns	−0.29 **												
NH_4-N	ns	ns	0.27 **	0.27 **	**0.61 ****	−0.30 **	−0.41 **	0.47 **	0.33 **	0.16 *	ns											
Nmin	ns	ns	ns	ns	0.15 *	ns	−0.20 **	ns	0.19 **	−0.23 **	**0.95 ****	0.33 **										
P	0.14 *	−0.14 *	−0.19 **	−0.36 **	**−0.77 ****	0.45 **	0.45 **	**−0.78 ****	−0.36 **	−0.42 **	0.41 **	−0.47 **	0.24 **									
EC	−0.15 *	0.15 *	0.27 **	0.26 **	0.20 **	ns	−0.29 **	ns	0.24 **	ns	0.39 **	0.24 **	0.45 **	ns								
pH	−0.27 **	0.27 **	0.25 **	**0.52 ****	0.29 **	ns	−0.38 **	0.44 **	0.33 **	ns	−0.28 **	ns	−0.23 **	−0.34 **	0.23 **							
Cmic	−0.27 **	0.27 **	0.38 **	0.36 **	**0.68 ****	−0.47 **	−0.33 **	**0.60 ****	**0.59 ****	ns	0.16 *	**0.57 ****	0.33 **	**−0.52 ****	0.19 **	ns						
Cmic/Corg	ns	ns	0.20 **	ns	0.25 **	−0.28 **	ns	ns	0.24 **	−0.16 *	0.36 **	0.31 **	0.43 **	ns	ns	−0.19 **	**0.75 ****					
Nmic	−0.22 **	0.22 **	0.43 **	ns	**0.55 ****	−0.42 **	−0.23 **	0.39 **	0.45 **	ns	0.28 **	0.51 **	0.43 **	−0.35 **	0.16 *	ns	**0.72 ****	**0.56 ****				
Nmic/TN	ns	ns	0.27 **	−0.15 *	0.39 **	−0.34 **	−0.13 *	0.16 *	ns	0.35 **	0.22 **	0.42 **	0.34 **	−0.25 **	ns	−0.19 **	**0.53 ****	**0.52 ****	**0.85 ****			
Cmic/Nmic	ns	ns	−0.19 **	0.24 **	ns	ns	ns	ns	ns	ns	−0.15 *	−0.14 *	−0.18 **	ns	ns	0.13 *	ns	ns	**−0.52 ****	**−0.53 ****		
Cact	−0.28 **	0.28 **	0.21 **	**0.60 ****	**0.52 ****	−0.29 **	−0.32 **	**0.64 ****	**0.62 ****	ns	ns	0.35 **	ns	−0.47 **	ns	0.18 **	**0.56 ****	0.20 **	0.30 **	ns	0.19 **	
Cact/Corg	0.18 **	−0.18 **	−0.13 *	ns	−0.28 **	ns	0.22 **	−0.40 **	ns	−0.37 **	0.25 **	−0.13 *	0.20 **	0.33 **	ns	−0.27 **	ns	0.28 **	−0.16 *	−0.19 **	0.20 **	0.41 **

ns: not significant, * p < 005, ** p < 0.001. Bold text indicates a significant correlation with r > 0.5.

3.2. Distinction of Land Uses Based on Physical, Chemical, and Biological Soil Quality Indicators

Discriminant analysis (DA) was used to identify the key SQIs that differentiate land uses. Using land use as a grouping parameter, DA produced four significant functions, explaining 100% of the total variability (Table 3).

Table 3. Standardized coefficients and properties of the discriminant analysis for physical, chemical, and biological SQIs.

Soil Quality Indicator	Function 1	Function 2	Function 3	Function 4
BD	0.511	0.307	−0.353	**0.734**
moist	−0.146	0.244	−0.486	0.186
WHC	0.356	0.754	−0.073	0.020
clay	0.690	0.595	0.495	0.306
silt	−0.318	0.857	0.229	0.417
Corg	0.511	0.406	−0.638	−0.266
TN	0.281	0.371	0.410	−0.358
C/N	−0.284	0.483	−0.130	0.123
NO_3-N	0.397	**−0.952**	−0.161	0.687
NH_4-N	0.402	−0.180	0.081	0.215
P	**−1.096**	0.767	0.510	−0.118
EC	0.438	0.094	0.074	0.234
pH	−0.161	0.105	−0.224	−0.063
Cmic	**−1.053**	−0.515	−1.229	0.546
Cmic/Corg	0.849	0.523	**1.306**	−0.572
Nmic	−0.217	0.491	0.067	0.182
Nmic/Ntot	0.599	−0.584	0.575	−0.656
Cmic/Nmic	0.235	−0.030	0.182	−0.093
Cact	0.474	**−1.087**	**1.581**	−0.456
Cact/Corg	−0.786	0.682	**−1.442**	0.245
Cmic/Nmic	8.01	7.02	−0.353	**0.734**
Cact	411.27	291.64	−0.486	0.186
Eigenvalues	41.596	5.727	2.362	1.744
Commulative variation %	80.9	92.0	96.6	100.0
Sig.	<0.001	<0.001	<0.001	<0.001

Bold text indicates the most significant coefficient for each function.

The first two functions accounted for 80.9% and 11.1% of the variability, respectively, with P, Cmic, Cact, and NO3-N being the most influential properties for land-use differentiation. Figure 4 shows the clear separation of land-use types, with crops grouped distinctly apart from forests and pastures.

While crops showed some overlap, particularly between double cultivation and wheats, the centroids for all land-use types remained distinct (Table 4).

Table 4. Pairwise comparison of land uses for existence significant difference of their centroids.

Land Use		Forest	Olive Trees	Wheats	Double Cultivation
Forest	F	357.378			
	Sig.	0.000			
Olive trees	F	207.950	57.095		
	Sig.	0.000	0.000		
Wheats	F	281.162	36.172	34.197	
	Sig.	0.000	0.000	0.000	
Double cultivation	F	70.362	260.748	165.884	230.242
	Sig.	0.000	0.000	0.000	0.000

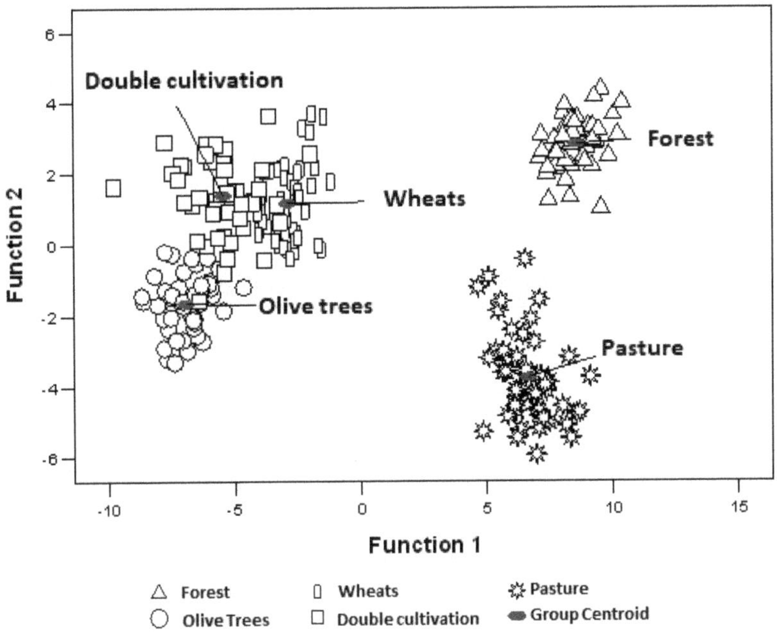

Figure 4. Scatter plot of land uses regarding the discriminant scores of the first two functions.

3.3. Factor Analysis

Principal component analysis (PCA) revealed six main factors with eigenvalues greater than 1, which explained 82.9% of the total variance of the data (Table 5).

Table 5. The first 10 factors of principal component analysis.

Component	Initial Eigenvalues		
	Total	% of Variance	Cumulative %
1	5.57	29.34	29.34
2	3.46	18.19	47.53
3	2.38	12.51	60.04
4	1.83	9.64	69.68
5	1.36	7.17	76.85
6	1.15	6.04	82.89
7	0.87	4.58	87.47
8	0.73	3.86	91.33
9	0.55	2.90	94.24
10	0.42	2.18	96.42

If there were no correlation between soil properties, the identification of factors would not be possible [52]. However, as significant correlations ($p < 0.05$) were found in 189 out of the 283 pairs of soil properties studied, PCA revealed that 6 main factors capture more variance than any individual original variable in the dataset (eigenvalues > 1), explaining 82.9% of the total variance. The amount of variance in each original variable captured by the extracted principal components (communalities) indicates that the six main factors explain > 90% of the variance in properties such as Corg, Nmic, Nmic/TN, Cmic, Nmin, Vp, and BD, and >80% of the variance for TN, P, Cmic/Corg, $N0_3$-N, and C/N. The six main factors, however, explain < 70% of the total variability in the properties of soil moist and NH_4-N (Table 6).

Table 6. "Weights" and percentage of explained variance of the SQI of the six factors.

Soil Properties	Factors						Communalities
	1	2	3	4	5	6	
TN	**0.82**	0.16	0.10	0.19	−0.26	−0.09	0.82
Corg	**0.82**	0.08	−0.10	0.22	0.44	−0.05	0.93
WHC	**0.80**	−0.18	0.08	0.30	0.08	0.13	0.79
Cact	**0.73**	0.30	0.00	0.15	−0.07	0.36	0.78
P	−0.61	−0.24	0.34	0.05	−0.52	0.02	0.81
Soil moist	0.54	0.11	0.05	0.01	−0.16	−0.50	0.58
NH_4-N	0.49	0.41	0.22	−0.22	0.35	−0.14	0.65
Cmic/Corg	0.07	**0.85**	0.16	−0.02	−0.21	0.13	0.82
Nmic/Ntot	−0.07	**0.78**	0.10	0.04	0.32	−0.45	0.93
Nmic	0.29	**0.75**	0.15	0.12	0.06	−0.52	0.95
Cmic	0.56	**0.74**	0.11	0.12	0.13	0.04	0.90
Nmin	−0.01	0.39	**0.87**	−0.05	−0.12	−0.03	0.93
NO_3-N	−0.18	0.28	**0.85**	0.01	−0.24	0.02	0.89
EC	0.29	−0.15	**0.76**	0.07	0.05	−0.14	0.70
VP	0.21	0.05	0.01	**0.96**	0.08	−0.04	0.98
BD	−0.21	−0.05	−0.01	**−0.96**	−0.08	0.04	0.98
C/N	−0.05	0.08	−0.15	0.12	**0.88**	0.12	0.84
Cact/Corg	−0.06	0.22	0.07	−0.13	−0.61	**0.52**	0.73

Bold text indicates the most significant coefficient (>0.7) for each Factor.

These factors were named based on the soil properties they most strongly correlated with. The first factor, *"organic matter"*, was strongly correlated ($r^2 > 0.70$) to TN, Corg, and Cact, as well as WHC. The second factor, *"microbial biomass"*, was associated with microbial biomass indicators like Cmic/Corg, Nmic/TN, Nmic, and Cmic. The third factor, *"nutrients"*, was strongly associated with Nmin, NO_3-N, and EC, which expresses the concentration of soil nutrients in the soil solution. The fourth factor, *"compaction"*, was strongly associated with the bulk density and porosity of the soil, which are key indexes of soil compaction. The sixth factor, *"C/N ratio"*, and seventh factor, *"active carbon"*, were determined solely by the C/N and Cact/Corg, respectively.

Each factor's performance was evaluated across land uses using analysis of variance (ANOVA), which showed that all factors except for *"active carbon"* varied significantly by land use (Table 7).

Table 7. Effect of land use on the factors score derived from principal component analysis.

	Factor Score					ANOVA	
Factors	Forest	Olive Trees	Wheats	Double Cultiv.	Pasture	F	Sig.
Organic Matter	1.53 a	−0.80 d	−0.38 c	−0.46 c	0.10 b	121.24	0.00
Microbial Biomass	−0.36 c	−0.41 cd	0.43 b	−0.79 d	1.12 a	52.76	0.00
Nutrients	−0.47 c	−0.96 d	0.31 b	0.78 a	0.33 b	37.99	0.00
Compaction	0.23 b	0.04 b	0.99 a	−0.74 c	−0.51 c	34.41	0.00
C/N	0.63 a	−0.33 b	−0.59 b	−0.31 b	0.60 a	21.04	0.00
Active Carbon	−0.11 a	0.13 a	−0.01 a	0.02 a	−0.03 a	0.35	0.84

Values for each factor with the same letter do not differ significantly for $p < 0.05$ according to LSD post hoc test.

The ANOVA results show that, except for "active carbon", all other soil quality factors exhibited significant variation across the different land uses (Table 7). The *"organic matter"* factor showed positive scores for forest and pasture soils but negative scores for other land uses, with the highest value in the forest and the lowest in crops, particularly in olive trees. The *"microbial biomass"* factor exhibited positive scores in pasture and crop soils, but negative scores were observed in forest, olive grove, and double-cropping systems, with pasture scoring the highest and double-cropping the lowest. In the same way, the "nutrients" factor exhibited higher scores for double-cropping, crops, and pasture. Forests and olive groves, on the contrary, recorded negative scores. The "*compaction*" factor revealed positive scores for crops, forests, and olive groves, but negative scores for double-cropping and pasture, with the highest score in crops and the lowest in double-cropping. The "*C/N ratio*" factor presented the highest positive scores in forest and pasture, with lower and negative scores in wheat and olive tree soils (Table 7).

Discriminant analysis (DA) was conducted using land use as the grouping variable to identify the factors that most significantly differentiated between land uses. DA generated four discriminant functions that explained 100% of the total variability ($p < 0.001$, Table 8).

Table 8. Eigenvalues and percentages of the explained variance of the four functions when discriminating land uses by soil quality factors.

Functions	Eigenvalue	% of Variance	Cumulative %
1	7.903	75.9	75.9
2	1.237	11.9	87.8
3	0.924	8.9	96.7
4	0.347	3.3	100.0

The first function explained 75.9% of the total variance, followed by the second with 11.9%, the third with 8.9%, and the fourth with 3.3% (Table 8).

$$Y1 = 2.50(\text{organic matter}) + 0.41(\text{soil nutrients}) + 0.10(\text{soil compaction}) + 1.43(C/N) - 0.20(\text{active carbon}) \quad (1)$$

DA (Table 9), recognized "*organic matter*" and "*C/N ratio*" factors as the most influential in distinguishing between land uses (Equation (1)). The discrimination coefficient for the "*organic matter*" factor was approximately five times greater than that for the "*soil nutrients*" and "*microbial biomass*" factors, and over ten times higher than those for the "*active carbon*" and "*soil compaction*" factors. Similarly, the discrimination coefficient for the "*C/N ratio*" was roughly three times higher than that of the "*nutrients* " and "*microbial biomass*" factors, and more than seven times higher than the coefficients for "*active carbon*" and "*compaction*".

Table 9. Unstandardized coefficients of discrete functions of soil quality factors.

Factos	Functions			
	1	2	3	4
1 Organic matter	2.504	−0.150	−0.117	0.385
2 Microbial biomass	0.416	1.069	0.756	−0.346
3 Nutrients	−0.460	0.903	−0.406	0.799
4 Compaction	0.104	−0.383	1.045	0.527
5 C/N	1.431	0.264	−0.221	−0.366
6 Active carbon	−0.203	−0.046	0.003	−0.082
(Constant)	0.000	0.000	0.000	0.000

3.4. Identification of "Minimum Set of Soil Quality Indicators"

Discriminant analysis (DA) was conducted again on the "organic matter" factor, which includes seven soil properties, to identify the most influential properties in differentiating land uses. This analysis produced four functions explaining 100% of the total variability ($p < 0.001$, Table 10), with the first function accounting for 84.2%, the second 12.8%, the third 1.9%, and the fourth 1.0% of the variance.

The linear discriminant function for the first factor is given in Equation (2) (Table 11). The analysis revealed that TN was the most important SQI for distinguishing land uses, with a discriminant coefficient several times larger than that of any other soil property (Equation (2)).

$$Y2 = 0.324(\text{Corg}) + 2.0(\text{TN}) + 0.19(\text{WHC}) + 0.004(\text{moist}) + 0.136(NH_4\text{-N}) - 0.234(\text{P}) - 0.003(\text{Cact}) \quad (2)$$

Table 10. Eigenvalues and percentages of the explained variance of the four functions when discriminating land uses by organic matter factor.

Function	Eigenvalue	% of Variance	Cumulative %
1	13.646	84.2	84.2
2	2.078	12.8	97.0
3	0.313	1.9	99.0
4	0.169	1.0	100.0

Table 11. Unstandardized coefficients of discrete functions of the organic matter factor.

SQI	Functions			
	1	2	3	4
Corg	0.324	−0.098	−0.237	0.271
TN	2.000	−0.099	3.187	3.318
WHC	0.019	0.169	0.008	−0.039
moist	0.004	0.035	−0.036	−0.064
NH_4-N	0.136	−0.021	0.312	−0.116
P	−0.234	0.043	0.031	0.089
Cact	−0.003	−0.010	0.000	0.004
(Constant)	−3.201	−6.596	−2.564	−5.163

In conclusion, the results of PCA, DA, and the derived equations clearly show that the soil properties of TN and C/N ratio play a pivotal role in differentiating between land uses. These two indicators can be considered highly sensitive soil properties and are recommended for monitoring soil quality changes in relation to land use in the study area.

4. Discussion

In this study, the dataset used in the multivariate analyses incorporates the effects of both land use and season on the examined SQIs. Land use within the same climate and soil type influences soil function through management practices such as tillage, irrigation, fertilization, and biomass removal. Additionally, vegetation cover plays a role, as this determines the quantity and quality of plant residues entering the soil system. Although land use impacts individual soil quality indicators (SQIs), it has a more substantial effect on the overall set of indices (physical, chemical, and biological), grouping soil functions into distinct categories for each land use. For instance, soil functions in crop lands, while differing between crop types, are more pronounced compared to those in forests and pastures.

Seasonal variability of SQIs is a crucial factor in understanding soil function and has been investigated in the study area, specifically for soil microbial biomass properties [31]. In Mediterranean regions, the hot and dry summer conditions are likely to have a more pronounced impact on the temporal changes of many SQIs compared to other factors. For example, soil microbial biomass often shows a decline during the summer, followed by an increase in autumn as rainfall returns. This pattern is consistent across various land uses, despite differences in vegetation, management practices, or Corg levels in Mediterranean agroecosystems. However, the magnitude of these seasonal shifts—from spring to summer and summer to autumn—varies significantly depending on land use [31]. Furthermore, the timing of management practices, such as irrigation, soil cultivation, and fertilization, strongly influences the variability of SQIs, including nutrient availability, electrical conductivity (EC), and soil microbial biomass indices.

PCA highlighted NO_3-N and P as key indicators distinguishing soil functions across different land uses. The observed differences may be linked to soil management practices, such as the regular application of chemical fertilizers (e.g., 11-15-15 type) that increase inorganic phosphorus levels, as animal excretions in pastures that increase nitrate nitrogen, and the closed nutrient cycling in forests. Two other indicators, Cmic and Cact, representing labile organic carbon in soils [52–55], also distinguish soil functions among land uses. These indicators are influenced by management practices such as tillage and biomass removal in crops, animal excretions in pastures, and nutrient cycling in forests, which affect labile carbon pools in the soil. Crop residues serve as a crucial source of energy and nutrients for microbial proliferation, contributing to the formation of soil organic carbon. Labile/active organic carbon pools represent a small part of soil organic carbon but serve as sensitive indicators of soil biogeochemical processes under agricultural management [56] that must be further studied.

In the current study, six key soil quality factors were identified: *organic matter*, *microbial biomass*, *nutrients*, *compaction*, *C/N ratio*, and *Cact/Corg ratio*. Each of these factors plays a role in supporting one or more essential soil functions.

The *organic matter* factor, in particular, reflects both long-term and short-term changes associated with land use change [57]. Soil organic matter (SOM) underpins crucial ecosystem services, such as food production, climate regulation, water filtration, erosion control, nutrient cycling, and providing energy for soil organisms [58,59]. It is widely regarded as a vital indicator of soil quality for the Mediterranean agroecosystems [27,60–62]. SOM also plays a key role in enhancing the resilience and adaptability of soils to environmental pressures [1]. The loss of soil organic carbon, often observed during the conversion of natural ecosystems to agricultural systems [63,64], is associated with reduced inputs of organic materials, decreased natural protection of organic carbon due to tillage, shifts in soil moisture and temperature that accelerate decomposition rates, and increased soil erosion [65]. Tillage, in particular, involves the physical disruption of the upper soil layers, which reduces soil aggregation and influences the turnover of aggregates, thereby impacting the soil carbon balance [66]. Conservation tillage techniques, such as reduced tillage, have been shown to increase total organic carbon in the surface layer, promoting microaggregation, and enhancing aggregate stability [67]. These practices can be effective alternatives for improving soil quality by increasing organic matter in cultivated soils and have to be incorporated into Mediterranean agroecosystem management. Additionally, crop rotations that include legumes, along with the application of organic amendments like animal residues and organic waste, can enhance carbon storage in soils [68]. Finally, if a significant portion of the diminishing soil organic matter could be restored through appropriate management practices, it might even be possible to mitigate some of the annual increases in atmospheric CO_2 levels [69].

The *microbial biomass* factor governs ecological processes that drive carbon and nutrient cycles, making it a sensitive measure of soil management impacts. Soil microbial biomass has been extensively recognized as a critical soil quality indicator [70–72] and has been reported among the most important ecological indicators of soil quality in the Mediterranean ecosystem [30]. Microbial biomass serves as both a source of mobile nutrients and a key player in the cycling and transformation of organic matter and plant nutrients in the soil [73]. Understanding microbial properties—such as the quantity, diversity, and activity of microbial biomass—is crucial for gaining deeper insights into the factors that contribute to soil health [74]. As a property that can predict future shifts in the amount of total organic matter [75], soil microbial biomass monitoring is a valuable tool for understanding and anticipating long-term changes in soil conditions. Many management practices in agroecosystems have been linked to the reduction of soil organic matter, leading to declines in soil biological fertility and resilience [76]. This issue is particularly pronounced in the rainfed agricultural systems of Mediterranean climates, where high summer temperatures and the alternation of wet and dry soil conditions contribute to high annual mineralization rates of organic matter [77]. The variability of abiotic factors in the Mediterranean agroecosystems is more extreme compared to temperate regions [78], and thus the synchronization of soil fauna and flora activity with the dynamics of certain chemical soil properties, influenced by seasonal climate changes, is a defining characteristic of Mediterranean-type ecosystems [79].

The *nutrients* factor affects nutrient availability, while the *compaction* factor influences water retention, aeration, and soil physical, chemical, and biological properties.

The *C/N* and *Cact/Corg* factors, while unidimensional in factor analysis, represent complex and dynamics soil functions. The C/N ratio is a key indicator of the quality of organic substrates available for decomposition [80,81], while the Cact/Corg ratio reflects the mineralization dynamics of organic matter [41]. Together, these factors provide a comprehensive assessment of soil quality in Mediterranean agroecosystems.

The differentiation of the six quality factors based on land use reflects dynamic soil qualities [14] and assesses the impact of land use and management practices on soil quality. The *organic matter* and *compaction* factors have been previously recognized by other researchers [21,81]. In this study, four additional factors are identified: *microbial biomass*, *nutrients*, *C/N*, and *Cact/Corg*. Evaluating these soil quality factors identifies five of the six as significant for assessing changes in soil quality due to land use changes. The *Cact/Corg* factor appears less important, as it does not effectively express soil quality dynamics.

In general, soil quality factors exhibit similar behavior across different land uses, consistent with the dominant indicators that comprise them. The *organic matter* factor is significantly affected by the land use and associated soil management like cultivation practices. The *microbial biomass* factor is influenced by carbon and nitrogen incorporation into microbial biomass, while the *nutrients* factor is impacted by nitrogen availability (via fertilization or mineralization). Soil management practices affect the *compaction* factor, and the *C/N* factor is influenced by the quantity and composition of plant residues. Among these, the *organic matter* and *C/N* factors, which reflect the quantity and quality of soil organic matter, appear to be the most crucial determinants of soil quality in Mediterranean agroecosystems. Changes in soil quality, resulting from land use and management practices, are reflected in all components of each factor [82].

The impact of land use on soil function is closely linked to the intensity of land management. While different crops generally have a similar characteristic effect on overall soil function, they also exhibit differences due to variations in the intensity and type of cultivation practices associated with each crop. Specifically, in this study, land use influences the content of soil organic matter, its quality, and the active pools of organic carbon. It seems that the quality of organic matter and the characteristics of active carbon pools, rather than the overall concentration of soil organic matter, play a dominant role in determining nutrient availability in soils.

In this study, TN is highlighted as critical for determining shifts in soil quality within the soil *organic matter* factor. Its significance as a fundamental property for soil quality is noted in numerous studies [83–85] due to its incorporation of a large portion of the information related to interacting soil parameters. In this study, forest soils show the highest TN stocks, followed by grasslands and croplands, similar to findings in other studies [85,86]. TN was significantly correlated with soil moisture, clay, Corg, Cmic, and Cact, indicating its influence on both labile and stable forms of soil organic matter. TN is a SQI that incorporates soil organic matter dynamics, and furthermore, is a significant and direct contributor to plant nitrogen nutrition, even in agricultural contexts [87]. TN has been reported by Zhao et al. [88] for other types of climatic zones as a sensitive SQI among different land uses.

Soil C/N ratio emerged as a second crucial factor for assessing changes in soil quality. Its significance lies in its ability to reflect the dynamics of organic matter decomposition, which plays a pivotal role in overall soil quality. Soil C/N ratio has long been recognized as a key indicator of organic matter quality and nitrogen mineralization-immobilization processes [89]. Shifts in soil C/N stoichiometry are known to significantly influence carbon dynamics in agroecosystems [80]. Microorganisms use labile carbon as an energy source to produce extracellular enzymes, facilitating nitrogen extraction from soil organic matter (SOM) and leading to SOM mineralization [90,91]. Thus, the C/N ratio, though often underestimated, plays a fundamental role in regulating soil organic matter decomposition, indirectly impacting soil quality. The C/N ratio also serves as a common proxy for organic matter stability [92], offering insights into soil quality changes. While interpreting shifts in C/N ratios in bulk soils is complex, especially in response to land use or climate change, it is essential for understanding potential soil organic carbon (SOC) sequestration or losses, as well as nutrient cycling and availability in agroecosystems. This makes the C/N ratio a valuable indicator for tracking soil quality changes in agricultural systems.

From a set of 23 physical, chemical, and biological SQIs, this study identifies soil properties of total nitrogen and C/N ratio as sensitive key indicators, capturing most of the variability across all 23 SQIs among land uses. These two indicators provide valuable insights into soil quality changes resulting from land use changes or the application of specific management practices. The proposed indicators are sufficient for assessing long-term soil quality changes.

Although this research focuses on a specific region of the Mediterranean, the findings related to soil quality factors are likely applicable to other regions with Mediterranean-type climates, where moisture and temperature patterns are the primary drivers of fundamental soil functions. Many of the soil quality indicators used in this study demonstrate similar behavior in response to land use across regions with Mediterranean climates, such as California [93,94]. This study's examination of how soil quality changes with land use change and cultivation practices in Mediterranean ecosystems underscores the need for further research. The significant role of the quantity and quality of soil organic matter, along with its components, in soil functions and nutrient availability highlights the importance of investigating organic matter dynamics using indicators that best reflect the parameters being studied. Additional research on sensitive indicators that can accurately predict changes in organic matter, such as active carbon and microbial biomass properties, would be particularly useful for the early detection of soil quality degradation.

As mentioned, the comparative evaluation of soil quality with changes in land use or different cultivation practices can be effectively conducted using a limited number of indicators. However, soil quality cannot be accurately assessed through one or two properties alone, as it is determined by a combination of physical, chemical, and biological characteristics. The cost and time required for comprehensive soil analysis make it impractical for farmers to regularly assess and monitor changes in soil quality. Therefore, using the selected indicators of TN and C/N that capture condensed information of a broader set of physical, chemical, and biological SQIs can serve as a practical tool for evaluating the sustainability of soil resources. These indicators allow producers and land managers to quickly and efficiently track changes in soil quality following land use change or the adoption of new farming practices, facilitating the re-evaluation of management strategies.

5. Conclusions

The concept of soil quality, which encompasses the holistic relationships and functions of physical, chemical, and biological soil properties, aligns with the sustainable management of non-renewable soil resources. In this study, the influence of land use on soil function and overall SQIs is emphasized, particularly under the unique conditions imposed by different land uses incorporating any seasonal variation of the SQIs in the Mediterranean agroecosystems. Agricultural use is shown to significantly affect soil parameters, differentiating them from the natural ecosystem of the forest, although considerable differences are observed even within agricultural systems. Cultivation practices appear to have a consistent impact on soil parameters, with olive trees, wheat, and double-cultivation systems grouped in a distinct manner, indicating that land cultivation affects soil function and differentiates it from other agricultural uses such as forest and pasture.

The study identifies five key factors that depict soil function: *"organic matter"*, *"microbial biomass"*, *"nutrients"*, the *"C/N ratio"*, and *"compaction"*. These factors influence one or more soil functions, explaining 82.9% of the total variability in the dataset of 23 physical, chemical, and biological SQIs, and can be used to comparatively reflect changes in soil quality due to land use change. Soil properties of TN and C/N ratio, which determine the quantity and quality of soil organic matter, emerge as particularly sensitive indicators of soil quality changes in Mediterranean agroecosystems. By monitoring TN and the C/N ratio, valuable insights into soil quality can be obtained, incorporating valuable evidence that can be derived by a bigger set of physical, chemical, and biological indicators. The comparison of soils based on these two indicators reveals the impact of land use changes and management practices on soil quality.

The proposed indicators, covering complex information about changes in soil quality, can serve as practical tools for assessing the sustainability of soil resources. Regular monitoring of TN and C/N soil properties enables producers and land management entities to rapidly and accurately determine shifts in soil quality following land use changes or the adoption of new cultivation practices, providing an opportunity to reassess and optimize management strategies.

Author Contributions: Conceptualization, E.E. and C.G.; methodology, E.E.; writing—original draft preparation, E.E.; writing—review and editing, E.E. and C.G.; supervision, C.G. All authors have read and agreed to the published version of the manuscript.

Funding: This research received no external funding.

Institutional Review Board Statement: Not applicable.

Informed Consent Statement: Not applicable.

Data Availability Statement: The data presented in this study are available on request from the corresponding author.

Conflicts of Interest: The authors declare no conflicts of interest.

References

1. Doran, J.W.; Parkin, T.B. Defining and assessing soil quality. In *Defining Soil Quality for Sustainable Environment*; Doran, J.W., Ed.; SSSA Special Publication: Madison, WI, USA, 1994; pp. 3–23.
2. Larson, W.E.; Pierce, F.J. Conservation and enhancement of soil quality. In *Evaluation for Sustainable Land Management in the Developing World*; Dumanski, J., Pushparajah, E., Latham, M., Myers, R., Eds.; International Board for Soil Research and Management, Bangkok, Thailand Technical Papers; Proceeding International Workshop: Chiang Rai, Thailand, 1991; Volume 2, pp. 15–21.
3. Karlen, D.L.; Mausbach, M.J.; Doran, J.W.; Cline, R.G.; Harris, R.F.; Schumann, G.F. Soil quality: A concept, definition and framework for evaluation. *Soil Sci. Soc. Am. J.* **1997**, *61*, 4–10. [CrossRef]
4. Sojka, R.E.; Upchurch, D.R. Reservations regarding the soil quality concept. *Soil Sci. Soc. Am. J.* **1999**, *63*, 1039–1054. [CrossRef]
5. Singer, M.J.; Ewing, S. Soil Quality. In *Handbook of Soil Science*; Sumner, M.E., Ed.; CRC Press: Boca Raton, FL, USA, 2000.
6. Sojka, R.E.; Upchurch, D.R.; Borlaug, N.E. Quality soil management or soil quality management: Performance versus semantics. In *Advances in Agronomy*; Sparks, D.L., Ed.; Academic Press: New York, NY, USA, 2003; Volume 79, pp. 1–68.
7. Doran, J.W. Soil health and global sustainability: Translating science into practice. *Agric. Ecosyst. Environ.* **2002**, *88*, 119–127. [CrossRef]
8. Lehmann, J.; Bossio, D.A.; Kögel-Knabner, I.; Rillig, M.C. The concept and future prospects of soil health. *Nat. Rev. Earth Environ.* **2020**, *1*, 544–553. [CrossRef] [PubMed]
9. Karlen, D.L.; Ditzler, C.A.; Andrews, S.S. Soil quality: Why and how? *Geoderma* **2003**, *114*, 145–156. [CrossRef]
10. Napoletano, P.; Colombo, C.; Di Iorio, E.; Memoli, V.; Panico, S.C.; Ruggiero, A.G.; Santorufo, L.; Maisto, G.; De Marco, A. Integrated Approach for Quality Assessment of Technosols in Experimental Mesocosms. *Sustainability* **2021**, *13*, 9101. [CrossRef]
11. Napoletano, P.; Barbarisi, C.; Maselli, V.; Rippa, D.; Arena, C.; Volpe, M.G.; Colombo, C.; Fulgione, D.; De Marco, A. Quantifying the Immediate Response of Soil to Wild Boar (*Sus scrofa* L.) Grubbing in Mediterranean Olive Orchards. *Soil Syst.* **2023**, *7*, 38. [CrossRef]
12. Lal, R. Restoring Soil Quality to Mitigate Soil Degradation. *Sustainability* **2015**, *7*, 5875–5895. [CrossRef]
13. Andrews, S.; Karlen, D.L.; Cambardella, C.A. The soil management-assessment framework: A quantitative soil quality evaluation method. *Soil Sci. Soc. Am. J.* **2004**, *68*, 1945–1962. [CrossRef]
14. Seybold, C.A.; Mausbach, M.J.; Karlen, D.L.; Rogers, H.H. Quantification of soil quality. In *Soil Processes and the Carbon Cycle*; Lal, R., Ed.; CRC Press LLC: Boca Raton, FL, USA, 1998; pp. 387–404.
15. James, F.C.; McCulloch, C.E. Multivariate analysis in ecology and systematics: Panacea or pandora's box? *Annu. Rev. Ecol. Syst.* **1990**, *21*, 129–166. [CrossRef]
16. Sena, M.M.; Frighetto, R.T.S.; Valarini, P.J.; Tokeshi, H.; Poppi, R.J. Discrimination of management effects on soil parameters by using principal component analysis: A multivariate analysis case study. *Soil Tillage Res.* **2002**, *67*, 171–181. [CrossRef]
17. Wang, Z.; Chang, A.C.; Wu, L.; Crowley, D. Assessing the soil quality of long-term reclaimed wastewater-irrigated cropland. *Geoderma* **2003**, *114*, 261–278. [CrossRef]
18. Shukla, M.K.; Lal, R.; Ebinger, M. Soil quality indicators for reclaimed minesoils in southeastern Ohio. *Soil Sci.* **2004**, *169*, 133–142. [CrossRef]

19. Rezaei, S.A.; Gilkes, R.J.; Andrews, S.S. Andrews A minimum data set for assessing soil quality in rangelands. *Geoderma* **2006**, *136*, 229–234. [CrossRef]
20. Peng, B.; Hong, H.; Xue, X.; Jin, D. On the measurement of socioeconomic benefits of integrated coastal management (ICM): Application to Xiamen. *China Ocean Coast. Manag.* **2006**, *49*, 93–109. [CrossRef]
21. Brejda, J.J.; Moorman, T.B.; Karlen, D.L.; Dao, T.H. Identification of regional soil quality factors and indicators: I. Central and southern high plains. *Soil Sci. Soc. Am. J.* **2000**, *64*, 2115–2124. [CrossRef]
22. Zhou, Y.; Ma, H.; Xie, Y.; Jia, X.; Su, T.; Li, J.; Shen, Y. Assessment of soil quality indexes for different land use types in typical steppe in the loess hilly area. *China Ecol. Indic.* **2020**, *118*, 106743. [CrossRef]
23. Clémence, S.; Coll, P.; Lardo, E.; Metay, A.; Villenave, C.; Marsden, C.; Blanchart, E.; Hinsinger, P.; Le Cadre, E. The soil quality concept as a framework to assess management practices in vulnerable agroecosystems: A case study in mediterranean vineyards. *Ecol. Indic.* **2016**, *61*, 456–465. [CrossRef]
24. Zornoza, R.; Mataix-Solera, J.; Guerrero, C.; Arcenegui, V.; Mataix-Beneyto, J.; Gómez, I. Validating the effectiveness and sensitivity of two soil quality índices based on natural forest soils under Mediterranean conditions. *Soil Biol. Biochem.* **2008**, *40*, 2079–2087. [CrossRef]
25. Armenise, E.; Redmile-Gordon, M.A.; Stellacci, A.M.; Ciccarese, A.; Rubino, P. Developing a soil quality index to compare soil fitness for agricultural use under different managements in the Mediterranean environment. *Soil Tillage Res.* **2013**, *130*, 91–98. [CrossRef]
26. Herencia, F. Soil quality indicators in response to long-term cover crop management in a Mediterranean organic olive system. *Biol. Agric. Hortic.* **2018**, *34*, 211–231. [CrossRef]
27. Farina, R.; Piccini, C.; Di Bene, C.; Fornasier, F.; Francaviglia, R.; Pennelli, B.; Vanino, S.; Russo, M.; Cerasuolo, M.; Troccoli, A. No tillage and leguminous cover crop improve soil quality in a typical rainfed Mediterranean system. *Can. J. Soil Sci.* **2024**, *104*, 257–272. [CrossRef]
28. Reis, A.; Dindaroğlu, T. Evaluating dynamic soil quality by the soil management assessment framework (SMAF) in the watershed scale in a semi-arid Mediterranean ecosystem in Turkey. *Geoderma Reg.* **2024**, *38*, e00829. [CrossRef]
29. Sánchez-Navarro, A.; Gil-Vázquez, J.M.; Delgado-Iniesta, M.J.; Marín-Sanleandro, P.; Blanco-Bernardeau, A.; Ortiz-Silla, R. Establishing an index and identification of limiting parameters for characterizing soil quality in Mediterranean ecosystems. *Catena* **2015**, *131*, 35–45. [CrossRef]
30. Babur, E.; Dindaroglu, T. Seasonal changes of soil organic carbon and microbial biomass carbon in different forest ecosystems. In *Environmental Factors Affecting Human Health*; Uher, I., Ed.; Intech Open: Rijeka, Croatia, 2020. [CrossRef]
31. Evangelou, E.; Tsadilas, C.; Giourga, C. Soil Microbial Carbon and Nitrogen as Affected by Land Use and Seasonal Patterns in a Mediterranean agro-ecosystem. *Commun. Soil Sci.* **2018**, *52*, 222–234. [CrossRef]
32. Yaalon, D.H. Soils in the Mediterranean region: What makes them different? *Catena* **1997**, *28*, 157–656. [CrossRef]
33. Marathianou, M.; Kosmas, C.; Gerontidis, S.; Detsis, V. Land-use evolution and degradationin Lesvos (Greece): A historical approach. *Land Degrad. Dev.* **2000**, *11*, 63–73. [CrossRef]
34. Giourga, H.; Margaris, N.S.; Vokou, D. Effects of grazing pressure on succession process and productivity of old fields on Mediterranean islands. *Environ. Manag.* **1998**, *2*, 89–596.
35. Kosmas, C.; Gerontidis, S.; Marathianou, M. The effect of land use change on soils and vegetation over various lithological formations on Lesvos (Greece). *Catena* **2000**, *40*, 51–68. [CrossRef]
36. Koulouri, M.; Giourga, C. Land abandonment and slope gradient as key factors of soil erosion in Mediterranean terraced lands. *Catena* **2007**, *69*, 274–281. [CrossRef]
37. Kosmas, C.; Briassoulis, H.; Gerontidis, S.; Detsis, B.; Marathianou, M.; Dalakou, B. *Medalus III: Lesvos, Soil–Vegetation Interactions*; ENV4-CT95-0119; Research Project: Brussels, Belgium, 1999.
38. Mandylas, C.; Kardakara, N. *Conservation and Promotion of Wetlands of Kalloni Bay, Lesvos Island, Part A*; Special Environmental Study; Ministry of Environment and Public Works: Athens, Greece, 1988. (In Greek)
39. El Behairy, R.A.; El Arwash, H.M.; El Baroudy, A.A.; Ibrahim, M.M.; Mohamed, E.S.; Kucher, D.E.; Shokr, M.S. How Can Soil Quality Be Accurately and Quickly Studied? A Review. *Agronomy* **2024**, *14*, 1682. [CrossRef]
40. Brady, N.C.; Weil, R.R. *The Nature and Properties of Soil*, 15th ed.; Prentice Hall: Upper Saddle River, NJ, USA, 2002.
41. Weil, R.R.; Islam, K.R.; Stine, M.A.; Gruver, J.B.; Samson-Liebig, S.E. Estimating active carbon for soil quality assessment: A simplified method for laboratory and field use. *Am. J. Altern. Agric.* **2003**, *18*, 3–17.
42. Gardner, W.H. Water content. In *Methods of Soil Analysis: Part 1—Physical and Mineralogical Methods*, 2nd ed.; Klute, A., Ed.; SSSA: Madison, WI, USA, 1986; pp. 493–544. [CrossRef]
43. Bouyoukos, G.H. A recalibration of the hydrometer method for making mechanical analysis of soils. *Agron. J.* **1951**, *43*, 434–438. [CrossRef]
44. Nelson, D.W.; Sommers, L.E. Total carbon, organic carbon, and organic matter. In *Methods of Soil Analysis: Part 2. Chemical and Microbiological Properties, Agronomy Monographs 9*, 2nd ed.; Page, A.L., Miller, R.H., Keeney, D.R., Eds.; ASA and SSSA: Madison, WI, USA, 1982.
45. Bremner, J.M.; Mulvaney, C.S. Nitrogen-Total. In *Methods of Soil Analysis. Part 2. Chemical and Microbiological Properties*, 2nd ed.; Page, A.L., Miller, R.H., Keeney, D.R., Eds.; Number 9 in the Series Agronomy; American Society of Agronomy and Soil Science Society of America, Inc.: Madison, WI, USA, 1982; pp. 595–624.

46. Keeney, D.R.; Nelson, D.W. Nitrogen-inorganic forms. In *Methods of Soil Analysis: Part 2. Chemical and Microbiological Properties, Agronomy Monographs 9*, 2nd ed.; Page, A.L., Miller, R.H., Keeney, D.R., Eds.; ASA and SSSA: Madison, WI, USA, 1982; pp. 643–698.
47. Olsen, S.R.; Sommers, L.E. Phosphorus. In *Methods of Soil Analysis: Part 2. Chemical and Microbiological Properties, Agronomy Monographs 9*, 2nd ed.; Page, A.L., Miller, R.H., Keeney, D.R., Eds.; ASA and SSSA: Madison, WI, USA, 1982; pp. 403–430.
48. Doran, J.W.; Jones, A.J.; Smith, J.L.; Doran, J.W. Measurement and use of pH and electrical conductivity for soil quality analysis. In *Methods for Assessing Soil Quality*; Doran, J.W., Jones, A.J., Eds.; SSSA Special Publication: Madison, WI, USA, 1996.
49. Vance, E.D.; Brookes, P.C.; Jenkinson, D.S. An extraction method for measuring soil microbial biomass. *Soil Biol. Biochem.* **1987**, *19*, 703–707. [CrossRef]
50. Brooks, P.; Landman, A.; Jenkinson, D. Chloroform fumigation and the release of soil nitrogen: A rapid direct extraction method to measure microbial biomass nitrogen in soil. *Soil Biol. Biochem.* **1985**, *17*, 837–842. [CrossRef]
51. Johnson, R.A.; Wichern, D.W. *Applied Multivariate Statistical Analysis*; Prentice-Hall: Englewood Cliffs, NJ, USA, 1992.
52. Gregorich, E.G.; Carter, M.R.; Angers, D.A. Towards a minimum data set to assess soil organic matter quality in agricultural soils. *Can. J. Soil Sci.* **1994**, *74*, 367–385. [CrossRef]
53. Blair, N.; Faulkner, R.D.; Till, A.R.; Crocker, G.J. Long-term management impacts on soil C, N and physical fertility. Part III. Tamworth crop rotation experiment. *Soil Till. Res.* **2006**, *91*, 48–56. [CrossRef]
54. Zou, J.; Wu, J.; Osborne, B.; Luo, Y. The response of ecosystem carbon and nitrogen pools to experimental warming in grasslands: A meta-analysis. *J. Plant Ecol.* **2022**, *15*, 733–742. [CrossRef]
55. Haynes, R.J. Labile organic matter fractions as central components of the quality of agricultural soils: An overview. *Adv. Agron.* **2005**, *85*, 221–268. [CrossRef]
56. Xie, N.; An, T.; Liang, X.; Schaeffer, S.; Sun, L.; Fu, S.; Wang, J. Interaction between maize residues and initial soil carbon status on soil labile organic carbon pools. *Appl. Soil Ecol.* **2024**, *202*, 105482. [CrossRef]
57. Lal, R. Digging deeper: A holistic perspective of factors affecting soil organic carbon sequestration in agroecosystems. *Glob. Chang. Biol.* **2018**, *24*, 3285–3301. [CrossRef] [PubMed]
58. Adhikari, K.; Hartemink, A.E. Hartemink Linking soils to ecosystem services—A global review. *Geoderma* **2016**, *262*, 101–111. [CrossRef]
59. Lorenz, K.; Lal, R.; Lal, R. Ehlers. Soil organic carbon stock as an indicator for monitoring land and soil degradation in relation to United Nations' Sustainable Development Goals. *Land Degrad. Dev.* **2019**, *30* (Suppl. S1), 824–838. [CrossRef]
60. Durán, G.A.; Sacristán, D.; Farrús, E.; Vadell, J. Towards defining soil quality of Mediterranean calcareous agricultural soils: Reference values and potential core indicator set. *Int. Soil Water Conserv. Res.* **2024**, *12*, 145–155. [CrossRef]
61. Aguilera-Huertas, J.; Parras-Alcántara, L.; González-Rosado, M.; Lozano-García, B. Intercropping in rainfed Mediterranean olive groves contributes to improving soil quality and soil organic carbon storage. *Agric. Ecosyst. Environ.* **2023**, *361*, 108826. [CrossRef]
62. Soto, R.L.; de Vente, J.; Padilla, M.C. Learning from farmers' experiences with participatory monitoring and evaluation of regenerative agriculture based on visual soil assessment. *J. Rural. Stud.* **2021**, *88*, 192–204. [CrossRef]
63. Mann, L.K. Changes in soil carbon storage after cultivation. *Soil Sci.* **1986**, *142*, 279–288. [CrossRef]
64. Davidson, E.A.; Ackerman, I.L. Changes in soil carbon inventories following cultivation of previously untilled soils. *Biogeochemistry* **1993**, *20*, 161–193. [CrossRef]
65. Tan, Z.X.; Lal, R. Carbon sequestration potential estimates with changes in land use and tillage practice in Ohio, USA. *Agric. Ecosyst. Environ.* **2005**, *111*, 140–152. [CrossRef]
66. Paustian, K.; Six, J.; Elliott, E.T.; Hunt, H.W. Management options for reducing CO_2 emissions from agricultural soils. *Biogeochemistry* **2000**, *48*, 147–163. [CrossRef]
67. Ogieriakhi, M.O.; Woodward, R.T. Understanding why farmers adopt soil conservation tillage: A systematic review. *Soil Secur.* **2022**, *9*, 100077. [CrossRef]
68. Hu, Q.; Zhang, Y.; Cao, W.; Yang, Y.; Hu, Y.; He, T.; Li, Z.; Wang, P.; Chen, X.; Chen, J.; et al. Legume cover crops sequester more soil organic carbon than non-legume cover crops by stimulating microbial transformations. *Geoderma* **2024**, *450*, 117024. [CrossRef]
69. Lal, R. Soil carbon sequestration to mitigate climate change. *Geoderma* **2004**, *123*, 1–22. [CrossRef]
70. Jenkinson, D.S.; Ladd, J.N. Microbial biomass. In *Soil: Measurement and Turnover*; Paul, E.A., Ladd, J.N., Eds.; Marcel Dekker, Inc.: New York, NY, USA; Basel, Switzerland, 1981; Volume 5, pp. 415–471.
71. Brookes, P.C. Measurement, properties and role of the soil microbial biomass in organic matter dynamics and the maintenamce of soil fertility. In *Integrated Plant Nutrition Systems, Report of an Expert Consultation*; Dudal, R., Roy, R.N., Eds.; FAO: Rome, Italy, 1995; pp. 113–128.
72. Moore, J.M.; Klose, S.; Tabatabai, M.A. Soil microbial biomass carbon and nitrogen as affected by cropping systems. *Biol. Fertil. Soil* **2000**, *31*, 200–210. [CrossRef]
73. Kaur, B.; Gupta, S.R.; Singh, G. Soil carbon, microbial activity and nitrogen availability in agroforesty systems on moderately alkaline soils in northern India. *Appl. Soil Ecol.* **2000**, *15*, 283–294. [CrossRef]
74. Hill, G.T.; Mitkowski, N.A.; Aldrich-Wolfe, L.; Emele, L.R.; Jurkonie, D.D.; Ficke, A.; Maldonado-Ramirez, S.; Lynch, S.T.; Nelson, E.B. Methods for assessing the composition and diversity of soil microbial communities. *Appl. Soil Ecol.* **2000**, *15*, 25–36. [CrossRef]
75. Brooks, P. The Soil Microbial Biomass: Concept, Measurement and Applications in Soil Ecosystem Research. *Microbes Environ.* **2001**, *16*, 131–140. [CrossRef]

76. Lal, R. Sustainable land use systems and soil resilience. In *Soil Resilience and Sustainable Land Use*; Greenland, J., Szabolcs, I., Eds.; CAB International: Wallingford, UK, 1994.
77. Wu, J.; Brookes, P.C. The proportional mineralisation of microbial biomass and organic matter caused by air-drying and rewetting of a grassland soil. *Soil Biol. Biochem.* **2005**, *37*, 507–515. [CrossRef]
78. Stamou, G.P. *Arthropods of Mediterranean-Type Ecosystems*; Springer: Berlin/Heidelberg, Germany, 1998.
79. Papatheodorou, E.M. Vegetation Dynamics and Nutrient Cycling (N and P) in a Degraded Evergreen Sclerophyllous Ecosystem at Mt. Hortiatis. Ph.D. Thesis, Aristotle University Thessaloniki, Thessaloniki, Greece, 1996.
80. Chen, S.; Huang, J.; Guo, R.; Ma, H.; Guo, J.; Ling, N.; Xu, Q.; Wang, M.; Shen, Q.; Guo, S. Soil net carbon balance depends on soil C: N: P stoichiometry. *Soil Tillage Res.* **2025**, *245*, 106298. [CrossRef]
81. de Lima, A.C.R.; Hoogmoed, W.; Brussaard, L. Soil quality assessment in rice production systems: Establishing a minimum data set. *J. Environ. Qual.* **2008**, *37*, 623–630. [CrossRef]
82. Brejda, J.J.; Moorman, T.B. Identification and Interpretation of Regional Soil Quality Factors for the Central High Plains of the Midwestern USA. In Proceedings of the Sustaining the Global Farm, Selected Papers from the 10th International Soil Conservation Organization Meeting, West Lafayette, IN, USA, 24–29 May 2001; pp. 535–540.
83. Govaerts, B.; Mezzalama, M.; Sayre, K.D.; Crossa, J.; Nicol, J.M.; Deckers, J. Long-term consequences of tillage, residue management, and crop rotation on maize/wheat root rot and nematode populations in subtropical highlands. *Appl. Soil Ecol.* **2006**, *32*, 305–315. [CrossRef]
84. Sparling, G.P.; Schipper, L.A.; Hill, R.; Bettjeman, W. Soil quality monitoring in New Zealand: Practical lessons from a six-year trial. *Agric. Ecosyst. Environ.* **2004**, *104*, 523–534. [CrossRef]
85. Cristian, P.; Dumitru, S.; Vizitiu, O.; Mocanu, V.; Lăcătusu, A.-R.; Ion, S.; Domnariu, H. Soil organic carbon and total nitrogen stocks related to land use and basic environmental properties—Assessment of soil carbon sequestration potential in different ecosystems. *Catena* **2024**, *246*, 108435.
86. Chatterjee, N.; Nair, P.K.R.; Chakraborty, S.; Nair, V.D. Changes in soil carbon stocks across the Forest-Agroforest-Agriculture/Pasture continuum in various agroecological regions: A meta-analysis. *Agric. Ecosyst. Environ.* **2018**, *266*, 55–67. [CrossRef]
87. Farzadfar, S.; Knight, J.D.; Congreves, K.A. Soil organic nitrogen: An overlooked but potentially significant contribution to crop nutrition. *Plant Soil* **2021**, *462*, 7–23. [CrossRef]
88. Zhao, X.; Tong, M.; He, Y.; Han, X.; Wang, L. A comprehensive, locally adapted soil quality indexing under different land uses in a typical watershed of the eastern Qinghai-Tibet Plateau. *Ecol. Indic.* **2021**, *125*, 107445. [CrossRef]
89. Thomsen, I.K.; Petersen, B.M.; Bruun, S.; Jensen, L.S.; Christensen, B.T. Estimating soil C loss potentials from the C to N ratio. *Soil Biol. Biochem.* **2008**, *40*, 849–852. [CrossRef]
90. Li, C.; Xiao, C.; Li, M.; Xu, L.; He, N. The quality and quantity of SOM determines the mineralization of recently added labile C and priming of native SOM in grazed grasslands. *Geoderma* **2023**, *432*, 116385. [CrossRef]
91. Wu, X.; Nguyen-Sy, T.; Sun, Z.; Wantanabe, T.; Tawaraya, K.; Hu, R.; Cheng, W. Soil Organic Matter Dynamics as Affected by Land Use Change from Rice Paddy to Wetland. *Wetlands* **2020**, *40*, 2199–2207. [CrossRef]
92. Amorim HC, S.; Araujo, M.A.; Lal, R.; Zinn, Y.L. What C:N ratios in soil particle-size fractions really say: N is preferentially sorbed by clays over organic C. *Catena* **2023**, *230*, 107230. [CrossRef]
93. Mitchell, J.P.; Shrestha, A.; Mathesius, K.; Scow, K.M.; Southard, R.J.; Haney, R.L.; Schmidt, R.; Munk, D.S.; Horwath, W.R. Cover cropping and no-tillage improve soil health in an arid irrigated cropping system in California's San Joaquin Valley, USA. *Soil Tillage Res.* **2017**, *165*, 325–335. [CrossRef]
94. Steenwerth, K.L.; Jackson, L.E.; Calderón, F.J.; Stromberg, M.R.; Scow, K.M. Soil microbial community composition and land use history in cultivated and grassland ecosystems of coastal California. *Soil Biol. Biochem.* **2002**, *34*, 1599–1611. [CrossRef]

sustainability

MDPI

Article

Chemical and Physical Aspects of Soil Health Resulting from Long-Term No-Till Management

Edyta Hewelke [1,*], Lilla Mielnik [2], Jerzy Weber [3], Aneta Perzanowska [4], Elżbieta Jamroz [3], Dariusz Gozdowski [4] and Paweł Szacki [5]

1 Institute of Environmental Engineering, Warsaw University of Life Sciences, Nowoursynowska 166, 02-787 Warsaw, Poland
2 Department of Bioengineering, West Pomeranian University of Technology in Szczecin, Papieża Pawła VI 3, 71-459 Szczecin, Poland; lilla.mielnik@zut.edu.pl
3 Institute of Soil Science, Plant Nutrition and Environmental Protection, Wrocław University of Environmental and Life Sciences, Grunwaldzka 53, 50-375 Wrocław, Poland; jerzyweber@gmail.com (J.W.); elzbieta.jamroz@upwr.edu.pl (E.J.)
4 Institute of Agriculture, Warsaw University of Life Sciences, Nowoursynowska 166, 02-787 Warsaw, Poland; aneta_perzanowska@sggw.edu.pl (A.P.); dariusz_gozdowski@sggw.edu.pl (D.G.)
5 The Prof. Marian Górski Experimental Station in Skierniewice, Warsaw University of Life Sciences, Nowoursynowska 166, 02-787 Warsaw, Poland; pawel_szacki@sggw.edu.pl
* Correspondence: edyta_hewelke@sggw.edu.pl

Abstract: The aim of this study was to compare the long-term effects of conventional tillage (CT) and no-till (NT) systems on the main soil properties that determine soil health. The research was conducted in a field experiment established in 1975 in Chylice, central Poland, at the WULS-SGGW Experimental Station Skierniewice. Soil samples collected from 0–10 and 10–20 cm of the mollic horizon of the Phaeozem were analysed for total organic carbon (TOC) content, fractional composition of SOM and spectroscopic properties of humin, soil structural stability, soil water retention characteristics and soil water repellency (SWR). The results showed that NT practice almost doubled the TOC in the 0–10 cm layer. However, optical parameters of humin indicated that NT management promoted the formation of humin with a lower molecular weight and lower degree of condensation of aromatic structures. In the NT 0–10 cm layer, a significant increase in the number of water-resistant macroaggregates was found. In the 0–10 cm layer, the water capacity increased by 9%, 18%, 22% and 26% compared to CT at (certain soil suction) pF values of 0.0, 2.0, 3.0 and 4.2, respectively. SWR occurs regardless of the cultivation method at a soil moisture equivalent to pF 4.2, and the greatest range of SWR was found in the NT 0–10 cm layer.

Keywords: soil organic carbon; humin; soil structure; soil water retention; soil water repellency; spectroscopic properties; fluorescence properties

1. Introduction

Soil is a key component of the terrestrial biosphere, not only for its production of food and fibre, but also for its role in maintaining environmental quality. For this reason, the provision of high-quality soil is a major concern for many researchers, not just soil scientists. The metaphor of "soil health", which refers to the ability of a soil to provide ecosystem services, was widely adopted by the scientific community in the 1990s [1]. However, the concept of applying this metaphor to soil is much older. The earliest mention of "soil health" comes from a 1910 thesis at Iowa State University by H.A. Wallace, who wrote about the importance of humus in maintaining soil health [2]. This first concept of soil health was based on the physical and chemical properties of the soil and its fertility, and over time has been complemented by the biological properties of the soil.

Today, "soil health" is defined as "the continued capacity of soil to function as a vital living ecosystem that sustains plants, animals, and humans" [3], while its longer version is "the capacity of soil to function as a vital living system, within ecosystem and land-use boundaries, to sustain plant and animal productivity, maintain or enhance water and air quality, and promote plant and animal health" [4]. A soil is considered healthy if it provides comparable or better ecosystem services than undisturbed reference soils of a similar type in the same region [5]. The sustainable management of agricultural soils fits into the current European research needs on the health of agricultural soils and soil resilience to climate change [6,7]. However, soil health is neither a readily quantifiable nor measurable object, so there are no simple indicators for it [8].

In addition to soil properties being limited by the parent rock type, soil organic matter (SOM) content and quality, moisture regime, and chemical and biological properties, they are influenced by anthropogenic activities [9]. This includes preferred cropping practices and intensive land use management, which can have positive or negative effects on soil health. No-tillage (NT) is considered an effective way to improve soil health in both the short and long term, with the magnitude of benefits depending on specific practices and environmental conditions [10–12].

NT practices have been observed to improve soil's physical, chemical and biological properties. The longer the duration of NT, the better the effect. Improvements include increased soil organic carbon (SOC) content, especially in surface layers [13,14], but little is known about changes in the soil's organic matter characteristics, including those of humic substances. In particular, the humin fraction is the least known component of the SOM, although this fraction may represent an important part of the pool responsible for carbon sequestration [15–18]. Changes in the directions of SOM transformation, including the humin fraction, can be accurately determined using advanced techniques such as UV–Vis supported by fluorescence [19–21].

Long-term NT significantly increases soil aggregation and aggregate stability, especially in the surface layers [10–12], which improves soil structure and reduces erosion [10,13]. Conventional tillage (CT), in contrast to NT, contributes to an increase in soil bulk density, particularly in the upper soil layers, which can lead to soil compaction problems [10]. NT has also been shown to increase soil infiltration and water storage, improving soil moisture retention [10,13,22,23]. In Tahad's et al. [24] review of agricultural practices, NT reduced the amount of irrigation water applied by 1/8 to 1/4 compared to CT. Thus, NT increases water use efficiency and, consequently, overall net returns. The NT system also increased the soil water repellency (SWR), contributing to soil degradation by changing some soil processes (e.g., carbon sequestration and soil erosion). Soil structure had a direct influence on SWR [25–27], and was a result of the interactions between the pore structure and hydrophobic substances [26]. Further studies into the mechanisms controlling SWR require more than a quantification of TOC [27].

The aim of this study was to assess the effects of almost 50 years of NT vs. CT treatments on soil's physical and chemical properties, including changes in the molecular structure of the rarely studied humin fraction of SOM, with particular attention to differences between the 0–10 cm and 10–20 cm layers.

2. Materials and Methods

2.1. Long-Term Tillage Experiment

The 47-year field experiment in Chylice (central Poland, 52.098° N, 20.548° E) was established in 1975 at the WULS-SGGW Experimental Station in Skierniewice. The experiment is the longest-running soil cultivation type of experiment in Poland and one of the oldest in Europe. The area has an average annual temperature of 9.3 °C and an average annual rainfall of 600 mm, classifying its climate as warm temperate (Cfb) [28]. In the decade 2011–2020, the average annual precipitation was 604.1 mm (min 476 mm in 2019, and max 725 mm in 2016) and the average annual temperature was 9.4 °C (min 8.8° in 2012, and max 10.5° in 2019), respectively. The groundwater level ranges from 70 cm in April to 170–200 cm in the summer months.

The experiment was carried out on Phaeozem derived from sandy loam (sand 70%, silt 17%, clay 13%) on 20 m^2 plots. A randomised block design was used with two contrasting tillage systems, i.e., traditional full inversion tillage with mouldboard plough (CT), and NT with direct drilling. Both tillage systems used the same crop rotation and mineral fertiliser regime. Crop rotation was dominated by cereals, but from 2020 onwards, a monoculture of maize was grown. The average fertilisation level over many years was 100–170 kg N adapted to the needs of crops (in legumes 20 kg), 30.6 kg P and 87.2 kg K per hectare. Cereals dominated crop rotation. In 2022 in CT and NT systems, the same crop (winter wheat in 2021), corn variety (Gallery), sowing date, fertilisation and level of chemical plant protection were used. On 28 April 2022, NPK mineral fertilisers were applied in the following doses: 21 kg of N in ammonium and 92 kg in amide forms, 30.6 kg of P and 87.2 kg of K in potassium chloride per hectare. Mineral fertilisers in the CT were mixed with the soil using a cultivator. On 29 April 2022, corn was sown in 88 thousand seeds per hectare with simultaneous under-sowing NP mineral fertilisers: 21 kg of N in ammonium form and 30,6 kg of P per hectare. A mixture of nicosulfuron, sulcotrione and terbuthylazine was sprayed after the emergence of maize plants. Three plots of each tillage system were selected (Figure 1) and in mid-season 2022, soil samples were collected from 0–10 cm and 10–20 cm soil layers at 10 points in each block plot and averaged by gently mixing.

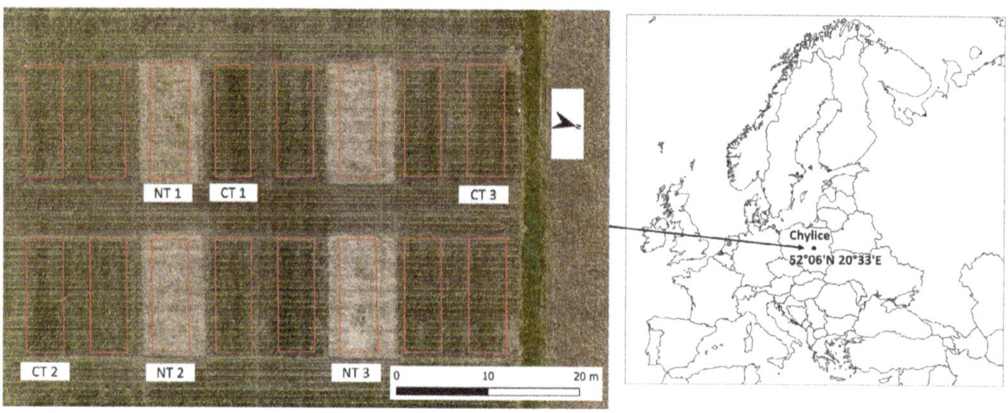

Figure 1. Localisation and the scheme of the long-term tillage experiment.

2.2. Basic Chemical Properties

Soil pH was measured potentiometrically in a 1:2.5 suspension of soil and 1 M KCl. Total organic carbon (TOC) and total nitrogen (TN) were determined using the Enviro TOC+N analyser (Elementar; Langenselbold, Germany). Cation exchange capacity (CEC) was calculated as the sum of H+ ions determined in 1 M KCl, and exchangeable base cations extracted with 1 M NH4Ac, and measured by atomic emission spectroscopy (K+, Na+ and Ca^{2+}) and atomic absorption spectroscopy (Mg^{2+}). Plant available nutrients (P, K, Mg) were determined by the Mehlich 3 method [29].

2.3. The Fractional Composition of SOM

The fractional composition of SOM was determined using the modified method of Swift [30]. H_2SO_4 was used to decalcify the soil, which allowed us to obtain a low-molecular-weight fraction called fulvic fraction (FF). The C content of the supernatant (FF) was determined using the Enviro TOC+N analyser (Elementar; Langenselbold, Germany). An exhaustive alkaline extraction with 0.1 M NaOH was then performed until the supernatant was light-coloured, in which the C content was determined as total extractable carbon (TEC). This fraction contained humic acids (HAs) and fulvic acids (FAs). In the next step, the alkaline supernatant was acidified to pH = 2, which caused HA precipitation. The HA was separated from the FA by centrifugation. Finally, hot 0.02 M NaOH was used to dissolve the precipitate and the C concentration in the whole volume was determined as HA. The content of the FA fraction was calculated as the balance between the TEC and HA fractions (FA = TEC − HA). The content of the humin fraction was calculated according to the formula HUM = TOC − (TEC + FF).

2.4. Spectroscopic Analysis

The spectroscopic properties of the humin isolated from the 0–10 cm layers were determined by means of UV–Vis spectroscopy and fluorescence analysis. The UV–Vis spectra were recorded in the wavelength range of 200–700 nm using a Jasco UV-VIS-NIR 770 spectrometer (Jasco-Global, Tokyo, Japan).

Fluorescence spectra were recorded using a Hitachi F 7000 spectrofluorometer (Hitachi, Tokyo, Japan). Emission spectra were recorded at several fixed excitation wavelengths ex. Synchronous scan spectra were measured in the range of 220–620 nm, keeping the scan difference constant, $\Delta\lambda = \lambda em - \lambda ex = 20$ nm. The monochromators of the excitation and emission slits were 5 nm and 10 nm, respectively, and the scan speed was a 240 nm min^{-1}.

Humin was dissolved in a mixture of DMSO and 98% H_2SO_4 (94% and 6% v/v, respectively) to obtain a carbon concentration of 10 mg dm^{-3}. Prior to optical analysis, the samples were filtered through a syringe filter with a pore size of 0.45 µm.

2.5. Bulk Density and Structural Stability

Soil bulk density was determined by the gravimetric method in undisturbed samples in three replications for each plot and layer. Structural stability was determined in 30 g samples of dry soil aggregates placed on top of a set of sieves in cylindrical containers filled with distilled water in Baksheev's apparatus [31].

Mean weight diameter of water-resistant aggregates (*MWDw*) was calculated based on the share of weighted mean diameters of aggregates of all size-fraction classes in the soil according to the formula described by Elliott [32]:

$$MWDw = \sum_{n=1}^{n} \overline{x}iwi \quad (1)$$

where MWD = mean weight diameter (mm) = mean diameter of each size faction,
wi = proportion of total sample weight,
n = number of size fractions.

2.6. Soil Water Retention Characteristic

The undisturbed, standard soil samples (100 cm^3) were collected in three replicates from each block plot design in the two layers (36 samples in total) for the determination of soil water retention characteristics (SWRCs). Retention curves were measured in the laboratory using reference methods [33]. Moisture content values in the range of pF = 0 to pF = 2 were determined on a sand table, while the amounts of water at the pF: 2.3, 2.7, 3.0 and 4.2 were measured in pressure chambers. The plant available water capacity (AWC) was calculated as the difference between the field capacity pF = 2.0 and the permanent wilting point (pF = 4.2), the easily available water (EAW) was calculated as the difference between pF = 2.0 and 3.0 and the difficultly available water (DAW) for the plant was calculated as the difference between pF = 3.0 and pF = 4.2.

2.7. Soil Water Repellency Assessment

SWR was assessed by the most widespread method, the water drop penetration time (WDPT) test [34]. The WDPT test was performed at different moisture contents that had been adjusted by equilibrating the undisturbed soil samples at characteristic pF values in triplicate, for each block plot and layer. A detailed description has been presented in previous work by Hewelke [35]. The classification of SWR proposed by Dekker and Jungerius [36] was used to evaluate the test results.

2.8. Statistical Analysis

Statistical analysis was performed using analysis of variance and the Fisher procedure of multiple comparisons, $\alpha = 0.05$. The Kruskal–Wallis test was used for WDPT test values that did not follow a normal distribution. Principal component analysis (PCA) was used to investigate the multivariate relationships between the examined soil variables studied and the tillage systems. All analyses were carried out using the data analysis programme Statistica version 13 [37].

3. Results and Discussion

3.1. Basic Chemical Properties

Compared to CT, NT can have a positive effect on most soil parameters, with changes usually limited to the top 10 cm [12]. The basic chemical properties of the soils studied are shown in Table 1. A global meta-analysis of soil pH responses to NT by Zhao et al. [38] suggested that changes in organic matter decomposition under undisturbed soil could lead to higher H+ concentration, thereby lowering soil pH. Our research indicated that long-term NT can reduce the pH of a surface layer of soil from 6.19 to 5.33. No such effect was observed in the 10–20 cm layer.

Table 1. Basic chemical properties.

Variant	pH (KCl)	TOC	TN	TOC/TN	CEC	P	K	Mg
		g kg^{-1}			cmol (+) kg^{-1}	mg kg^{-1}		
0–10 cm layer								
CT	6.19 b	9.87 a	1.00 a	10.50 a	8.21 a	128.33 a	125.33 a	107.67 a
NT	5.33 a	17.60 b	1.63 b	10.83 a	9.31 a	196.67 b	218.00 b	114.67 a
10–20 cm layer								
CT	5.38 a	12.13 a	1.07 a	11.30 a	7.56 a	121.33 a	121.67 a	100.67 a
NT	5.54 a	11.43 a	1.20 a	9.73 a	10.61 a	121.67 a	109.00 a	120.00 a

NT resulted in a significant increase in TOC and TN, but only in the top layer of the soil, indicating the beneficial effect of NT on SOM content [39–41]. Increased soil C levels under NT compared to CT are a result of a 1.5 times slower C turnover, leading to a stabilisation of C within microaggregates [42]. Our results showed that compared to CT (9.87 g kg^{-1}), almost 50 years of continuous use of NT increased TOC by 78%, indicating that this type of practice maintains soil health [43] and minimises the risk of soil degradation [44].

Despite an increase in TOC content of almost 80% in the surface soil layer, no effect of NT on CEC was observed in either the surface or deeper soil layers. However, there was a significant increase in the content of plant-available forms of potassium and phosphorus, in the upper 0–10 cm layer.

3.2. The Fractional Composition of SOM

Several studies have demonstrated the positive effects of NT on soil organic carbon stocks, but little is known about the effects of this soil management on the characteristics of accumulated SOM. CT generally reduces the aggregation and content of particulate organic matter (POM). Six et al. [42] found that C concentrations in fine intra-aggregate POM were on average 51% lower under CT than under NT. Results obtained by Aduhene-Chinbuah et al. [45] in a 19-year field experiment showed an increase in carbon, nitrogen and phosphorus in POM fractions at 0–7.5 cm depth in the NT system, leading the authors to suggest that this system must be highly effective in improving soil fertility.

There is a lack of comprehensive research on the effect of NT on changes in the fractional composition of humic substances. Our results showed that this type of soil management had a significant effect on the FF fraction in the 0–10 cm soil layer (Table 2). The content of this highly mobile fraction, consisting mainly of low molecular weight organic matter, was significantly lower in the NT system (8.77% of TOC) than in the CT (11.65% of TOC). Szajdak et al. [46] found a 42 to 59% higher concentration of HA in NT soils, whereas the concentration of FA was 54% higher in conventionally cultivated soils. In contrast, Wulanningtyas et al. [47] showed negative effects of NT in combination with fallow, hairy vetch and rye on the ratio of humic acids to fulvic acids. Our research did not show a clear effect of NT on the humification process. A higher, but not statistically significant, proportion of HA (51.55% of TOC for CT vs. 46.25% of TOC for NT) and a higher HA/FA ratio (3.46 for CT vs. 2.50 for NT) were found in both the 0–10 cm and 10–20 cm layers. This may suggest that CT favours the humification process and the formation of organic matter with highly reactive organic fractions to a greater extent than NT practice. However, this hypothesis requires more in-depth research on a larger number of samples.

Table 2. Fractional composition of SOM.

Variant	TOC g kg^{-1}	FF	HA	FA	HUM	HUMg kg^{-1}	HA/FA
		% of TOC					
			0–10 cm layer				
CT	9.87 a	11.65 b	51.66 a	15.65 a	21.05 a	2.08 a	3.46 a
NT	17.60 b	8.77 a	46.25 a	18.64 a	26.34 a	4.63 b	2.50 a
			10–20 cm layer				
CT	12.13 a	10.04 ab	48.84 a	15.65 a	25.47 a	3.09 a	3.28 a
NT	11.45 a	11.22 b	41.98 a	21.10 a	25.70 a	2.92 a	2.03 a

TOC—total organic carbon, FF—low molecular fraction, HA—humic acids, FA—fulvic acids, HUM—humin; HA and FA% of TOC and HA/FA value were calculated based on values from individual replications and, because of that, are not exactly equal to ratio calculated based on the means of HA and FA% of TOC and HA and FA values.

Also noteworthy is the quantitatively significant increase in the content of the humin fraction in the NT system (from 2.08 g kg^{-1} to 4.63 g kg^{-1} for CT and NT, respectively, in the 0–10 cm layer). This is of environmental importance, as the increase in this fraction, considered the most resistant to decomposition, has a significant effect on the increase in carbon sequestration. However, it should be noted that this is the effect of the increase in TOC content.

3.3. *Spectroscopic Properties of Humin*

In the study of various organic substances, including SOM, their optical properties are increasingly used. Thanks to the high sensitivity of fluorescence methods, they allow the identification of even small differences in their structure, which may indicate the directions of incipient changes in their transformation. Sometimes they are not advanced enough to be detected by other methods.

3.3.1. UV–Vis Analysis

The UV–Vis spectra of the investigated humin showed a specific profile with a prominent double maximum in the short wavelength range, at 260 nm and 280 nm (Figure 2). In this range, humin isolated from objects with CT showed a much higher absorption capacity. Light absorption by organic matter in this wavelength range increases with increasing degree of condensation of aromatic rings, a higher ratio of carbon in the aromatic core of the molecule to carbon in aliphatic chains and higher molecular weight [48]. On the other hand, at higher wavelength values, corresponding to the Vis range, the absorption efficiency is due to acceptor–donor complexes formed as a result of internal and external molecular aggregation [48,49]. In this range, the analysed humin did not show significant differences.

Differences in soil management and their influence on the transformation of fresh organic matter are also reflected in the absorption coefficient values often used in the literature (Table 3). They are calculated for the absorption values at individual wavelengths. Lower E_{260}:E_{280} and E_{465}:E_{665} values for humin from CT indicated their larger size and molecular weight [50] and increased highly complex aromatic structures and alkyl substitution [51], indicating a more advanced stage of the transformation [52]. Changes in the values of the E_{280}:E_{365} and E_{280}:E_{470} coefficients of the discussed humin samples also indicated different dynamics of the transformation processes of SOM, as well as the influx of fresh mass [53,54]. Higher values of these indices for humin from CT indicated its higher degree of humification.

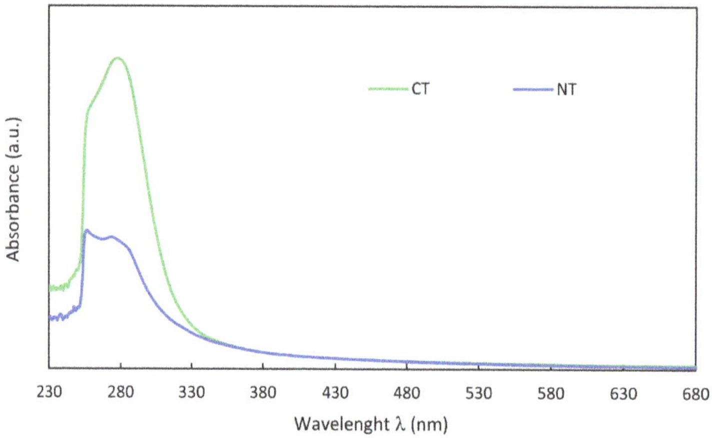

Figure 2. UV–Vis spectra of investigated humin (0–10 cm layer).

Table 3. UV–Vis and fluorescence spectroscopic analysis and correspondent indices.

Variant	UV–Vis Spectroscopy				Fluorescence Spectroscopy			
	E_{260}:E_{280}	E_{465}:E_{665}	E_{280}:E_{365}	E_{280}:E_{470}	IFl_{330}:IFl_{390}	IFl_{330}:IFl_{470}	HIX	A_{440}
CT	0.77	2.37	12.90	35.48	1.88	5.48	1.76	30.04
NT	1.04	3.40	6.61	14.79	2.67	7.69	1.39	25.24

3.3.2. Fluorescence

Differences in SOM transformation processes in NT and CT soils were also visible in the synchronous scan fluorescence spectra (Figure 3), mainly in the shortwave region (280–340 nm). Humin from NT showed a greater ability to emit fluorescence. This band is attributed to the presence of fluorophores of lower molecular complexity. In particular, the band at 280–300 nm is due to the presence of aromatic amino acids, such as tryptophan, and other substances with strongly coupled aliphatic structures [55,56]. According to Duarte et al. [57], in the synchronous scan fluorescence spectra, the band at about 280 nm can be attributed to structural fragments derived from lignin. Its intensity increases towards smaller sizes and lower molecular complexity.

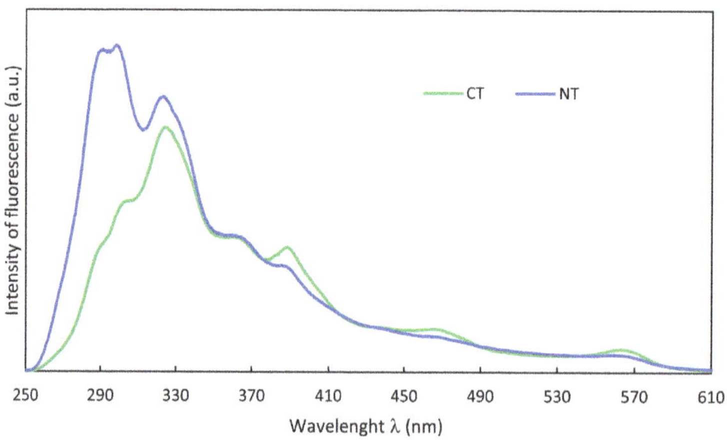

Figure 3. Synchronous scan fluorescence spectra.

The humin fractions tested showed very weak fluorescence in the 340–600 nm range, with CT humin showing slightly higher fluorescence. Peaks in the 420–600 nm range indicate the presence of high molecular weight polycyclic aromatic compounds, while increased fluorescence in the 340–420 nm range indicates the presence of large amounts of simple, dissociated phenolic and quinone groups [18,56,58,59]. It is assumed that the increase in humification leads to a higher fluorescence intensity in the longer wavelength range. This is related to the increase in the number of highly substituted aromatic rings and/or highly conjugated unsaturated systems [54,60]. The observed low fluorescence in the long wavelength part of the tested humin may indicate the degradation of high molecular weight components and the formation of smaller fractions.

The calculated fluorescence coefficients (Table 3) are related to the degree of humification and to the degree of condensation of the aromatic group in humin. They indicate a similar effect of the cultivation method on the humin properties as the UV–Vis data. Higher values of IFl_{330}:IFl_{390} and IFl_{330}:IFl_{470} for NT indicate a lower degree of humification of these samples. This reflects the relationship between relatively simple fluorophores and the number of strongly coupled and condensed aromatic nuclei [54,61]. Similarly, lower HIX and A_{440} values for NT confirm a lower degree of internal transformation of the humin molecule, thus indicating a lower "packing" of this humin. According to Fuentes et al. [62], higher HIX values may be associated with a lower share of oxygen-containing functional groups.

3.4. Bulk Density and Structural Stability

In our investigation, NT caused a significant reduction in bulk density for the 0–10 cm layer, with values of 1.56 g cm^{-3} for NT and 1.65 g m^{-3} for CT, respectively. There were no differences in the 10–20 cm layer. Similar results were reported by Blanco-Canqui and Ruis [12], although they noted in their review that NT can have different effects on bulk density, depending on soil texture.

The 0–10 cm layer showed significantly more large water-resistant macroaggregates with fraction sizes of 10–7, 7–5, 5–3 and 3–1 mm in NT compared to CT (Figure 4a,b). Their proportion was 204, 302 and 244% higher in NT, respectively. In contrast, there were significantly more water-stable macroaggregates of smaller diameters (1–0.5 mm and 0.5–0.25 mm) and microaggregates (<0.25 mm) in CT (by 39.5, 42.6 and 45.3%). This varied share of water-resistant fractions of different sizes significantly influenced *MWDw*—the average diameter of the water-resistant aggregate in NT was almost twice that in CT (Figure 4c). The obtained results clearly indicate that the soil structure of CT was unstable and easily disrupted in the water environment into smaller macroaggregates and microaggregates compared to NT. Zheng et al. [63] also reported the enhancement of soil macroaggregate stability in the 0–10 cm surface soil layer under long-term NT conditions. The improvement of the surface soil structure in NT was also confirmed in different regions and soils [64–66]. The use of NT was associated with the cessation of periodic mechanical destruction of soil aggregates and the accumulation soil organic matter, which increases the dominance of stabilisation processes over destabilisation processes and enhances the process of the cementation of soil particles into stable structures by organic binding agents [67,68].

3.5. Soil Water Retention Characteristic

SWRCs were significantly affected by the tillage treatments, especially in the 0–10 cm layer for all soil water pressures (Figure 5a–d). The NT treatment in 0–10 cm significantly increased the SWRCs at all soil water pressures compared to the CT. In the 0–10 cm layer, under the influence of 47 years of NT, the water capacity increased by 9%, 18%, 22% and 26% compared to CT at pF values of 0.0 (Figure 5a), 2.0 (Figure 5b), 3.0 (Figure 5c) and 4.2 (Figure 5d), respectively. It is noteworthy that differences were obtained at pressures corresponding to the permanent wilting point in the 0–10 cm layer, compared to the 10–20 cm layers, where no significant differences were found. Layer 0–10 cm can retain 50% more water in the NT system compared to CT and 25% more compared to the 10–20 cm layer in both treatments. Similar observations were made by De Vita et al. [69], who concluded that NT performed better with limited rainfall during the durum wheat growing season. Higher moisture contents obtained for NT under certain conditions indicate the water-saving effect due to low soil structure disturbance, as documented for Scandinavia [70], the North American drylands [71] as well as for Germany [72].

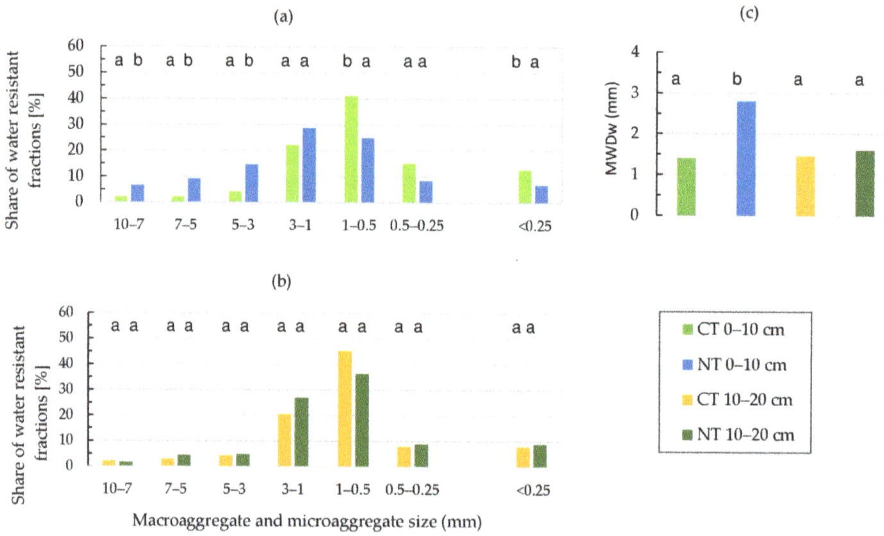

Figure 4. Water-stable soil macroaggregate and microaggregate size distribution for CT and NT treatment in 0–10 cm (**a**) and 10–20 cm (**b**) layers. Mean weight diameter of the water-stable macroaggregates (*MWDw*) (**c**). Different letters indicate significant differences (according to analysis of variance and Fisher's procedure, $p < 0.05$).

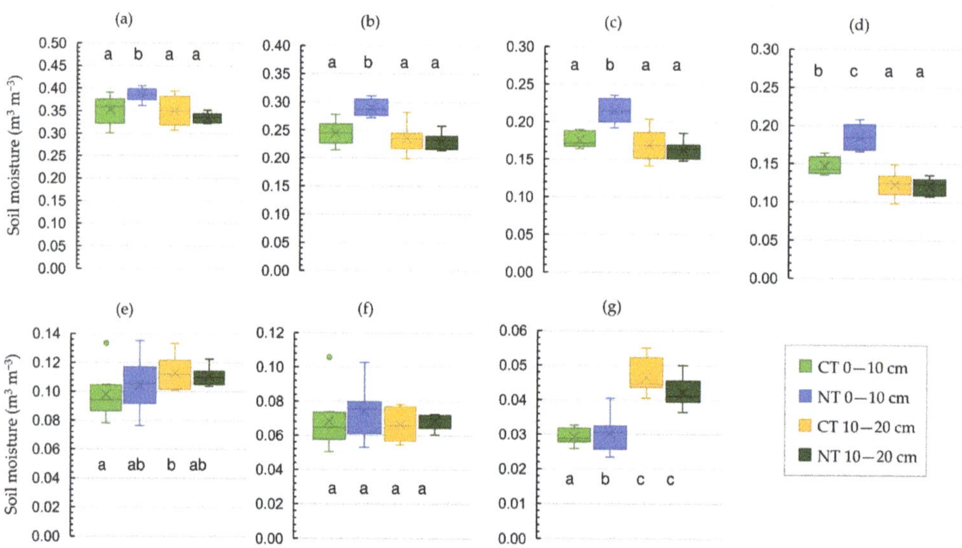

Figure 5. Soil water retention (box plots) measured at pF 0.0 (**a**), pF 2.0 (**b**), pF 3.0 (**c**) and pF 4.2 (**d**). Water available (boxplots and outliers) to the plant (**e**), including easily available (**f**) and difficultly available (**g**). Mean value $n = 9$; different letters indicate significant differences (according to analysis of variance and Fisher's procedure, $p < 0.05$) for the analysed CT and NT, and for the layers 0–10 and 10–20 cm.

Research conducted by Jabro and Stevens [73] during a four-year tillage study showed that soils under NT management had significantly higher plant AWC than soils under CT practices. This long-term study provided critical information on the effect of NT and CT on the range of water that is difficult for plants to access (Figure 5g), which may not effect plant development (Figure 5e,f), but may affect SOM transformation processes.

3.6. Soil Water Repellency

The results indicated by the WDPT test medians for dry soil, i.e., potential SWR and actual SWR obtained at different soil moisture levels, assess it as the hydrophilic, wettable class of SWR [32]. Hydrophobicity was obtained at a single soil moisture content of pF 4.2 (Figure 6) for different system managements and layers, allowing it to be classified as SWR class 2, i.e., strongly repellent. As drought becomes more common, SWR is expected to become more common [74]. Bianco-Canqui [75] found an increase in SWR of 1.5 to 40 times in NT compared to CT management. Fifteen years of NT practice increased the water repellency index compared to the CT, as a consequence of the interactions between the hydrophobic substance and pore structure [26]. In our case, the significant difference and the lowest value of SWR were found for 10–20 cm of the NT system, which is beneficial for water infiltration. However, the widest range of values was found for NT 0–10 cm, from wettable to highly hydrophobic, suggesting that hydrophobic components accumulate in the unmixed 0–10 cm layer. The importance of the wetting history, i.e., the strength and duration of soil drying and wetting [76,77], is suggested to be the key for the effectiveness of SOM stabilisation by SWR. Bianco-Canqui and Ruis [12] and Behrends Kraemer et al. [78] proposed SWR as a driver to maintaining or improving the structural quality of the soil. Zhang et al. [79] stressed the SWR protection of aggregates is mainly related to the reduction in the initial wetting rate, which diminishes the build-up of air pressure in soil pores and reduces the slaking stress. The environmental implications of SWR occurrence are negatively perceived, i.e., slow water infiltration, increased runoff, reduced water storage and thus plant growth [80–82]. Increased SWR may cause significant restrictions and conditions for agricultural production, land use and environmental protection [83].

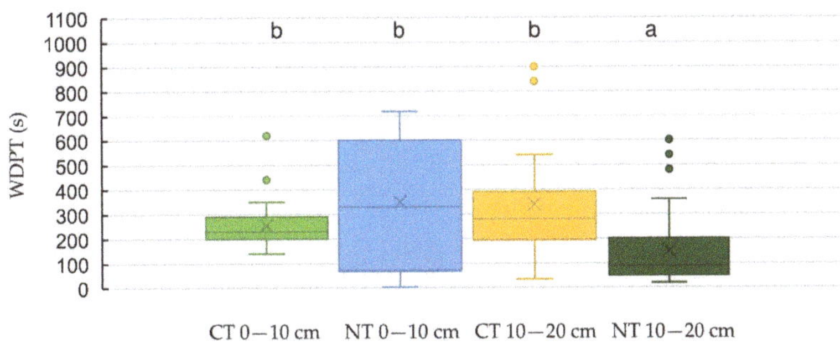

Figure 6. Soil water repellency, indicated by Water Drop Penetration Time (WDPT) test (boxplots and outliers) at pF = 4.2 for different system managements and layers. Median value $n = 45$, different letters indicate significant differences determined with the Kruskal–Wallis test ($\alpha < 0.05$).

3.7. Soil Health Assessment

The results of the PCA are presented in Figure 7. The first two principal components (PC1 and PC2) explain about 81% of the total variability of the data set of soil properties, which was used for the analysis. This means that most of the variability is explained by these two PCs. Both treatments (NT and CT) for the soil layer 10–20 cm were very similar according to all the soil properties used for the PCA. These two treatments, NT (10–20) and CT (10–20), have high DAW, BD and AWC, while simultaneously having a low p = F0, pF = 4.2, SWR and sand content. For the 0–10 cm layer, very large differences were observed between NT and CT. CT (0–10) was characterised by high HA (C% in TOC), pH (KCl), parts < 0.25 and low HUM (C% in TOC).

NT (0–10) was characterised by high HUM (g kg^{-1}), TOC, TN, parts 10–7, 3–7, pF = 0, PF = 4.2 and SWR and simultaneously by low BD and FF (C% in TOC).

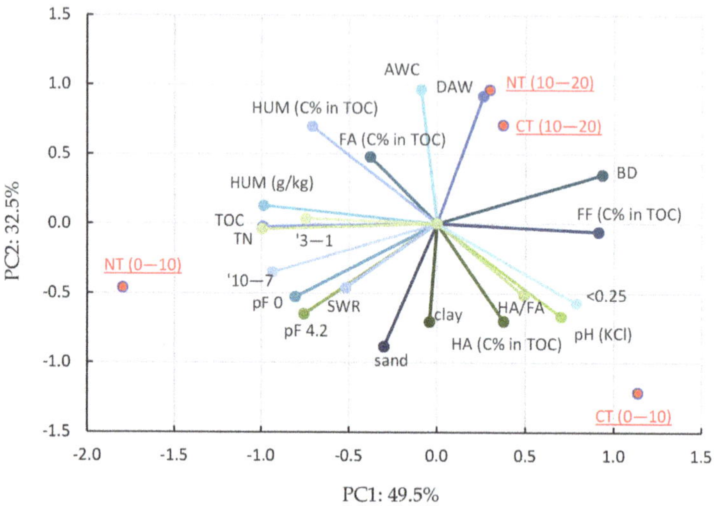

Figure 7. Biplot based on the results of PCA showing multivariate differences between treatments (NT 0–10—no tillage, CT—conventional tillage) and soil layers (and 10–20 cm) based on the data-characterised soil properties.

4. Conclusions

Our long-term study showed that a 47-year application of NT, compared to CT, resulted in a doubling of the TOC content and an improvement in the soil structure and water regime. The fractional composition of humic substances, considered as a percentage of TOC, did not change significantly, indicating that, despite a significantly slowed mineralisation of SOM, the directions of its transformation were similar in both treatments. As a result, the humin content, quantified in g kg^{-1}, doubled under NT. This is ecologically important because the increase in this fraction, which is considered to be the most resistant to decomposition, has a significant effect on carbon sequestration.

The UV–Vis and fluorescence properties of the humin studied showed the different dynamics of the SOM transformation processes occurring under the cultivation methods discussed. The spectroscopic properties of the humin fraction formed under NT conditions indicated its smaller size and lower molecular weight, suggesting a lower degree of humification of these substances formed under NT cultivation conditions.

NT also induced changes in physical properties, but these were confined to the 0–10 cm layer only. There was a significant improvement in soil structure and water holding capacity. A particularly positive effect on water retention was observed at pF values of 3.0 and 4.2, where soil water retention increased by 22% and 26%, respectively. However, it should be stressed that SWR occurs at low soil moisture levels, regardless of the cultivation method. The identified wide range of its values in 0–10 cm NT may lead to unfavourable phenomena related to the irregular wetting front after dry periods. Increasing the SOM content and water retention of the soils under NT conditions may be a good way to contribute to climate change mitigation and ensure food security. An integrated sustainable approach linked to soil health is needed for a long-term strategy and recognition of resilience to climate change.

Author Contributions: Conceptualisation: E.H. and J.W.; methodology: E.H., L.M., A.P. and E.J.; formal analysis: E.H., L.M., J.W., A.P., E.J. and D.G. writing—original draft preparation: E.H., J.W., L.M., A.P. and E.J. writing—review and editing, E.H., J.W., L.M., A.P., E.J., D.G. and P.S. All authors have read and agreed to the published version of the manuscript.

Funding: This research was supported by the EJP SOIL programme NCBR project EJPSOIL/I/78/SOMPACS/2022.

Data Availability Statement: Data are available on request from the authors. The data are not publicly available due to the ongoing project, which will be completed in 2025.

Acknowledgments: The authors would like to thank Przemysław Chłopek for his help in collecting soil samples.

Conflicts of Interest: The authors declare no conflicts of interest.

References

1. Powlson, D.S. Soil Health—Useful Terminology for Communication or Meaningless Concept? Or Both? *Front. Agric. Sci. Eng.* **2020**, *7*, 246. [CrossRef]
2. Brevik, E.C. *A Brief History of the Soil Health Concept*; Soil Science Society of America: Madison, WI, USA, 2019.
3. USDA-NRCS Soil Health. *Soil Health*; Natural Resources Conservation Service, U.S. Department of Agriculture: Washington, DC, USA, 2019. Available online: https://www.nrcs.usda.gov/wps/portal/nrcs/main/soils/health/ (accessed on 25 September 2024).
4. Doran, J.W.; Zeiss, M.R. Soil Health and Sustainability: Managing the Biotic Component of Soil Quality. *Appl. Soil Ecol.* **2000**, *15*, 3–11. [CrossRef]
5. Guo, M. Soil Health Assessment and Management: Recent Development in Science and Practices. *Soil Syst.* **2021**, *5*, 61. [CrossRef]
6. EJP Soil. 2020. Available online: https://ejpsoil.eu/ (accessed on 25 September 2024).
7. Smreczak, B.; Hewelke, E.; Kowalik, M.; Ukalska-Jaruga, A.; Weber, J. Rational Management of Agricultural Soils under Climate Change. *Soil Sci. Annu.* **2024**, *75*, 193074.
8. Harris, J.A.; Evans, D.L.; Mooney, S.J. A New Theory for Soil Health. *Eur. J. Soil Sci.* **2022**, *73*, e13292. [CrossRef]
9. Purakayastha, T.J.; Trivedi, A.; Biswas, S.; Kumar, D. Soil Health and Climate Change. In *Climate Change Impacts on Soil-Plant-Atmosphere Continuum*; Advances in Global Change Research; Pathak, H., Chatterjee, D., Saha, S., Das, B., Eds.; Springer Nature: Singapore, 2024; Volume 78, pp. 107–137. ISBN 978-981-9979-34-9.
10. Mitchell, J.P.; Shrestha, A.; Mathesius, K.; Scow, K.M.; Southard, R.J.; Haney, R.L.; Schmidt, R.; Munk, D.S.; Horwath, W.R. Cover Cropping and No-Tillage Improve Soil Health in an Arid Irrigated Cropping System in California's San Joaquin Valley, USA. *Soil Tillage Res.* **2017**, *165*, 325–335. [CrossRef]
11. Page, K.L.; Dang, Y.P.; Dalal, R.C.; Reeves, S.; Thomas, G.; Wang, W.; Thompson, J.P. Changes in Soil Water Storage with No-Tillage and Crop Residue Retention on a Vertisol: Impact on Productivity and Profitability over a 50 Year Period. *Soil Tillage Res.* **2019**, *194*, 104319. [CrossRef]
12. Blanco-Canqui, H.; Ruis, S.J. No-Tillage and Soil Physical Environment. *Geoderma* **2018**, *326*, 164–200. [CrossRef]
13. Toth, M.; Stumpp, C.; Klik, A.; Strauss, P.; Mehdi-Schulz, B.; Liebhard, G.; Strohmeier, S. Long-Term Effects of Tillage Systems on Soil Health of a Silt Loam in Lower Austria. *Soil Tillage Res.* **2024**, *241*, 106120. [CrossRef]
14. Sangotayo, A.O.; Chellappa, J.; Sekaran, U.; Bansal, S.; Angmo, P.; Jasa, P.; Kumar, S.; Iqbal, J. Long-term Conservation and Conventional Tillage Systems Impact Physical and Biochemical Soil Health Indicators in a Corn–Soybean Rotation. *Soil Sci. Soc. Am. J.* **2023**, *87*, 1056–1071. [CrossRef]
15. Ćwieląg-Piasecka, I. Soil Organic Matter Composition and pH as Factors Affecting Retention of Carbaryl, Carbofuran and Metolachlor in Soil. *Molecules* **2023**, *28*, 5552. [CrossRef] [PubMed]
16. Hayes, M.H.B.; Mylotte, R.; Swift, R.S. Humin. In *Advances in Agronomy*; Elsevier: Amsterdam, The Netherlands, 2017; Volume 143, pp. 47–138. ISBN 978-0-12-812421-5.
17. Hayes, M.H.B.; Swift, R.S. Vindication of Humic Substances as a Key Component of Organic Matter in Soil and Water. In *Advances in Agronomy*; Elsevier: Amsterdam, The Netherlands, 2020; Volume 163, pp. 1–37. ISBN 978-0-12-820769-7.
18. Weber, J.; Jerzykiewicz, M.; Ukalska-Jaruga, A.; Ćwieląg-Piasecka, I.; Jamroz, E.; Kocowicz, A.; Debicka, M.; Bekier, J.; Mielnik, L.; Bejger, R.; et al. Properties of Humin Isolated from Polish Arable Soils: The Most Recalcitrant Fraction of Soil Organic Matter That Prevent Soil Degradation. *Land Degrad. Dev.* **2024**, *35*, 2425–2436. [CrossRef]
19. Debska, B.; Spychaj-Fabisiak, E.; Szulc, W.; Gaj, R.; Banach-Szott, M. EPR Spectroscopy as a Tool to Characterize the Maturity Degree of Humic Acids. *Materials* **2021**, *14*, 3410. [CrossRef]
20. Jiménez-González, M.A.; Álvarez, A.M.; Carral, P.; Almendros, G. Influence of Soil Forming Factors on the Molecular Structure of Soil Organic Matter and Carbon Levels. *Catena* **2020**, *189*, 104501. [CrossRef]
21. Machado, W.; Franchini, J.C.; De Fátima Guimarães, M.; Filho, J.T. Spectroscopic Characterization of Humic and Fulvic Acids in Soil Aggregates, Brazil. *Heliyon* **2020**, *6*, e04078. [CrossRef] [PubMed]
22. Reynolds, W.D.; Drury, C.F.; Yang, X.M.; Fox, C.A.; Tan, C.S.; Zhang, T.Q. Land Management Effects on the Near-Surface Physical Quality of a Clay Loam Soil. *Soil Tillage Res.* **2007**, *96*, 316–330. [CrossRef]
23. Fernández, F.G.; Sorensen, B.A.; Villamil, M.B. A Comparison of Soil Properties after Five Years of No-Till and Strip-Till. *Agron. J.* **2015**, *107*, 1339–1346. [CrossRef]
24. Tahat, M.M.; Alananbeh, K.M.; Othman, Y.A.; Leskovar, D.I. Soil Health and Sustainable Agriculture. *Sustainability* **2020**, *12*, 4859. [CrossRef]

25. Li, S.; Liu, X.; Wu, X.; Lu, J.; Abdelrhman, A.A.; Liang, G. Factors Governing Soil Hydrological Function under Long-term Tillage Practices: Insight into Soil Water Repellency. *Soil Sci. Soc. Am. J.* **2023**, *87*, 781–796. [CrossRef]
26. Li, S.; Lu, J.; Liang, G.; Wu, X.; Zhang, M.; Plougonven, E.; Wang, Y.; Gao, L.; Abdelrhman, A.A.; Song, X.; et al. Factors Governing Soil Water Repellency under Tillage Management: The Role of Pore Structure and Hydrophobic Substances. *Land Degrad. Dev.* **2021**, *32*, 1046–1059. [CrossRef]
27. Urbanek, E.; Hallett, P.; Feeney, D.; Horn, R. Water Repellency and Distribution of Hydrophilic and Hydrophobic Compounds in Soil Aggregates from Different Tillage Systems. *Geoderma* **2007**, *140*, 147–155. [CrossRef]
28. Kottek, M.; Grieser, J.; Beck, C.; Rudolf, B.; Rubel, F. World Map of the Köppen-Geiger Climate Classification Updated. *Meteorol. Z.* **2006**, *15*, 259–263. [CrossRef] [PubMed]
29. Mehlich, A. Mehlich 3 Soil Test Extractant: A Modification of Mehlich 2 Extractant. *Commun. Soil Sci. Plant Anal.* **1984**, *15*, 1409–1416. [CrossRef]
30. Swift, R.S. *Methods of Soil Analysis. Part 3. Chemical Methods*; Soil Science Society of America: Madison, WI, USA, 1996.
31. Suwara, I.; Pawlak-Zaręba, K.; Gozdowski, D.; Perzanowska, A. Physical Properties of Soil after 54 Years of Long-Term Fertilization and Crop Rotation. *Plant Soil Environ.* **2016**, *62*, 389–394. [CrossRef]
32. Elliott, E.T. Aggregate Structure and Carbon, Nitrogen, and Phosphorus in Native and Cultivated Soils. *Soil Sci. Soc. Am. J.* **1986**, *50*, 627–633. [CrossRef]
33. Klute, A. Water Retention: Laboratory Methods. In *SSSA Book Series*; Klute, A., Ed.; Soil Science Society of America, American Society of Agronomy: Madison, WI, USA, 2018; pp. 635–662. ISBN 978-0-89118-864-3.
34. Doerr, S.H.; Shakesby, R.A.; Walsh, R.P.D. Soil Water Repellency: Its Causes, Characteristics and Hydro-Geomorphological Significance. *Earth-Sci. Rev.* **2000**, *51*, 33–65. [CrossRef]
35. Hewelke, E. Influence of Abandoning Agricultural Land Use on Hydrophysical Properties of Sandy Soil. *Water* **2019**, *11*, 525. [CrossRef]
36. Dekker, L.W.; Jungerius, P.D. Water Repellency in the Dunes with Special Reference to the Netherlands. *Catena* **1990**, *18*, 173–183.
37. TIBCO Statistica (Data Analysis Software System). Version 13; TIBCO Software Inc.: Santa Clara, CA, USA, 2017.
38. Zhao, X.; He, C.; Liu, W.; Liu, W.; Liu, Q.; Bai, W.; Li, L.; Lal, R.; Zhang, H. Responses of Soil pH to No-till and the Factors Affecting It: A Global Meta-analysis. *Glob. Change Biol.* **2022**, *28*, 154–166. [CrossRef]
39. Arshad, M.A.; Schnitzer, M.; Angers, D.A.; Ripmeester, J.A. Effects of till vs No-till on the Quality of Soil Organic Matter. *Soil Biol. Biochem.* **1990**, *22*, 595–599. [CrossRef]
40. Jayaraman, S.; Naorem, A.; Dalal, R.C.; Sinha, N.K.; Srinivasa Rao, C.; Lal, R.; Kundu, S.; Prasad, J.V.N.S.; Singh, A.K. No-till Farming and Climate Change Mitigation: Lessons Learnt from Long-Term No-till Experiments and Future Perspectives. In *Advances in Agronomy*; Elsevier: Amsterdam, The Netherlands, 2024; Volume 187, pp. 21–107. ISBN 978-0-443-29530-0.
41. Mondal, S.; Chakraborty, D.; Paul, R.K.; Mondal, A.; Ladha, J. No-till is More of Sustaining the Soil than a Climate Change Mitigation Option. *Agric. Ecosyst. Environ.* **2023**, *352*, 108498. [CrossRef]
42. Six, J.; Elliott, E.T.; Paustian, K. Aggregate and Soil Organic Matter Dynamics under Conventional and No-Tillage Systems. *Soil Sci. Soc. Am. J.* **1999**, *63*, 1350–1358. [CrossRef]
43. Nunes, M.R.; Van Es, H.M.; Schindelbeck, R.; Ristow, A.J.; Ryan, M. No-till and Cropping System Diversification Improve Soil Health and Crop Yield. *Geoderma* **2018**, *328*, 30–43. [CrossRef]
44. Hussain, S.; Sharma, V.; Arya, V.M.; Sharma, K.R.; Rao, C.S. Total Organic and Inorganic Carbon in Soils under Different Land Use/Land Cover Systems in the Foothill Himalayas. *Catena* **2019**, *182*, 104104. [CrossRef]
45. Aduhene-Chinbuah, J.; Sugihara, S.; Komatsuzaki, M.; Nishizawa, T.; Tanaka, H. No Tillage Increases SOM in Labile Fraction but Not Stable Fraction of Andosols from a Long-Term Experiment in Japan. *Agronomy* **2022**, *12*, 479. [CrossRef]
46. Szajdak, L.; Jezierski, A.; Cabrera, M.L. Impact of Conventional and No-Tillage Management on Soil Amino Acids, Stable and Transient Radicals and Properties of Humic and Fulvic Acids. *Org. Geochem.* **2003**, *34*, 693–700. [CrossRef]
47. Wulanningtyas, H.S.; Gong, Y.; Li, P.; Sakagami, N.; Nishiwaki, J.; Komatsuzaki, M. A Cover Crop and No-Tillage System for Enhancing Soil Health by Increasing Soil Organic Matter in Soybean Cultivation. *Soil Tillage Res.* **2021**, *205*, 104749. [CrossRef]
48. Enev, V.; Doskočil, L.; Kubíková, L.; Klučáková, M. The Medium-Term Effect of Natural Compost on the Spectroscopic Properties of Humic Acids of Czech Soils. *J. Agric. Sci.* **2018**, *156*, 877–887. [CrossRef]
49. Ukalska-Jaruga, A.; Bejger, R.; Smreczak, B.; Podlasiński, M. Sorption of Organic Contaminants by Stable Organic Matter Fraction in Soil. *Molecules* **2023**, *28*, 429. [CrossRef]
50. Chen, Y.; Senesi, N.; Schnitzer, M. Information Provided on Humic Substances by E4/E6 Ratios. *Soil Sci. Soc. Am. J.* **1977**, *41*, 352–358. [CrossRef]
51. Bensa, A.; Miloš, B.; Jurković Balog, N.; Torić, I. Spectroscopic Characterization of Humic Substances of Anthropogenic Soils Derived from Terra Rossa. *Agrofor* **2016**, *1*, 5–11. [CrossRef]
52. Fuentes, M.; Baigorri, R.; Garcia-Mina, J.M. Maturation in Composting Process, an Incipient Humification-like Step as Multivariate Statistical Analysis of Spectroscopic Data Shows. *Environ. Res.* **2020**, *189*, 109981. [CrossRef] [PubMed]
53. Mielnik, L.; Hewelke, E.; Weber, J.; Oktaba, L.; Jonczak, J.; Podlasiński, M. Changes in the Soil Hydrophobicity and Structure of Humic Substances in Sandy Soil Taken out of Cultivation. *Agric. Ecosyst. Environ.* **2021**, *319*, 107554. [CrossRef]
54. Nieweś, D.; Marecka, K.; Huculak-Mączka, M. Application of Alkaline Deep Eutectic Solvents as a Green Alternative to the Traditional Extractants for the Isolation of Humic Substances. *ACS Omega* **2024**, *9*, 25265–25276. [CrossRef] [PubMed]

55. Coble, P.G. Characterization of Marine and Terrestrial DOM in Seawater Using Excitation-Emission Matrix Spectroscopy. *Mar. Chem.* **1996**, *51*, 325–346. [CrossRef]
56. Peuravuori, J.; Koivikko, R.; Pihlaja, K. Characterization, Differentiation and Classification of Aquatic Humic Matter Separated with Different Sorbents: Synchronous Scanning Fluorescence Spectroscopy. *Water Res.* **2002**, *36*, 4552–4562. [CrossRef]
57. Duarte, R.M.B.O.; Santos, E.B.H.; Duarte, A.C. Spectroscopic Characteristics of Ultrafiltration Fractions of Fulvic and Humic Acids Isolated from an Eucalyptus Bleached Kraft Pulp Mill Effluent. *Water Res.* **2003**, *37*, 4073–4080. [CrossRef]
58. Rodríguez, F.J.; Schlenger, P.; García-Valverde, M.A. Comprehensive structural evaluation of humic substances using several fluorescence techniques before and after ozonation. Part I: Structural characterization of humic substances. *Sci. Total Environ.* **2014**, *476–477*, 718–730. [CrossRef]
59. Chen, J.; LeBoeuf, E.J.; Dai, S.; Gu, B. Fluorescence Spectroscopic Studies of Natural Organic Matter Fractions. *Chemosphere* **2003**, *50*, 639–647. [CrossRef]
60. Kalbitz, K.; Geyer, W.; Geyer, S. Spectroscopic Properties of Dissolved Humic Substances? A Reflection of Land Use History in a Fen Area. *Biogeochemistry* **1999**, *47*, 219–238. [CrossRef]
61. Weber, J.; Mielnik, L.; Leinweber, P.; Hewelke, E.; Kocowicz, A.; Jamroz, E.; Podlasiński, M. The Influence of Different, Long-Term Fertilizations on the Chemical and Spectroscopic Properties of Soil Organic Matter. *Agronomy* **2024**, *14*, 837. [CrossRef]
62. Fuentes, M.; González-Gaitano, G.; García-Mina, J.M. The Usefulness of UV–Visible and Fluorescence Spectroscopies to Study the Chemical Nature of Humic Substances from Soils and Composts. *Org. Geochem.* **2006**, *37*, 1949–1959. [CrossRef]
63. Zheng, H.; Liu, W.; Zheng, J.; Luo, Y.; Li, R.; Wang, H.; Qi, H. Effect of Long-Term Tillage on Soil Aggregates and Aggregate-Associated Carbon in Black Soil of Northeast China. *PLoS ONE* **2018**, *13*, e0199523. [CrossRef] [PubMed]
64. Blanco-Canqui, H. Enhancing Soil Aggregation in No-Till Farming Systems. In *No-Till Farming Systems for Sustainable Agriculture*; Dang, Y.P., Dalal, R.C., Menzies, N.W., Eds.; Springer International Publishing: Cham, Switzerland, 2020; pp. 233–249. ISBN 978-3-030-46408-0.
65. Li, P.; Ying, D.; Li, J.; Deng, J.; Li, C.; Tian, S.; Zhao, G.; Wu, C.; Jiao, J.; Jiang, M.; et al. Global-Scale No-Tillage Impacts on Soil Aggregates and Associated Carbon and Nitrogen Concentrations in Croplands: A Meta-Analysis. *Sci. Total Environ.* **2023**, *881*, 163570. [CrossRef]
66. Sithole, N.J.; Magwaza, L.S.; Thibaud, G.R. Long-Term Impact of No-till Conservation Agriculture and N-Fertilizer on Soil Aggregate Stability, Infiltration and Distribution of C in Different Size Fractions. *Soil Tillage Res.* **2019**, *190*, 147–156. [CrossRef]
67. Six, J.; Elliott, E.T.; Paustian, K. Soil Macroaggregate Turnover and Microaggregate Formation: A Mechanism for C Sequestration under No-Tillage Agriculture. *Soil Biol. Biochem.* **2000**, *32*, 2099–2103. [CrossRef]
68. Oades, J.M. The Role of Biology in the Formation, Stabilization and Degradation of Soil Structure. *Geoderma* **1993**, *56*, 377–400. [CrossRef]
69. Devita, P.; Dipaolo, E.; Fecondo, G.; Difonzo, N.; Pisante, M. No-Tillage and Conventional Tillage Effects on Durum Wheat Yield, Grain Quality and Soil Moisture Content in Southern Italy. *Soil Tillage Res.* **2007**, *92*, 69–78. [CrossRef]
70. Rasmussen, K.J. Impact of Ploughless Soil Tillage on Yield and Soil Quality: A Scandinavian Review. *Soil Tillage Res.* **1999**, *53*, 3–14. [CrossRef]
71. Fuentes, J.P.; Flury, M.; Huggins, D.R.; Bezdicek, D.F. Soil Water and Nitrogen Dynamics in Dryland Cropping Systems of Washington State, USA. *Soil Tillage Res.* **2003**, *71*, 33–47. [CrossRef]
72. Gruber, S.; Möhring, J.; Claupein, W. On the Way towards Conservation Tillage-Soil Moisture and Mineral Nitrogen in a Long-Term Field Experiment in Germany. *Soil Tillage Res.* **2011**, *115–116*, 80–87. [CrossRef]
73. Jabro, J.D.; Stevens, W.B. Soil-Water Characteristic Curves and Their Estimated Hydraulic Parameters in No-Tilled and Conventionally Tilled Soils. *Soil Tillage Res.* **2022**, *219*, 105342. [CrossRef]
74. Davis, P.; Park, D.M. Food Security in a Changing Climate Starts with Managing Soil Water Repellency. *Geoderma* **2024**, *447*, 116922. [CrossRef]
75. Blanco-Canqui, H. Does No-till Farming Induce Water Repellency to Soils? *Soil Use Manag.* **2011**, *27*, 2–9. [CrossRef]
76. Lamparter, A.; Bachmann, J.; Goebel, M.-O.; Woche, S.K. Carbon Mineralization in Soil: Impact of Wetting–Drying, Aggregation and Water Repellency. *Geoderma* **2009**, *150*, 324–333. [CrossRef]
77. Goebel, M.-O.; Bachmann, J.; Reichstein, M.; Janssens, I.A.; Guggenberger, G. Soil Water Repellency and Its Implications for Organic Matter Decomposition—Is There a Link to Extreme Climatic Events?: Extreme Events and Soil Water Repellency. *Glob. Change Biol.* **2011**, *17*, 2640–2656. [CrossRef]
78. Behrends Kraemer, F.; Hallett, P.D.; Morrás, H.; Garibaldi, L.; Cosentino, D.; Duval, M.; Galantini, J. Soil Stabilisation by Water Repellency under No-till Management for Soils with Contrasting Mineralogy and Carbon Quality. *Geoderma* **2019**, *355*, 113902. [CrossRef]
79. Zhang, B.; Yao, S.-H.; Hu, F. Microbial Biomass Dynamics and Soil Wettability as Affected by the Intensity and Frequency of Wetting and Drying during Straw Decomposition. *Eur. J. Soil Sci.* **2007**, *58*, 1482–1492. [CrossRef]
80. Coelho, C.O.A.; Laouina, A.; Regaya, K.; Ferreira, A.J.D.; Carvalho, T.M.M.; Chaker, M.; Naafa, R.; Naciri, R.; Boulet, A.K.; Keizer, J.J. The Impact of Soil Water Repellency on Soil Hydrological and Erosional Processes under Eucalyptus and Evergreen Quercus Forests in the Western Mediterranean. *Soil Res.* **2005**, *43*, 309. [CrossRef]
81. Lichner, L.; Hallett, P.D.; Feeney, D.S.; Ďugová, O.; Šír, M.; Tesař, M. Field Measurement of Soil Water Repellency and Its Impact on Water Flow under Different Vegetation. *Biologia* **2007**, *62*, 537–541. [CrossRef]

82. Hewelke, E.; Szatylowicz, J.; Gnatowski, T.; Oleszczyk, R. Spatial Variability in Soil Moisture Content under Preferential Flow in Hydrophobic Organic Soil. *Rocz. Ochr. Sr.* **2014**, *16*, 580–607.
83. Keesstra, S.; Nunes, J.P.; Saco, P.; Parsons, T.; Poeppl, R.; Masselink, R.; Cerdà, A. The Way Forward: Can Connectivity Be Useful to Design Better Measuring and Modelling Schemes for Water and Sediment Dynamics? *Sci. Total Environ.* **2018**, *644*, 1557–1572. [CrossRef] [PubMed]

sustainability

MDPI

Article

The Influence of Rural Urbanization on the Change in Soil Organic Matter of Farmland in Northeast China

Xiaodong Wang [1], Long Fei [1], Yu An [2], Xiaohui Liu [2] and Mei Zhang [1,*]

1 College of Geographical Sciences, Changchun Normal University, Changchun 130032, China; wangxd219@nenu.edu.cn (X.W.); feilong@ccsfu.edu.cn (L.F.)
2 Northeast Institute of Geography and Agroecology, Chinese Academy of Sciences, Changchun 130102, China; anyu@iga.ac.cn (Y.A.); liuxh@iga.ac.cn (X.L.)
* Correspondence: zhangmei@ccsfu.edu.cn

Abstract: Studying the impact of urbanization on changes in the soil organic matter (SOM) content of farmland plays an important role in determining the influence mechanism of urbanization regarding regional environmental change. Taking the farmland in Yushu City, northeast China, as the research area, in May 2019, 68,393 sample plots (each plot: 60 m × 60 m) were set up in farmland and sampled to measure the SOM content of each plot while combining image data from the same period in the study area (resolution: 60 m). This investigation was based on 17 levels divided by the size of areas occupied by residences, using residential areas as the center and a radius of 60 m. Through a gradually buffered extrapolation method combined with mathematical functions, the influence of rural urbanization on the changes in SOM content was revealed. These results showed that the slope of the linear function between the SOM content and the residential area level was greater than zero and that with the continuous advancement of urbanization, the SOM content had an increasing trend. When urbanization advanced to the stage of larger cities, large-scale mechanized production led to land degradation. When urbanization advanced to the stage of towns, intensive cultivation was beneficial for land restoration. The findings of this study provide a reference basis for future studies of the relationship between rural urbanization and agricultural mechanization around the world.

Keywords: cultivated land; mathematical function; rural towns process; SOM

1. Introduction

Much of the rural population has shifted to urban areas, and methods of agricultural production have also changed in various regions in the world with global urbanization. Meanwhile, the content of soil organic matter (SOM) is also constantly changing in farmland [1]. In this context, it is important to study the relationship between the SOM content and the process of rural urbanization, in order to reveal the reasons for the continuous decline in farmland quality and to maintain sustainable agricultural development [2,3]. By applying the PARAFAC model, one such study revealed that the SOM content surrounding suburbs decreased with the urbanization of Beijing, China, by 2015 [4]. Elsewhere, with the advancement of rural urbanization in the Heilongjiang Reclamation Area of China in 2014, based on data from 113 agricultural pastures, it was revealed that agricultural mechanization was extensively used, and fertilizer was excessively applied, which caused soil compaction, soil degradation, and a rapid reduction in SOM [5]. Likewise, in Harbin, China, the rural urbanization was found to have changed the soil properties, reduced the SOM content by 59 g/kg to 38 g/kg, and produced severe soil degradation [6]. This phenomenon is not unique to China; in the Nile Delta, the urban area increased from 452 km^2 in 1972 to 2644 km^2 in 2017, and over the same period, the loss of SOM (cultivation layer: 0–75 cm) from farmland increased from 25,000 Mg/km^2 to 141,000 Mg/km^2 [7]. Similarly, in the Tombel region of southwestern Cameroon, the SOM content of farmland decreased from 2.77 ± 1.09 kg/m^2 to 2.16 ± 0.93 kg/m^2, while the urban area expanded by 83% from 1985 to 2017 [8]. The results of these studies have demonstrated that as urbanization has steadily advanced, SOM has continually decreased, and farmland degradation has gradually intensified.

The influence of changes in the methods of agricultural production on the SOM content has great significance for sustainable agricultural development during the process of rural urbanization [9,10]. Rural urbanization has expanded the scale of centralized agricultural centralized management, and the mechanization rate has also increased in China. At the same time, the agricultural environment has declined as the SOM has decreased on farmland [11]. For instance, in Yunnan Province, China, mechanization led to a decrease in farmland fertility, an SOM decrease, and accelerated nutrient loss, while the land occupation of urban construction increased by 49.8% from 1989 to 2018, paired with increases in agricultural product and fertilizer application [12]. Similarly, in Shanghai, China, changes in the crop structure and farming system have led to a decrease in the SOM content of farmland with rapid urbanization, which has brought about the conversion of the soil carbon sink to a carbon source since 2004 [13]. Elsewhere, in West Java, Indonesia, a large amount of chemical fertilizer has been applied to inhibit microbial activity, and the SOM content has reduced in tandem, creating an urgent need to alter the islanders' methods of land cultivation [14]. These cases exemplify how, with the urbanization of rural areas and the transformation of methods of agricultural production, the SOM content of farmland has decreased. However, there are other cases where the SOM content of farmland has shown an increasing trend with the application of modern technology in certain areas of the world. For instance, there was one report that minimal tillage (MT) and no tillage (NT) techniques significantly reduced SOM loss by 17% and 63%, respectively, compared to traditional tillage [15]. Furthermore, in soybean growing areas in Europe, low-input organic agriculture enhanced the SOM content of farmland compared to traditional mechanization and maintained sustainable agricultural development [16]. Therefore, it can be said that agricultural mechanization has a very complex impact on the SOM of farmland, while the process of rural urbanization has a profound impact. Nevertheless, existing research on the impact of urbanization on the SOM content has mainly concentrated on certain specific stages of the urbanization process. There is a lack of research on the continuous changes in SOM content during the process of continuous urbanization in rural areas.

With the continuous advancement of urbanization, the SOM content is also constantly changing. Accordingly, it seems necessary to distinguish at which stage of urbanization the SOM content is highest. Different levels of urbanization resulted in differences in the cultivation methods of surrounding farmland [17], and further clarification is needed of how differences in agricultural production have affected soil quality. Based on this, we took the farmland distribution in Yushu City, China, as the research area, and we used buffer analysis based on a field investigation in order to study the relationship between different residential levels and SOM content, understand the changes in SOM content during the process of urbanization, and identify the interrelationships between the urbanization process, agricultural production, and SOM content. Specifically, the main scientific problem addressed in this study is to determine the effectiveness of the following methods:

(1) Understanding the change in the SOM content of farmland with an increase in residential area level.

(2) Revealing the variation characteristics of SOM content extending outward from residential areas within each residential area level.

(3) Analyzing the impacts of alteration in the methods of agricultural production on the change in the SOM content with the advancement of rural urbanization. We can provide suitable research methods for future studies that will support the effective protection and utilization of farmland resources and establishing sustainable development models suitable for local agriculture.

2. Materials and Methods

2.1. Study Area

Northeast China is the main grain producing area nationally, with the highest yield in China [18]. However, due to its the prolonged cultivation, the quality of the farmland has significantly decreased [19]. Meanwhile, the urbanization level in Northeast China has increased (e.g., from 27.61% in 1952 to 63.15% in 2019). The advancement of urbanization has led to continuous improvements in agricultural mechanization and great changes in the methods of farming. As such, this is an ideal area for studying the impacts of rural urbanization on SOM changes on farmland [20]. The study area is located at Yushu City and belongs to the center of the Northeast China Plain (126°01′44″–127°05′09″ E, 46°30′57″–45°15′02″ N), covering an area of 4712 km^2 [21] (Figure 1). The soil type is mainly black soil in the study area [22], matching the most typical black soil distribution in Northeast China. Its high grain yield and fertile soil have made the area a national commodity grain base in China [23,24].

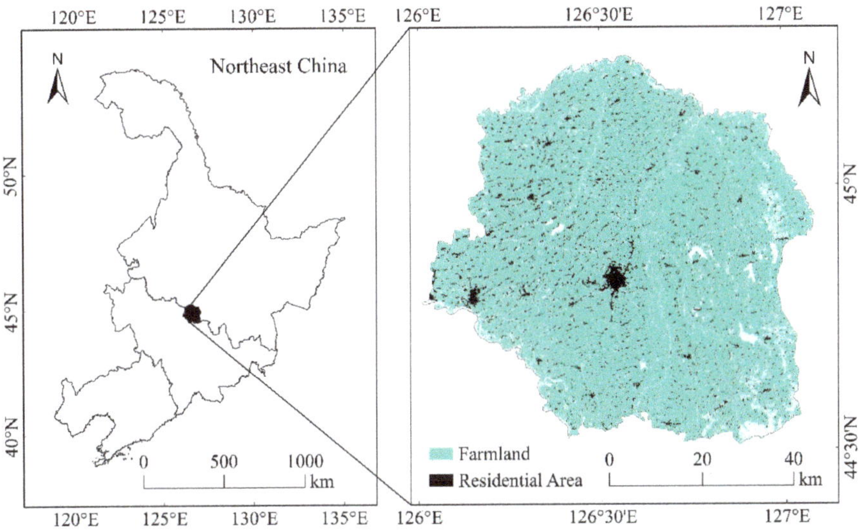

Figure 1. The distribution of residential areas and croplands in the study area.

2.2. Data Collection and Processing

In May 2019, soil sampling was conducted in the field. At this point, frozen soil had just melted, and spring plowing had not yet begun, so it was convenient for field sampling. Each plot measured 60 m × 60 m, matching the minimum resolution of the imagery we accessed. Firstly, GPS was used to measure the longitude and latitude at the center of a plot. Plots were set up in all field parcels of Yushu City to ensure the coverage of all farmlands distributed in our study area. Then, five samples (each sample: 0.5 kg) were taken at the four corners and center points of each plot from the cultivated soil layer. Last, these five samples were mixed into one sample, and we took 1 kg of that mixture as our single one soil sample for each plot. All soil samples were sealed and brought to the laboratory. After drying them and removing impurities, the soil samples were weighed, ground, and passed through a 0.1 mm sieve. The SOM content of each soil sample was determined using the potassium dichromate sulfuric acid method [25]. Finally, the SOM content of each plot in the study area (totaling 68,393) was obtained. The SOM content ranged from 1.9% to 3.9%, with a mean of 2.54% and a standard deviation (STD) of 0.25% among 68,393 plots. The Mesoscale Landsat 8 Operational Land Imager (OLI) was used to collect the remote sensing data. The imaging period was May 2019, which was consistent with the soil sampling time. The data were then subjected to data importation (import), multi-band image blending (utilities), image cropping (subset), and geometric correction for images (geometric correction). All residential areas and farmland were classified in the study area using the random forest classification method in the ENVI environment. After comparing the longitude and latitude of each plot in its image with the data from our field investigations, an SOM content was assigned to each plot to form a new data layer using the ArcGIS model.

2.3. Data Analysis

2.3.1. Buffer Analysis

In the new data layer, according to the Sturges formula, all residential areas were evenly divided into 15 levels based on land occupation scale, with a difference of 145,000 square meters between adjacent levels [26–28]. The large land area of Yushu Development Zone (only one) and the location of the municipal government (urban area) were not included in the formula calculation. Instead, the two areas were separately classified as the 16th (municipal government) and the 17th (development zone) level, and so a total of 17 levels of residential areas were determined. Because the 14th level had no residential areas, it was not included in the following analysis (Table 1).

Table 1. Classification of residential areas and function selection of each level.

Level	Level Range (Ten Thousand m^2)	X_1	X_2 (%)	X_3 (m)	X_4 (%)	X_5 (m)
1	0–14.5	−0.0308	2.55	900	2.37	2700
2	14.5–29	0.0006	2.56	9660	2.48	5340
3	29–43.4	−0.0007	2.61	60	2.45	12,900
4	43.4–57.9	−0.0072	2.58	7620	2.21	20,880
5	57.9–72.4	−0.0021	2.57	60	2.47	26,160
6	72.4–86.9	−0.0016	2.59	12,840	2.47	26,400
7	86.9–101.4	−0.0012	2.67	41,460	2.44	28,320
8	101.4–115.9	0.0034	2.66	21,240	2.47	5220
9	115.9–130.3	−0.0018	2.66	5100	2.45	42,780
10	130.3–144.8	−0.0001	2.66	68,640	2.44	4620
11	144.8–159.3	0.0014	2.61	48,240	2.14	60
12	159.3–173.8	−0.0015	2.59	8760	2.46	33,060
13	173.8–188.3	−0.0039	2.58	15,840	2.17	46,380
15	202.8–217.2	0.0001	2.66	28,860	2.40	47,460
16	859.3	0.0004	2.59	38,520	2.34	60
17	2612.8	−0.0028	2.82	60	2.47	28,740

Note: X_1, the slope of the linear function between the distance (circled area and residential area) (independent variable) and the SOM mean (dependent variable) in the circled area; X_2, the maximum of function with the best fitting effect between the distance (circled area and residential area) (independent variable) and the SOM mean (dependent variable) in the circled area. X_3, the distance corresponding to the maximum of functions with the best fitting effect between the distance (circled area and residential area) (independent variable) and the SOM mean (dependent variable) in the circled area. X_4, the maximum of functions with the best fitting effect between the distance (circled area and residential area) (independent variable) and the SOM mean (dependent variable) in the circled area. X_5, the distance corresponding to the maximum of functions with the best fitting effect between the distance (circled area and residential area) (independent variable) and the SOM mean (dependent variable) in the circled area.

Within each level, buffer analysis was conducted on surrounding farmland, centered around residential areas and gradually expanded from the inside out in farmland. Because the minimum resolution of the plot was 60 m, the buffer radius was also chosen as 60 m. Using the 1st level of residential areas as an example, buffer analysis was introduced: first, we centered around the residential area of the 1st level, circled the area of a farmland with a radius of 60 m in the outer farmland, and the mean was calculated of the SOM content of farmland in the circled area. Then, we took the boundary of the circled area in the first round as the starting line, circled again a new farmland area with a radius of 60 m, and calculated the mean of the SOM content within the new circled area. Thus, this proceeded, with continuous buffer analysis outward until all farmland was covered in the study area.

At last, a series of the circled areas was formed, centered around the 1st level of residential area. Following this, buffer analysis was performed on the 2nd level of residential area to form another series of the circled areas, and so on. Each level formed a corresponding circled areas series, with a total of 16 series, meaning that each level of residential area was a series of circled areas.

2.3.2. Function Fitting and Step-By-Step Elimination

The 1st (linear) to the 6th functions were performed, using the residential area level as the independent variable and the SOM mean of the residential area level as the dependent variable. In addition to the linear function, the 6th function ($R^2 = 0.5457$, $p < 0.05$) (Figure 2) was selected as appropriate to reveal the relationship between the level of residential area and the SOM content according to the change characteristics of the 6th function curve. At each residential area level, the distance (circled area and residential area) was taken as the independent variable and the SOM mean in the circled area as the dependent variable; the linear function was performed, and we calculated the slope of each function (Table 1). Then, using these slopes as the dependent variable and the residential area level as the independent variable, six functions were fitted from the linear to the 6th function. Finally, since the 4th function had the best fitting effect ($R^2 = 0.6489$, $p < 0.05$) (Figure 3), we selected the linear function and the 4th function with which to analyze the relationship between the SOM content and distance (circled area and residential area) at different residential area levels.

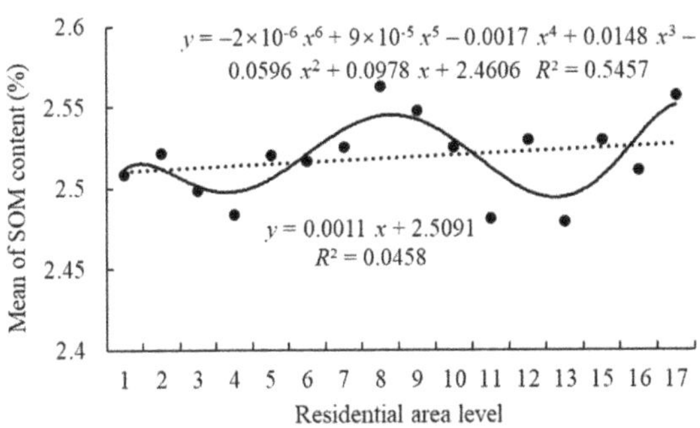

Figure 2. The sixth function with the best fitting effect between residential area level (independent variable) and the mean of SOM content in all circled areas (dependent variable).

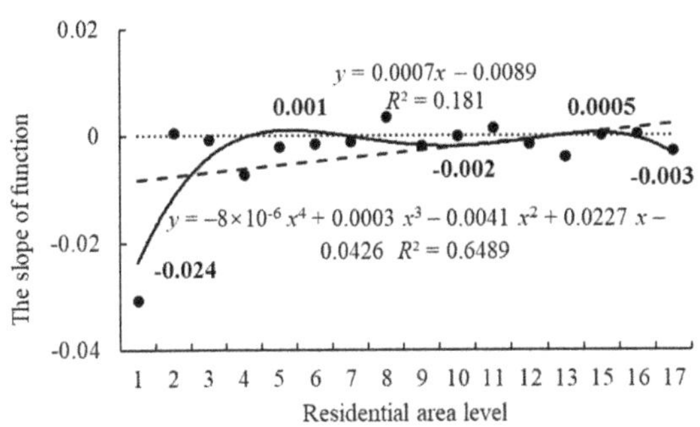

Figure 3. The 4th function with the best fitting effect between residential area level (independent variable) and the slope (dependent variable) of the function; the function referred to the distance (circled area and residential area) as the independent variable and the SOM mean in the circled area as the dependent variable. The dotted line was 0 for the slope of the function; the dashed line was the linear function (k = 0.007) for the slope of the function.

At each level of residential area, the distance (circled area and residential area) was taken as the independent variable, the SOM mean in the circled area was taken as the dependent variable, and the six functions (from the 1st to the 6th) were fitted separately. The function was selected with the biggest R^2, and we calculated the SOM maximum of the function curve. In addition, the distance corresponding to the SOM maximum of the function was computed too (Table 1). Then, the SOM maximum of each function with the biggest R^2 for each level was taken as the dependent variable, the residential area level was taken as the independent variable, the linear function was performed, and we calculated the slope of the function (Figure 4). Using the same method, another linear function was also obtained, and the slope of the function was also calculated, where another function referred to the distance (circled area and residential area) as the independent variable and the SOM mean in the circled area as the dependent variable (Figure 4).

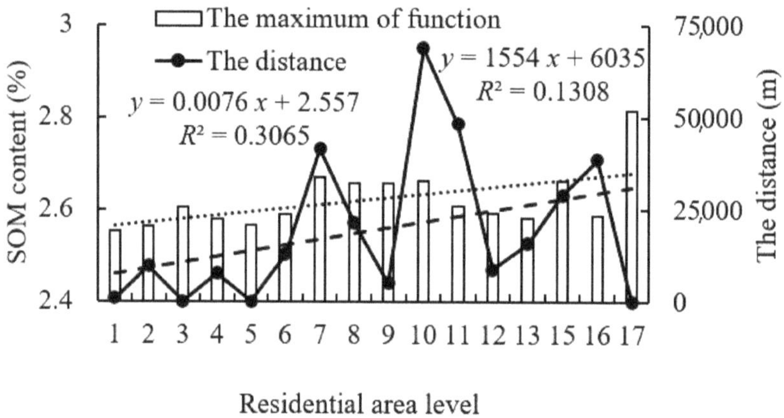

Figure 4. The linear function between residential area level (independent variable) and the SOM maximum of the function. The linear function between residential area level (independent variable) and the distance corresponding to the SOM maximum of the function (dependent variable). The function referred to the distance (circled area and residential area) as the independent variable and the SOM mean in the circled area as the dependent variable with the best fitting effect. The dotted line is the linear function (k = 0.0076) for the SOM maximum; the dashed line is the linear function (k = 1554) for the distance corresponding to the SOM maximum.

Finally, the eliminated correlation coefficient (R) for the SOM maximum and the distance corresponding to the SOM maximum was calculated by a step-by-step elimination analysis. In the first round of calculation, the pair of data numbered as 1 among the 16 pairs for SOM maximum and the distance corresponding to SOM maximum, which represented SOM maximum and the distance corresponding to SOM maximum of the residential area as 1, was eliminated, and the R of the remaining 15 pairs of the residential area levels was calculated. Then, the pair of the residential area level numbered as 1 was returned, while the pair numbered as 2 was removed, and the R of the remaining residential area levels was calculated again. Following this process, 16 Rs were obtained. From these, the level with the biggest R among the 16 was eliminated in the first round. The same elimination method was then used in the second round for the remaining 15 levels. The elimination process was considered complete after the second round since the R reached the level of significance ($p < 0.01$) in the second round. With the same method of step-by-step elimination, the Rs were calculated of SOM minimum and the distance corresponding to SOM minimum. It was only in step six that a significant positive R was achieved (eliminated R: 0.64, $p < 0.05$), and in step five that a significant negative R was achieved (eliminated R: −0.67, $p < 0.05$). In doing so, the relationship was revealed between the residential area level and SOM maximum based on the slope of the linear function and these step-by-step elimination processes. With the same methods, the SOM minimum and the distance corresponding to the SOM minimum were calculated for each level's function curve, using the SOM minimum as the dependent variable and the distance (circled area and residential areas) as the independent variable for the linear function. The impact of urbanization on the SOM content was revealed by the slope of the linear function and by the eliminated R between the SOM minimum and the distance corresponding to the SOM minimum.

3. Results

3.1. Urbanization Process

The rural population decreased from 1.04 million in 1991 to 0.73 million in 2018, a decrease of 29.8%. The urban population, meanwhile, increased from 0.13 million to 0.15 million, an increase of 8.3%, while the urbanization rate increased from 12.9% to 21.2% in Yushu City [29,30]. Alongside the processes of urbanization in the study area, the comprehensive input–output efficiency of farmland increased from 0.54 in 2006 to 1 in 2015 [31]. The intensity of soil development also continually increased, while the SOM content decreased by 2.46% to 5.46% compared to the initial background [32]. The results of our analysis of variance of the SOM content in different plots ($p < 0.01$) indicated that severe differential degradation occurred in Yushu City.

3.2. The Change in Means of SOM Contents among Different Residential Area Levels

The slope of the linear function was greater than 0 (k = 0.0112) (Figure 2) between the level of residential areas (independent variable) and the mean of SOM content in all circled areas (dependent variable), with an upward trend, indicating that as the level of residential areas increased, the SOM content had an increasing trend, and the urbanization was beneficial for soil recovery. The sixth function with the best fitting effect ($R^2 = 0.5457$) had a downward trend at a level of less than 3.79 (Figure 2), indicating that once farmland was cultivated, soil degradation immediately began. Between the levels of 3.79 and 8.30 (Figure 2), the curve showed an upward trend, indicating that with the concentration of the population towards the town, the intensity of farmland remediation increased, and soil was restored to a certain extent. However, between the levels of 8.30 and 12.71 (Figure 2), the curve had a downward trend, indicating that when urbanization reached this stage, the increase in the development intensity of farmland led to a degradation trend in soil. After the 12.71 level (Figure 2), the curve had an upward trend again, indicating that when the urbanization scale reached this stage, the economic and technological strengths were enhanced, which led to the use of advanced agricultural production methods, raising the improvement ability and quality of soil.

3.3. The Changes in SOM Content Extending Outward from Residential Areas

3.3.1. Linear Trend Change

The slope of the linear function was greater than 0 (k = 0.0007) (Figure 3) between the residential area level (independent variable) and the slope (dependent variable) of the function. The function refers to the distance (circled area and residential area) as the independent variable and the SOM mean in the circled area as the dependent variable, with an upward trend. As the residential area level increased, the SOM content increased with distance, indicating that the advancement of urbanization has continuously expanded the spatial scope of soil restoration. The variation in the sixth function with the best fitting effect between the residential area level (independent variable) and the slope (dependent variable) of the function showed that when the residential area was less than the 4.14 level, the slope was negative (from −0.024 to 0) (Figure 3), indicating that in the early stage of farmland cultivation, the farther away from the residential area, the more extensive the farmland development, and the more severe the soil degradation. Between the levels of 4.14 and 6.89, the slope was positive (from 0 to 0.001) (Figure 3), indicating that as the population concentrated towards the town, the farther away from the residential area, the higher the SOM content. The high-intensity use of farmland around the residential area led to rapid soil degradation near the town. At levels 6.89 to 12.75, the slope was negative (from −0.002 to 0) (Figure 3), indicating that when urbanization reached this stage, people began to pay attention to land degradation; however, land restoration was focused on the farmland near high-level towns, which the farmland near the high-level town could restore to a certain extent. Between levels 12.75 and 15.76, the slope was positive (from 0 to 0.0005) (Figure 3), which indicated that when urbanization was reached at this stage, the advanced management mode expanded the scope of farmland restoration to areas far away from towns. After the 15.76 level, the slope was negative (from −0.003 to 0) (Figure 3), indicating that when urbanization expanded to the urban stage, the surrounding farmland shifted towards intensive farming and that the increased farmland input significantly increased the SOM content.

3.3.2. The Change in SOM Maximum

The slope of the linear function was greater than 0 (k = 0.0076) (Figure 4) between the residential area level (independent variable) and the SOM maximum of the function (dependent variable), where the function referred to the distance (circled area and residential area) as the independent variable and the SOM mean in the circled area as the dependent variable. The slope of the linear function was 1554 (Figure 4) between the residential area level (independent variable) and the distance corresponding to the SOM maximum of the function (dependent variable) with the best fitting effect, which were all positive, indicating that as the residential area level increased, the SOM maximum continued to increase and the distance corresponding to the SOM maximum also continued to increase. Urbanization has continuously expanded the spatial scope of surrounding farmland to improve the SOM content. The SOM maximum continued to rise, and the distance corresponding to the SOM maximum continually moved away from residential areas. According to the step-by-step elimination analysis, except for the 17th and 8th levels in the first and second rounds, the R was 0.66 ($p < 0.01$) between the residential area level and the SOM maximum fitting curve of each level, indicating that except for these two levels, the SOM maximum continued to increase with the distance from the residential area in other residential area levels. This shows that advancements in urbanization could improve the soil quality and that the intensity and spatial scope of improvement constantly expand. The SOM maximum was 2.47% of the 17th level for the function with the best fitting effect between the distance (circled area and residential area) as the independent variable and the SOM mean in the circled area as the dependent variable (Figure 4). The distance corresponding to the SOM maximum of the function of the 17th level was 60 m (Figure 4), indicating that the farmland adjacent to the residential area was a production base for agricultural food. The result also indicated that only when urbanization was advanced to this level, a specialized agricultural production base would emerge in the surrounding areas of urban areas.

3.3.3. The Change in SOM Minimum

The slope of the linear function was negative (k = −0.0455) (Figure 5) between the residential area level (independent variable) and the SOM minimum of the function, where the function referred to the distance (circled area and residential area) as the independent variable and the SOM mean in the circled area as the dependent variable with the best fitting effect. The slope of the linear function was positive (k = 1198) (Figure 5) between the residential area level (independent variable) and the distance corresponding to the SOM minimum of the function with the best fitting effect (dependent variable), indicating that as the spatial scope of soil restoration continued to expand, the interference with soil also continued to expand. The slope (k = 1554) of the linear function for the distance corresponding to the SOM maximum of the function was greater than that (k = 1198) of the function for the distance corresponding to the SOM minimum of the function (Figures 4 and 5), indicating that as the level of the residential areas increased, the intensity of soil improvement was greater than the intensity of soil interference. As such, urbanization had a promoting effect on soil recovery. In addition, the R was 0.05 between the residential area level and the SOM minimum function of each level, which showed that there was a weak positive correlation. This revealed that urbanization is beneficial for soil restoration. The results of the step-by-step elimination analysis showed that a significant negative correlation was only reached in step 5, and a significant positive correlation was only reached in step 6. With the continuous advancement of urbanization, the contradiction between the improvement and degradation of soil was very complex, with the soil moving in the direction of recovery in a tortuous process.

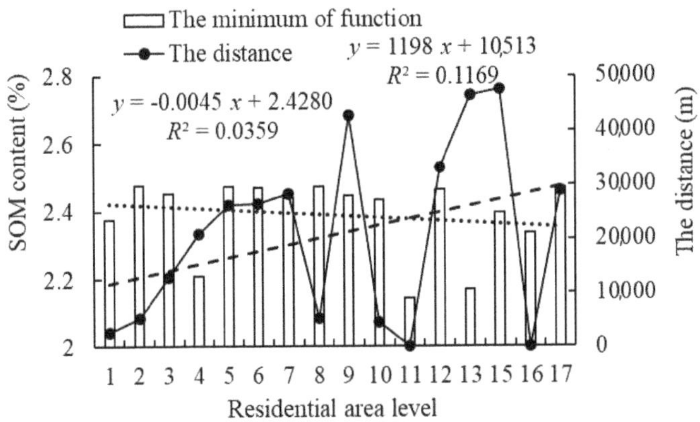

Figure 5. The linear function between the residential area level (independent variable) and the SOM minimum of the function. The linear function between the residential area level (independent variable) and the distance corresponding to the SOM minimum of the function (dependent variable), where the function referred to the distance (circled area and residential area) as the independent variable and the SOM mean in the circled area as the dependent variable with the best fitting effect. The dotted line is the linear function between the residential area level and the SOM minimum of the function; the dashed line is the linear function between the residential area level and the distance corresponding to the SOM minimum of the function.

4. Discussion

4.1. The Damage of Cultivation and Reclamation

Our results indicated that once the land was cultivated, soil degradation occurred immediately. Specifically, as land was cultivated, surface vegetation was destroyed, soil erosion increased, SOM loss accelerated, land degradation accelerated, and recovery was slow [33]. This phenomenon occurs around the world. For instance, in the semi-arid grassland area of Nigeria, after land reclamation, the SOM content rapidly decreased [34]. Elsewhere, in the northern region of Ethiopia, the SOM content decreased significantly after forest land reclamation, and the decrease in surface soil was greater than that in deep soil [35]. Meanwhile, in southern Belgium, when forest and grassland were converted into farmland, the SOM content decreased by 14.4 t C/ha and 23.4 t C/ha, respectively [36]. Therefore, we propose that the first development of farmland should be particularly cautious, and that it is necessary to monitor and maintain nutrients to prevent a rapid decline in soil quality and even irreversible situations after reclamation.

4.2. Non-Universal Application of Large-Scale Mechanized Production

The maximum and minimum of the best effect functions for the 13th level were 2.58% and 2.17% between the distance (circled area and residential area) (independent variable) and the SOM mean (dependent variable), which were the lowest in all residential area levels. When rural urbanization reached this stage (13th level), the intensity of soil improvement was the smallest, the interference was the greatest, and soil degradation was the most significant. This stage was a key stage in the transition from rural to urban areas, where a large number of farmers had left agricultural production and entered other industries in urban areas. People's production and lifestyle underwent fundamental changes [37]. With the concentration of the population towards cities, the number of people engaged in agricultural activities in rural areas has decreased; however, the demand for rural land output is higher so as to support more citizens [38]. Therefore, the production mode was mechanized, and in order to improve crop yield, the application of fertilizers and pesticides also rapidly increased [39]. With the popularization of mechanization, the application of fertilizers and pesticides became more extensive, ultimately leading to a significant decrease in SOM content in farmland [40]. The current production model of agricultural mechanization has been continuously popularized worldwide [41], but the agricultural production model of large machinery is not applicable to all regions of the world. In previous research, it was found that, in Eastern China, the coupling degree was between 0.30 and 0.50 between new urbanization and the agricultural ecological environment. At this time, a large number of farmers entered cities, the rural mechanization of production was fully promoted, and farmland degradation was the most severe [42]. Elsewhere, after six years of the implementation of mechanization in South Pampas, Argentina, the SOM content in farmland had decreased by 0.7% [43]. Meanwhile, in a province of Iran, the SOM elasticity was -1.58 under long-term agricultural mechanization, and the SOM elasticity was -1.58, posing a threat to farmland sustainability and making it impossible to sustain the agricultural mechanization of production in this Iranian province [44]. These findings demonstrate that as large-scale mechanized production spreads globally in rural areas, it is necessary to strengthen the monitoring of the SOM content. Once the SOM is rapidly reduced, measures should be taken to restore the production capacity of farmland to increase conservation support.

4.3. *The Suitability of Intensive Cultivation*

In the second round of the step-by-step elimination between the maximum and the distance corresponding to the maximum for all levels of residential areas, the eighth level was removed. Meanwhile, in the first round of the step-by-step elimination between the minimum and the distance corresponding to the minimum for all levels of the residential area, the eighth level was also removed. These findings revealed that the eighth level did not follow the rule of an increasing or decreasing SOM content as the distance to residential areas increased. The mean (2.56%) of SOM content of the eighth level was the highest, and the difference (0.19%) between the maximum (2.66%) and minimum (2.47%) was the smallest in all residential area levels. When rural urbanization advanced to this stage (town level) [45], the contrast between improvement and interference was very small, the extrapolation in the region tended to be consistent, and the spatial heterogeneity was the smallest in all levels of residential areas, indicating that cultivation methods for farmland were the most favorable for SOM restoration at this stage. We found that the process of rural urbanization was at the township level, and that the agricultural production occurred through intensive cultivation. At this stage, farmers had relatively high incomes and a strong dependence on their farmland, and they were willing to make the investments necessary for land restoration. Mechanization was mainly focused on miniaturization, which was convenient and flexible and caused little interference with the farmland. At this stage, farmers were likely to invest more resources into the land to ensure higher returns in the future, and the soil fertility was well maintained thanks to their higher investments. Intensive cultivation was the most effective means of farmland restoration in China, as well as globally, as intensive cultivation plays an important role in maintaining the SOM content in farmland [46]. For instance, in a case from the Czech Republic, traditional agriculture led to soil degradation, whereas organic or bio-based agricultural management practices based on intensive cultivation restored soil health and productivity. After more than 50 years of intensive cultivation, for the high SOM input generated by organic agriculture, the mean of SOM content increased by 5.51%, and soil fertility was restored [47]. Elsewhere, in the eastern region of Iran, due to the adoption of intensive farming, the SOM content increased by 0.3% from 2004 to 2018 [48]. Similarly, in southeastern Brazil, intensive management increased the soil carbon storage in farmland at a rate of 0.28 Mg/ha from 2010 to 2016, promoting an SOM increase while also playing a positive role in carbon sequestration [49]. Therefore, from the perspective of soil protection, if the process of rural urbanization in China is not advanced to a higher level and is instead maintained at the township level, this is beneficial for soil restoration. Currently, China's rural areas are undergoing urbanization through a process of merging villages and towns, with a drop in the rural population and an intensive cultivation trend of agricultural production. In Nancun township and Qingdao, Shandong Province, merging villages and towns activated the rural economy while maintaining effective land conservation, and intensive cultivation was a beneficial approach in maintaining the sustainable development of soil fertility [50]. In sum, based on our research results, intensive cultivation seems to be beneficial for increasing the SOM content. Furthermore, based on the literature, it seems that promoting the merging of villages and towns is an effective measure for soil protection in China.

5. Conclusions

With the continuous advancement of rural urbanization, the SOM content undergoes different changes in different stages of the process, which we determined according to the changes in linear functions or the maxima and minima of the best functions. With the continuous improvement of urbanization from low to high levels, the restoration and disturbance of soil expanded outward from residential areas, and their intensity increased during this expansion. Our specific key findings were as follows: (1) As the level of residential areas increased, the SOM had an increasing trend (k = 0.0112), and the maximum of SOM continuously increased with the distance away from residential areas (k = 1554). As such, the advancement of urbanization was beneficial for soil recovery. (2) At the first level of residential areas, the SOM mean was only 2.51%. This showed that once the land was cultivated, soil degradation immediately occurred. Meanwhile, at the 13th level of residential areas, the SOM mean was the lowest (2.48%). This demonstrated that when rural areas began to transition to larger cities, the popularization of large-scale mechanized production had the greatest impact on land disturbance and caused the most obvious degradation. (3) At the eighth level, the SOM mean was the highest (2.56%), and the difference between the maximum and minimum was the smallest (0.19%). This highlighted that when rural urbanization advanced to the town level, the improvement was great, and interference was low. At this time, intensive cultivation was conducive to soil restoration and protection.

Author Contributions: Conceptualization, X.W.; methodology, M.Z.; writing—original draft preparation, L.F.; investigation, Y.A.; writing—review and editing, X.L. All authors have read and agreed to the published version of the manuscript.

Funding: This research was funded by Joint Funds of the National Natural Science Foundation of China (No. U2243230); the National Natural Science Foundation of China (No. 42230516); Technology Development Program of Jilin Province (No. YDZJ202301ZYTS524); and Natural Science Foundation of Changchun Normal University (No. CSJJ2022009ZK).

Institutional Review Board Statement: Not applicable.

Informed Consent Statement: Not applicable.

Data Availability Statement: The data presented in this study are available from the author upon reasonable request.

Conflicts of Interest: The authors declares no conflicts of interest.

References

1. Zhang, Z.P.; Ding, J.L.; Zhu, C.M.; Chen, X.Y.; Wang, J.Z.; Han, L.J.; Ma, X.; Xu, D. Bivariate empirical mode decomposition of the spatial variation in the soil organic matter content: A case study from NW China. *Catena* **2021**, *206*, 105572. [CrossRef]
2. Liu, S.W.; Zhang, P.Y.; Li, J. Analyzing the urbanization driving force in Heilongjiang Reclamation Areas. *Geogr. Res.* **2013**, *32*, 2066–2078. (In Chinese with English Abstract)

3. Li, W.B.; Wang, D.Y.; Liu, S.H.; Zhu, Y.L. Measuring urbanization-occupation and internal conversion of peri-urban cultivated land to determine changes in the peri-urban agriculture of the black soil region. *Ecol. Indic.* **2019**, *102*, 328–337. [CrossRef]
4. Chai, L.W.; Huang, M.K.; Fan, H.; Wang, J.; Jiang, D.L.; Zhang, M.J.; Huang, Y. Urbanization altered regional soil organic matter quantity and quality: Insight from excitation emission matrix (EEM) and parallel factor analysis (PARAFAC). *Chemosphere* **2019**, *220*, 249–258. [CrossRef] [PubMed]
5. Ye, Y.J.; Qi, Q.W.; Jiang, L.L.; Zhang, A. Impact factors of grain output from farms in Heilongjiang reclamation area based on geographical detector. *Geogr. Res.* **2018**, *37*, 171–182. (In Chinese with English Abstract)
6. Wang, W.J.; Wang, Q.; Zhou, W.; Xiao, L.; Wang, H.M.; He, X.W. Glomalin changes in urban-rural gradients and their possible associations with forest characteristics and soil properties in Harbin City, Northeastern China. *J. Environ. Manag.* **2018**, *224*, 225–234. [CrossRef] [PubMed]
7. Abd-Elmabod, S.K.; Fitch, A.C.; Zhang, Z.H.; Ali, R.R.; Jones, L. Rapid urbanisation threatens fertile agricultural land and soil carbon in the Nile delta. *J. Environ. Manag.* **2019**, *252*, 109668. [CrossRef] [PubMed]
8. Nguemezi, C.; Tematio, P.; TSilatsa, F.B.; Yemefack, M. Spatial variation and temporal decline (1985–2017) of soil organic carbon stocks (SOCS) in relation to land use types in Tombel area, South-West Cameroon. *Soil. Till Res.* **2021**, *213*, 105114. [CrossRef]
9. Wu, K.N.; Han, C.J.; Sun, Z.Y.; Lv, Q.L.; Feng, X.W.; Fu, Q.L.; Li, L. Spatial and temporal distribution of soil fertility as affected by urbanization. *Chin. J. Soil. Sci.* **2007**, *38*, 242–246. (In Chinese with English Abstract)
10. Parenti, A.; Zegada-Lizarazu, W.; Pagani, E.; Monti, A. Soil organic carbon dynamics in multipurpose cropping systems. *Ind. Crop Prod.* **2022**, *187*, 115315. [CrossRef]
11. Zhang, Y.N.; Long, H.L. Agricultural production transition and its environmental effects: Research progress and prospect. *J. Nat. Resour.* **2022**, *7*, 1691–1706. [CrossRef]
12. Yang, C.L.; Li, C.F.; Tian, Y.F.; Zhou, H.; He, J.N.; Liu, J.P.; Ma, Y.Q. Situation of non-grain cropland and its eco-environmental effect in Yunnan Province. *Miner. Explor.* **2020**, *11*, 2573–2591.
13. Zhao, M.; Hu, J.; Tang, Q.H. Evaluation of soil carbon sinks in Shanghai. *Admin Tech. Environ. Monit.* **2012**, *24*, 17–22. (In Chinese with English Abstract)
14. Moeskops, B.; Sukristiyonubowo; Buchan, D.; Sleutel, S.; Herawaty, L.; Husen, E.; Wati, R.S.; Setyorini, D.; Neve, S.D. Soil microbial communities and activities under intensive organic and conventional vegetable farming in West Java, Indonesia. *Appl. Soil. Ecol.* **2010**, *45*, 112–120. [CrossRef]
15. Cillis, D.; Maestrini, B.; Pezzuolo, A.; Marinello, F.; Sartori, L. Modeling soil organic carbon and carbon dioxide emissions in different tillage systems supported by precision agriculture technologies under current climatic conditions. *Soil. Till Res.* **2018**, *183*, 51–59. [CrossRef]
16. Scharer, M.L.; Dietrich, L.; Kundel, D.; Mader, P.; Kahmen, A. Reduced plant water use can explain higher soil moisture in organic compared to conventional farming systems. *Agr. Ecosyst. Environ.* **2022**, *332*, 107915. [CrossRef]
17. Abebe, M.G. Impacts of urbanization on food security in Ethiopia. A review with empirical evidence. *J. Agr. Food Res.* **2024**, *15*, 100997. [CrossRef]
18. Zhang, M.; Chen, T.; Latifmanesh, H.; Feng, X.M.; Cao, T.H.; Qian, C.R.; Deng, A.X.; Song, Z.W.; Zhang, W.J. How plant density affects maize spike differentiation, kernel set, and grain yield formation in Northeast China? *J. Integr. Agr.* **2018**, *17*, 1745–1757. [CrossRef]
19. Li, H.Q.; Yao, Y.F.; Zhang, X.J.; Zhu, H.S.; Wei, X.R. Changes in soil physical and hydraulic properties following the conversion of forest to cropland in the black soil region of Northeast China. *Catena* **2021**, *198*, 104986. [CrossRef]
20. Yang, Q.F.; Zhang, P.Y.; Li, J.; Liu, W.X.; He, X.L. Development Level and Spatio-temporal Evolution of Agricultural Modernization in Northeast China. *Sci. Geogr. Sin.* **2022**, *9*, 1588–1599. (In Chinese with English Abstract)
21. Fang, Y.G.; Shi, K.J.; Niu, C.C. A comparison of the means and ends of rural construction land consolidation: Case studies of villagers' attitudes and behaviours in Changchun City, Jilin province, China. *J. Rural. Stud.* **2016**, *47*, 459–473. [CrossRef]
22. Yu, R.; Wang, Y.; Wang, C.X.; Wang, Q.C.; Cui, Z.W. Survey of heavy metal pollution and source identification of black soil in Zea mays L. cultivated region of Yushu city. *China* **2017**, *26*, 1788–1794. (In Chinese with English Abstract)
23. Wang, D.D.; Shi, X.Z.; Lu, X.X.; Wang, H.J.; Yu, D.S.; Sun, W.X.; Zhao, Y.C. Response of soil organic carbon spatial variability to the expansion of scale in the uplands of Northeast China. *Geoderma* **2010**, *154*, 302–310. [CrossRef]
24. Zhao, H.B.; Ma, Y.J.; Miao, C.H. Integrated assessment and obstacle factors of environmental carrying capacity of national strategic economic zones based on entropy and catastrophe progression methods: A case study in Changchun-Jilin-Tumenjiang development and opening-up pilot area. *Sci. Geogr. Sin.* **2015**, *35*, 1525–1532. (In Chinese with English Abstract)
25. Institute of Soil Science, Chinese Academy of Sciences. *Soil Physicochemical Analysis*; Routledge: Beijing, China, 1977.
26. Imbusch, G.F.; Yen, W.M. The mcCumber and sturge formula. *J. Lumin.* **2000**, *85*, 177–179. [CrossRef]
27. Zhang, L.J.; Wu, Y.; Hu, M.H.; Guo, W.Y. Automatic image analysis of episcleral hemangioma applied to the prognosis prediction of trabeculotomy in Sturge–Weber syndrome. *Displays* **2022**, *71*, 102118. [CrossRef]
28. Du, M.M.; Li, H.Y.; Jin, X.; Roshnianfard, A.; Zhou, H. Extracting micro-topographical features of farmland by using UAV remote sensing image. *Agri Eng.* **2023**, *13*, 77–81. (In Chinese with English Abstract)
29. Li, Q.Y.; Li, Y.F.; Gong, H.Y.; Wang, D.Y. Prediction of urban and rural construction lands supply and demand indicators under human-land Linking in Changchun and Jilin. *J. Northeast Normal Univ. (Nat. Sci. Ed.)* **2017**, *49*, 120–127. (In Chinese with English Abstract)

30. Liu, Z.M.; Xiu, C.L.; Wei, Y.; Fang, Z.X. Spatial Pattern of Population Change and Its Influencing Factors in Northeast China from 1990 to 2010. *Northwest Popul.* **2017**, *5*, 19–26. (In Chinese with English Abstract)
31. Yang, Y. Study on the Coupling Relationship between the Land Input-Output Efficiency and Urbanization Quality in Jilin Province. Ph.D. Thesis, Jilin University, Changchun, China, 2018.
32. Liu, C.; Liu, X.Y.; Liu, P. Analysis on the changes of organic matters and their influencing factors of typical black soil areas in Songliao plain. *Geol. Resour.* **2020**, *29*, 550–555. (In Chinese with English Abstract)
33. Wick, A.F.; Ingram, L.G.; Stahl, P.D. Aggregate and organic matter dynamics in reclaimed soils as indicated by stable carbon isotopes. *Soil. Biol. Biochem.* **2009**, *41*, 201–209. [CrossRef]
34. Xu, W.Q.; Chen, X.; Luo, G.P.; Lin, Q. Using the CENTURY model to assess the impact of land reclamation and management practices in oasis agriculture on the dynamics of soil organic carbon in the arid region of North-western China. *Ecol. Complex.* **2011**, *8*, 30–37. [CrossRef]
35. Okolo, C.C.; Gebresamuel, G.; Zenebe, A.; Haile, M.; Eze, P.N. Accumulation of organic carbon in various soil aggregate sizes under different land use systems in a semi-arid environment. *Agr. Ecosyst. Environ.* **2020**, *297*, 106924. [CrossRef]
36. Stevens, A.; Wesemael, B. Soil organic carbon stock in the Belgian Ardennes as affected by afforestation and deforestation from 1868 to 2005. *For. Ecol Manag.* **2008**, *256*, 1527–1539. [CrossRef]
37. Gu, S.Z.; Wu, Y.B.; Wang, J.R. Thinking and strategic countermeasures of cultivating the fifth-tier cities. *Chin. J. Pop. Sci.* **2016**, *6*, 2–10. (In Chinese with English Abstract)
38. Yang, S.D.; Yang, X.; Zhang, J.X.; Gao, X.; Zhou, J.; Wu, X.L. Assessing the impacts of rural depopulation and urbanization on vegetation cover: Based on land use and nighttime light data in China, 2000–2020. *Ecol. Indic.* **2024**, *159*, 111639. [CrossRef]
39. Liao, L.W.; Long, H.L.; Ma, E.P. Rural labor change and farmland use transition. *Econ. Geogr.* **2021**, *41*, 148–155. (In Chinese with English Abstract)
40. Cambou, A.; Chevallier, T.; Barthès, B.G.; Derrien, D.; Cannavo, P.; Bouchard, A.; Allory, V.; Schwartz, C.; Vidal, B.L. The impact of urbanization on soil organic carbon stocks and particle size and density fractions. *J. Soil. Sediment.* **2022**, *23*, 792–803. [CrossRef]
41. Loon, J.V.; Woltering, L.; Krupnik, T.J.; Baudron, F.; Boa, M.; Govaerts, B. Scaling agricultural mechanization services in smallholder farming systems: Case studies from sub-Saharan Africa, South Asia, and Latin America. *Agr. Syst.* **2020**, *180*, 102792. [CrossRef]
42. Cai, J.; Li, X.P.; Liu, L.J.; Chen, Y.Z.; Wang, X.W.; Lu, S.H. Coupling and coordinated development of new urbanization and agro-ecological environment in China. *Sci. Total Environ.* **2021**, *776*, 145837. [CrossRef]
43. Virginia, A.; Zamora, M.; Barbera, A.; Castro-Franco, M.; Domenech, M.; Geronimo, E.D.; Costa, J.L. Industrial agriculture and agroecological transition systems: A comparative analysis of productivity results, organic matter and glyphosate in soil. *Agr. Syst.* **2018**, *167*, 103–112. [CrossRef]
44. Amiri, Z.; Asgharipour, M.R.; Campbell, D.E.; Sabaghi, M.A. Comparison of the sustainability of mechanized and traditional rapeseed production systems using an emergy-based production function: A case study in Lorestan Province, Iran. *J. Clean. Prod.* **2020**, *258*, 120891. [CrossRef]
45. Fang, C.L.; Cui, X.G.; Liang, L.W. Theoretical analysis of urbanization and eco-environment coupling coil and coupler control. *Acta Geogr. Sin.* **2019**, *74*, 2529–2546. (In Chinese with English Abstract)
46. Pudelko, K.; Kolodziel, J.; Mankowski, J. Restoration of minesoil organic matter by cultivation of fiber hemp (*Cannabis sativa* L) on lignite post-mining areas. *Ind. Crop Prod.* **2021**, *171*, 113921. [CrossRef]
47. Juricova, A.; Chuman, T.; Zizala, D. Soil organic carbon content and stock change after half a century of intensive cultivation in a chernozem area. *Catena* **2022**, *211*, 105950. [CrossRef]
48. Maleki, S.; Karimi, A.; Zeraatpisheh, M.; Poozeshi, R.; Feizi, H. Long-term cultivation effects on soil properties variations in different landforms in an arid region of eastern Iran. *Catena* **2021**, *206*, 105465. [CrossRef]
49. Bieluczyk, W.; Piccolo, M.C.; Pereira, M.G.; Moraes, M.T.; Soltangheisi, A.; Bernardi, A.C.C.; Pezzopane, J.R.M.; Oliveira, P.P.A.; Moreira, M.Z.; Camargo, P.B.; et al. Integrated farming systems influence soil organic matter dynamics in southeastern Brazil. *Geoderma* **2020**, *371*, 114368. [CrossRef]
50. Gu, C.L.; Li, Y.; Han, S.S. Development and transition of small towns in rural China. *Habitat. Int.* **2015**, *50*, 110–119. [CrossRef]

sustainability

MDPI

Article

Internal Force Mechanism of Pisha Sandstone as a Soil Amendment to Improve Sandy Soil Structural Stability in Mu Us Sandy Land

Zhe Liu [1,2,3], Lin Zhou [4], Yang Zhang [1,2,3,*], Jichang Han [1,2,3], Yingying Sun [1,2,3], Ruiqing Zhang [1,2,3], Xuxiang Li [1] and Feinan Hu [4,*]

[1] Technology Innovation Center for Land Engineering and Human Settlements, Shaanxi Land Engineering Construction Group Co., Ltd. and Xi'an Jiaotong University, School of Human Settlements and Civil Engineering, Xi'an Jiaotong University, Xi'an 710049, China; liuzhe168@stu.xjtu.edu.cn (Z.L.); hanjc_sxdj@126.com (J.H.); sunyy526@163.com (Y.S.); zhangruiq2010000@163.com (R.Z.); xxli@mail.xjtu.edu.cn (X.L.)

[2] Shaanxi Provincial Land Engineering Construction Group Co., Ltd., Xi'an 710075, China

[3] Institute of Land Engineering and Technology, Shaanxi Provincial Land Engineering Construction Group Co., Ltd., Xi'an 710075, China

[4] State Key Laboratory of Soil Erosion and Dryland Farming on the Loess Plateau, Institute of Soil and Water Conservation, Chinese Academy of Sciences and Ministry of Water Resources, Yangling, Xianyang 712100, China; zhoulin@nwafu.edu.cn

* Correspondence: zhangyang_dj@163.com (Y.Z.); hufn@nwafu.edu.cn (F.H.); Tel./Fax: +86-29-8662-5019 (Y.Z. & F.H.)

Abstract: Compounding Pisha sandstone (PSS) with sandy soil in Mu Us Sandy Land is a viable agronomical measure to effectively reduce soil erosion and improve soil quality due to the complementary characters and structures of the two materials. Aggregate stability is an important indicator to assess sandy soil erosion resistance and quality, which could be largely affected by soil surface electrochemical properties and particle interaction forces. However, the effect of the compound ratio and particle interaction forces on the aggregate stability of compound soils with Pisha sandstone and sandy soil is still unclear. Therefore, in this study, the electrochemical properties, particle interaction forces, and their effects on the aggregate stability of PSS and sandy soil at five volume ratios (0:1, 1:5, 1:2, 1:1, and 1:0) were determined to clarify the internal force mechanism of PSS to increase sandy soil structural stability in a 10-year field experiment. Experiments were measured by a combined method for the determination of surface properties and aggregate water stability. A ten-year field study revealed that the incorporation of Pisha sandstone significantly enhanced the soil organic carbon (SOC) and cation exchange capacity (CEC) ($p < 0.05$), while the CEC value notably increased from 4.68 to 13.76 cmol·kg^{-1} ($p < 0.05$). The soil surface potential (absolute value) and the electric field intensity gradually decreased with the increase in the Pisha sandstone content. For the compound soil particle interaction force, the addition of Pisha sandstone enhanced the van der Waals attraction force, reduced the net repulsive force between compound soil particles, and promoted the agglomeration of aeolian sandy soil. The overall trend of the aggregate breaking strength of compound soils under different addition ratios of PSS was 1:0 > 1:1 > 0:1 > 1:5 > 1:2. When the Pisha sandstone content in the compound soils was <50%, the aggregate stability was mainly influenced by compound soil particle interaction forces, and the interaction force increase was the key reason for the aggregate breakdown. When the Pisha sandstone content in the compound soils was ≥50%, the aggregate stability was affected by the combined effects of the compound soil particle composition and particle interaction forces. These results indicate that PSS addition ratios and particle interaction force are important factors affecting the structural stability of compound soils, in which the volume ratio of PSS to sandy soil of 1:2 is the appropriate ratio. Our study provides some theoretical references for further understanding of the compound soil structure improvement and sandy soil erosion control in Mu Us Sandy Land.

Keywords: Pisha sandstone; compound soil; electrochemical properties; aggregate structural stability; particle interaction; Mu Us Sandy Land

1. Introduction

Mu Us Sandy Land, one of China's four major sandy lands, is in the southeastern part of Inner Mongolia and the northern loess Plateau of Shaanxi Province with an area of about 4×10^6 ha [1]. Pisha sandstone (PSS), also known as soft rock or feldspathic sandstone, is widely distributed throughout the region, covering more than 1.67×10^6 ha [2]. PSS is a distinct type of terrigenous clastic rock from the Mesozoic and Late Paleozoic periods (approximately 250 million years ago) and is composed of argillaceous sandstone, sand shale, and mudstone [3]. PSS exhibits a loosely bonded structure, minimal diagenesis, and relatively low compressive strength [4]. When dry, PSS is hard as a rock, but it disintegrates rapidly into mud when exposed to water [5]. Due to its limited usability and severe soil erosion, PSS is viewed by local residents as an environmental menace. Nevertheless, researchers have discovered that PSS has excellent hydrophilicity and expansibility, making it a natural water-retention agent and soil amendment [4,6]. Pisha sandstone is rich in clay–silt particles and montmorillonite, with large specific surface area, strong cation adsorption capacity, and outstanding colloidal properties, which promote the cementation and agglomeration of aeolian sandy soil in Mu Us Sandy Land [6–8]. Hence, researchers suggested using a combination of PSS and sandy soil to improve water retention ability and soil quality due to the complementary characters and structures of the two materials.

Incorporating PSS into sandy soil to create compound soils offers an effective approach for managing soil degradation and severe erosion, expanding arable land to support local agriculture and gradually establishing a new soil resource [7,9]. Several studies have demonstrated that the addition of PSS can significantly alter particle distribution and improve the texture of the compound soil [10,11]. Furthermore, the incorporation of PSS into sandy soil enhances its water and fertilizer retention capacity. Wang et al. (2013) reported a 2.7-fold increase in the water retention capacity of sandy land after PSS and sandy soil blending and that PSS is an effective measure to improve the water retention and fertilizer retention capacity of aeolian sandy soil, in which the volume ratio of PSS to aeolian sandy soil of 1:2 is the appropriate ratio [9]. As a soil amendment, PSS also contributes to the enhancement of the field soil organic matter (SOM) content, cation exchange capacity (CEC), and aggregate formation in sandy soil [7]. In practical applications, over 1600 ha of newly cultivated land resources have been established by incorporating PSS into sandy soils, and the addition of PSS has improved the utilization efficiency and productivity of aeolian sandy soil resources in Mu Us Sandy Land [9,12]. Despite extensive research efforts, the current focus is predominantly on macro aspects such as erosion resistance, hydraulic parameters, or productivity [11,13]. The microscopic particle interaction force mechanism underlying the impact of PSS addition on aggregate stability in compound soil remains incompletely understood.

The stability of aggregates is the primary and direct factor in assessing soil erosion and quality, as it is significantly affected by soil electrochemical properties and interaction forces among particles [14,15]. For example, Liu et al. (2020) demonstrated that inputting SOM changed electrochemical properties on the soil particle surface during vegetation succession, subsequently enhancing the van der Waals attraction between the particles and ultimately improving soil structure stability [16]. Similarly, Hu et al. (2021) observed that incorporating biochar as a soil amendment increased the CEC, specific surface area (SSA), and charge density of the soil surface, which significantly enhanced the molecular attraction, and weakened the repulsive force between soil particles and consequently improved aggregate stability [17]. Xu et al. (2015) reported that the release of small particles (<2 μm) from a compound soil sample (20% montmorillonite and 80% kaolinite) was four times higher than that of pure montmorillonite, indicating that the addition of fine particles could enhance soil aggregate stability [15]. Moreover, the difference of surface potentials or kaolinite contents led to a difference in soil aggregate stability, highlighting that the influence of soil particle composition ratios on surface electrochemical properties and aggregate stability is critical. In conclusion, it is reasonable to infer that electrochemical properties and particle internal forces among compound soil (PSS and sandy soil) particles can be altered with the addition of PSS, further profoundly impacting soil aggregate stability. However, there have been limited studies to date on how PSS addition influences sandy soil particle interaction forces by changing particle surface properties and its impacts on aggregate stability in compound soil. Further investigations in this area are crucial for elucidating the mechanisms by which PSS stabilizes compound soil aggregates.

Therefore, in this study, we collected soil samples from a ten-year field experiment with varying application rates of PSS in sandy soil (PSS: sandy soil = 0:1, 1:5, 1:2, 1:1, and 1:0, v/v) to assess soil aggregate stability. The primary objective is to quantitatively investigate the impact of PSS addition on soil interaction forces (SIFs) and soil aggregate stability while also elucidating the underlying mechanisms of compound soil aggregate stability and compound soil erosion control in Mu Us Sandy Land. Our study will provide some theoretical references for further understanding of the compound soil structure improvement and sandy soil erosion control in Mu Us Sandy Land.

2. Materials and Methods

2.1. Experimental Site and Design

The Pisha sandstone and sandy soil used in this study were obtained from Daji Han Village in Yuyang District, Yulin city, located within the Mu Us Sandy Land of China (109°28′ E, 38°27′ N) (Figure 1). Prior to field compounding, any remaining plant roots and boulders were removed from the samples. After natural drying, the samples were crushed and passed through a 5 mm sieve.

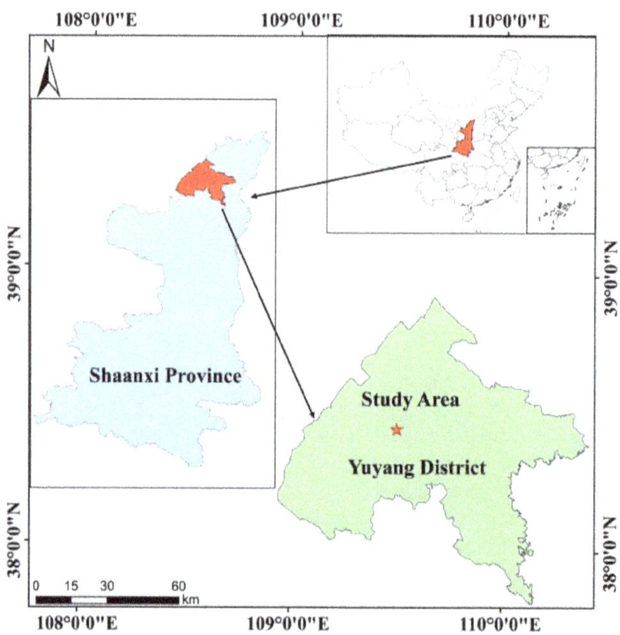

Figure 1. Location of the experimental site.

The field experiment was conducted in Fuping County (109°11′ E, 34°42′ N), Shaanxi Province, China. The area represents a typical semi-arid continental climate, and the mean annual rainfall is 473 mm. Rainfall distribution during the year is highly uneven, with approximately 59% occurring from July to September. The potential annual evaporation is nearly 1300 mm, more than twice the amount of the rainfall. The field experimental plots (2 × 2 m) were filled 30 cm deep by compound soil consisting of five volume ratios of PSS and sandy soil (0:1, 1:5, 1:2, 1:1, and 1:0, v/v), and the electrochemical properties, particle interaction forces, and their effects on aggregate stability were studied for five compound soil samples with different volume ratios of Pisha sandstone and sandy soil. The objective is to determine the internal force control mechanism of the addition of PSS at different ratios to improve the structure and quality of aeolian sandy soil. Corn and wheat were cultivated on the compound soil for ten years. Each group was replicated three times.

2.2. Determination of the Basic Properties of the Compound Soil

The compound soil samples were collected randomly from three locations in each plot using a soil auger at a depth of 0–20 cm. The test materials were naturally dried, followed by the removal of stones and plant roots, and then sieved through a 2 mm mesh. The soil organic carbon (SOC) content was measured by the $K_2Cr_2O_7$ oxidation method [18]. Soil pH was determined by a pH meter with a soil to water ratio of 1:2.5 [19]. The carbonate content was determined using a gas volume method [20]. The primary clay minerals in compound soils were determined by X-ray diffraction analysis (Ultima IV, manufactured by Rigaku Corporation of Japan, Tokyo, Japan). The primary clay minerals in PSS were montmorillonite (~7%), hydromica (~10%), kaolinite (~12%), and chlorite (~11%), whereas the primary clay minerals in sandy soil were hydromica (~8%) and kaolinite (~6%). Soil particle size distributions were determined using the pipette method and were classified into sand (2–0.02 mm), silt (0.02–0.002 mm), and clay (<0.002 mm) based on the International Soil Texture Classification System. CEC and SSA were measured according to the method proposed by Li et al. (2013), involving the saturation of soil samples with hydrogen ions and conducting ion exchange experiments with a mixture of $Ca(OH)_2$ and NaOH. The concentrations of Ca^{2+} and Na^+ were detected to calculate CEC, *SSA*, and surface charge density (σ_0) according to the double layer theory [21]. Specific procedures followed those in previous research [16]. The basic compound soil properties are presented in Table 1.

Table 1. Basic physicochemical properties and surface electrochemical properties of compound soils under different compound ratios.

v (PSS):v (SS)	Particle Size Distribution			pH	$CaCO_3$ ($g{\cdot}kg^{-1}$)	SOC ($g{\cdot}kg^{-1}$)	CEC ($cmol{\cdot}kg^{-1}$)	*SSA* ($m^{2.5}{\cdot}g^{-1}$)
	Sand (%)	Silt (%)	Clay (%)					
0:1	93.8 ± 0.25 a	2.33 ± 0.21 d	3.87 ± 0.22 a	8.39 ± 0.01 b	29.14 ± 2.49 d	3.28 ± 1.13 cd	4.68 ± 0.04 d	2.54 ± 0.31 d
1:5	88.15 ± 0.52 b	8.85 ± 0.71 c	3.00 ± 0.19 a	8.78 ± 0.05 a	40.84 ± 1.76 c	4.50 ± 0.86 bc	6.45 ± 0.07 c	7.47 ± 2.12 cd
1:2	82.82 ± 2.13 c	12.94 ± 1.74 b	4.24 ± 1.7 a	8.75 ± 0.06 a	44.59 ± 1.95 c	5.24 ± 0.52 ab	6.67 ± 0.08 c	10.96 ± 0.77 c
1:1	83.29 ± 0.61 c	14.13 ± 1.37 b	2.58 ± 0.82 a	8.72 ± 0.02 a	55.09 ± 2.26 b	6.59 ± 0.52 a	13.76 ± 0.22 b	28.71 ± 3.84 b
1:0	64.18 ± 1.81 d	31.59 ± 1.86 a	4.23 ± 0.07 a	8.21 ± 0.10 c	144.66 ± 5.16 a	2.22 ± 0.52 cd	17.91 ± 0.18 a	37.24 ± 4.90 a

Notes: v (PSS):v (SS) represents the combined treatment of the volume ratio of Pisha sandstone to sandy soil; SOC, soil organic carbon; CEC, cation exchange capacity; *SSA*: specific surface area. Different lowercase letters for the same index are significantly different at the 0.05 level ($p < 0.05$); Values are means ± standard deviation (n = 3).

2.3. Determination of Soil Surface Charge Properties

Compound soil surface charge properties include cation exchange capacity (CEC), specific surface area (SSA), surface charge density (σ_0), surface electric field strength (E), and surface potential (φ_0). These parameters were determined according to the combined method for the determination of soil surface properties [22]. The detailed steps were as follows. Firstly, due to the high content of $CaCO_3$ in soft rock, it was necessary to decalcify the soil samples. Briefly, differently treated crushed (<0.25 mm) samples were decalcified by washing with 0.5 mol L^{-1} HCl solution three times until no CO_2 was released in the suspension. Secondly, H^+-saturated samples were prepared by washing approximately 100 g soil four times with 500 mL of 0.1 mol L^{-1} HCl and then with deionized water repeatedly until the solution was free of Cl^{-1} in the suspension. The H^+-saturated soil samples were obtained, oven-dried at 60 °C, and sieved through a 0.25 mm sieve. Thirdly, certain amounts of H^+-saturated samples were weighed into 150 mL triangular bottles, and equal volumes (approximately 55 mL) of 0.01 mol L^{-1} $Ca(OH)_2$ and NaOH solution were added. Because pH affects the soil surface charge properties, after 24 h of shaking, the pH values of the differently treated suspension were adjusted to 7.0 with 1 mol L^{-1} HCl solution. The quantities of Ca^{2+} and Na^+ absorbed on soil particles were determined by measuring the activities and concentrations of Ca^{2+} and Na^+ in the supernatants using a flame photometer and an atomic absorption spectrometer, respectively.

Finally, the electrochemical properties of compound soil were calculated by introducing the measured data into the following Equations (1)–(5) [22,23].

$$\varphi_0 = \frac{2RT}{2(\beta_{Ca} - \beta_{Na})F} \ln \frac{\alpha^0_{Ca} N_{Na}}{\alpha^0_{Na} N_{Ca}} \quad (1)$$

$$\sigma_0 = \mathrm{sgn}(\varphi_0)\sqrt{\frac{\varepsilon RT}{2\pi}\left(\alpha^0_{Na} \exp\frac{\beta_{Na} F \varphi_0}{RT} + \alpha^0_{Ca} \exp\frac{2\beta_{Ca} F \varphi_0}{RT}\right)} \quad (2)$$

$$E = 4\pi\sigma_0/\varepsilon \quad (3)$$

$$SSA = \frac{N_{Na}\kappa}{m\alpha^0_{Na}} \exp\frac{\beta_{Na} F \varphi_0}{2RT} = \frac{N_{Ca}\kappa}{m\alpha^0_{Ca}} \exp\frac{\beta_{Ca} F \varphi_0}{RT} \quad (4)$$

$$CEC = 10^5 S\sigma/F \quad (5)$$

where φ_0 (mV) is the particle surface potential; σ_0 (C m^{-2}) is the surface charge density; R (J K^{-1} mol^{-1}) is the universal gas constant; T (K) is the absolute temperature; F (C mol^{-1}) is the Faraday constant; E (V m^{-1}) is the surface electric field strength; SSA (m^2 g^{-1}) is the specific surface area; CEC (cmol kg^{-1}) is the cation exchange capacity; Z is the charge of the cation; β_{Na} and β_{Ca} are the corresponding modification factors of Z for Na^+ and Ca^{2+}, respectively; ε is the dielectric constant for water (8.9×10^{-9} C^2 $J^{-1}m^{-1}$); κ (dm^{-1}) is the Debye–Hückel parameter; I (mol L^{-1}) is the ionic strength; $c^0{}_i$ (mol L^{-1}) is the equilibrium concentration of the cation (i = Ca^{2+}, Na^+) in the bulk solution; $\alpha^0{}_i$ (mol L^{-1}) is the activity of the cation (i = Ca^{2+}, Na^+) in the bulk solution; N_i (mol g^{-1}) is the total number of cations (i = Ca^{2+}, Na^+) adsorbed on the soil particle surface.

In this experiment, the calculation equations of compound soil electric field intensity distribution with distance are as follows:

$$\varphi(x) = \frac{4RT}{F} \tan h^{-1}\left(ae^{-kx}\right) \quad (6)$$

$$a = \tan h\left(\frac{ZF\varphi_0}{4RT}\right) \quad (7)$$

$$E(x) = \sqrt{\frac{8\pi RT}{\varepsilon}\left[c_0\left(e^{\frac{-ZF\varphi(x)}{RT}} - 1\right)\right]} \quad (8)$$

where φ(x) is the potential at x distance from the particle surface, V; x is the distance between two adjacent particles in the double electric layer, nm; a is the intermediate variable; E(x) is the electric field strength at x distance from the particle surface, V m^{-1}.

2.4. Quantification of Soil Particle Interaction Forces

The compound soil particle interaction forces (SIFs) include electrostatic, van der Waals, and hydration forces. The net pressure (P_{net}) of the compound soil interaction force is the sum of the electrostatic repulsive pressure (P_{ele}), hydration pressure (P_{hyd}), and van der Waals force (P_{vdW}) in this experiment. Here, based on the determination and calculation results of the surface charge properties of composite soils, the P_{net} can be calculated according to the following equations [17,24]:

$$P_{net} = P_{ele} + P_{hyd} + P_{vdW} \quad (9)$$

$$P_{ele} = \frac{2}{101} RTc_0 \left\{ \cos h\left[\frac{ZF\varphi(d/2)}{RT}\right] - 1 \right\} \quad (10)$$

$$P_{hyd} = 3.33 \times 10^4 exp^{-5.76\times10^9 d} \quad (11)$$

$$P_{vdW} = -\frac{A_{eff}}{0.6\pi}(10d)^{-3} \quad (12)$$

where R ($J\cdot K^{-1}\cdot mol^{-1}$) is the universal gas constant; T (K) represents absolute temperature; c_0 ($mol\cdot L^{-1}$) refers to the cation concentration of the equilibrium solution; Z is the cation valence; F ($C\ mol^{-1}$) is the Faraday constant; A_{eff} (J) is the effective Hamaker constant, which is usually between 10^{-21} and 10^{-19} J [25] and can be calculated using the soil water characteristic curve [26,27]. As PSS and sandy soil are not real soils in the strict sense, the A_{eff} for quartz [17,28] (main composition of sandy soil) instead of that for sandy soil was used, and the A_{eff} of PSS was measured, while that of other compound soils was interpolated based on the mixing ratio. d (dm) represents the distance between two adjacent particles, while $\varphi(d/2)$ (V) denotes the electric potential at the midpoint of the overlap region of their respective double layers.

2.5. Evaluation of Soil Aggregate Stability

In this experiment, the aggregate breaking strength was utilized for assessing the aggregate stability to characterize the structural stability of compound soil aggregates [17,24]; this strength is defined as the percentage of particles (diameters < 10 and <5 μm). An increase in the percentage of fine particles indicates a decrease in soil aggregate stability. In this study, NaCl solutions with varying concentrations (1, 10^{-1}, 10^{-2}, 10^{-3}, and 10^{-5} mol L^{-1}) were prepared in five cylinders (500 mL), and 20 g of 1–5 mm Na^+-saturated soil aggregate was slowly added to each cylinder (three replicates). Subsequently, the soil aggregate was submerged for 2 min, followed by careful inversion of the cylinders four times within the next 2 min. The mass percent of released particles (<10 and <5 μm) to total aggregate was determined using the pipette method according to Stokes law [28]. Throughout the experiment, strict adherence to proper techniques and precautions was maintained.

3. Results

3.1. Influences of PSS on Compound Soil Properties

Table 1 presents the fundamental properties of different compound soils. As can be seen from this table, sandy soil had the highest sand content while PSS has less sand and more clay than sandy soil. With the addition of PSS, the silt content in the compound soil increased significantly, while the sand content decreased ($p < 0.05$). The pH of the compound soils was slightly higher compared to PSS and sandy soil. The incorporation of PSS led to a significant increase in the $CaCO_3$ content ($p < 0.05$), and the $CaCO_3$ content of the composite soil at the ratio of 1:1 was nearly twice that of the sandy soil. The initial SOC content in PSS was only 2.22 $g\cdot kg^{-1}$; however, the addition of PSS substantially elevated the SOC levels in the compound soils. After ten years of cultivation, among the various ratios tested, the SOC content was highest in the compound soil with a ratio of 1:1, approximately twice that of sandy soil and three times that of pure PSS. The organic matter content of PSS and sandy soil in the combined treatment at 1:5, 1:2, and 1:1 increased by 37.32%, 59.78%, and 101.07%, respectively, compared with the control treatment without PSS. Across the different proportions, the CEC ranged from 4.68 cmol kg^{-1} to 17.91 cmol kg^{-1}, with an average of 9.89 cmol kg^{-1}. Furthermore, the SSA of the compound soil varied from 2.54

to 37.24 $m^2\ g^{-1}$, with an average of 17.38 $m^2\ g^{-1}$. Overall, the incorporation of PSS significantly enhanced the CEC and SSA ($p < 0.05$), while the CEC value notably increased from 4.68 to 13.76 $cmol \cdot kg^{-1}$ ($p < 0.05$).

3.2. Changes of Surface Potential and Electric Field Strength of Compound Soils

The surface electric field intensity and internal force of compound soil particles are affected by the composition of soil electrolytes and the concentration of the electrolyte solution [29]. In general, with the increase in electrolyte concentration, the electric field strength and internal force between soil particles show a decreasing trend, and the net attraction between soil particles increases. It was found that with the increase in the distance between soil particles, the electrostatic repulsive force, van der Waals attractive force, and hydration repulsive force between compound soil particles will gradually decrease [30,31]. The surface potential of soil particles at different electrolyte concentrations can be calculated by Equation (1), and the results are shown in Table 2. Under the same electrolyte concentration, the soil surface potential (absolute value) decreased with the increase in the soft rock content. For all the compound soils, the surface potential (absolute value) decreased with the increase in the electrolyte concentration. Specifically, when the electrolyte concentration varied from 10^{-5} to 1 mol L^{-1}, the absolute value of the surface potential for the 1:5, 1:2, and 1:1 groups decreased by 290.6, 288.9, and 287.6 mV, respectively.

Table 2. Surface potentials of soil particles under different electrolyte concentrations and compound ratios.

Electrolyte Concentration ($mol \cdot L^{-1}$)	Compound Soil Surface Potential (mV)				
	0:1	1:5	1:2	1:1	1:0
1	−470.3	−431.4	−413.4	−401.2	−401.4
0.1	−352.2	−313.3	−295.4	−283.2	−283.4
0.01	−293.2	−254.6	−236.9	−224.8	−225.0
0.001	−234.7	−196.7	−179.4	−167.7	−167.9
0.00001	−177.3	−140.8	−124.5	−113.6	−113.8

The electric field distribution curves around soil particles under different electrolyte concentrations can be further obtained based on the surface potential values and Equations (6)–(8). As shown in Figure 2, the electric field intensity decreased with the increase in the distance between particles. Similarly, the increase in electrolyte concentration also led to the reduction of the electric field intensity around soil particles, and the variation range of the surface electric field was sharply reduced. For example, the electric field action distance was only within 10 nm when the electrolyte concentration was 1 mol L^{-1}; however, with the dilution of the electrolyte concentration to 10^{-5} mol L^{-1}, the range of the electric field intensity increased rapidly, reaching more than 100 nm.

The electric field intensity values of the five compound soils at 10 nm and different concentrations are shown in Table 3. The electric field intensity gradually decreased with the addition of soft rock, while the difference of the electric field intensity of compound soils with the same concentration was small. However, for the same treatment, the change of electrolyte concentration resulted in an order of magnitude change in the electric field intensity. When the concentration was diluted from 1 to 10^{-2} mol L^{-1}, the intensity of the electric field jumped from double figures to 10^6 orders of magnitude, but when the concentration decreased from 10^{-2} to 10^{-5} mol L^{-1}, the electric field intensity was still within the range of 10^6 orders of magnitude. The above results indicate that an electrolyte concentration of 10^{-2} mol L^{-1} was the key concentration that affected the electric field intensity in this study.

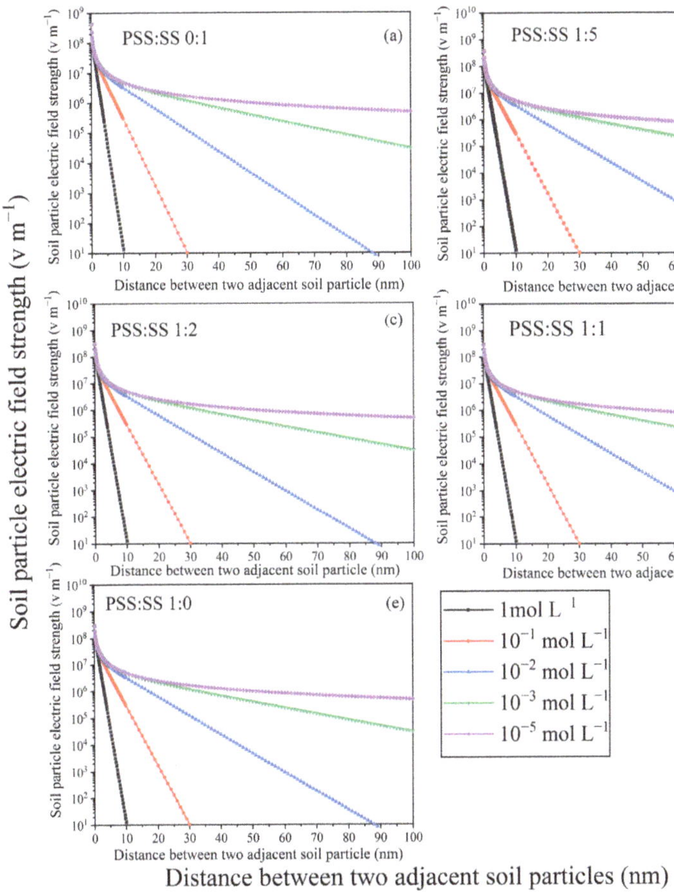

Figure 2. Distribution of compound soil electric field around soil particles under different electrolyte concentrations and compound ratios.

Table 3. Soil surface electric field strength at the distance of 10 nm from particle surfaces under different electrolyte concentrations and compound ratios.

Electrolyte Concentration ($mol \cdot L^{-1}$)	Compound Soil Electric Field Strength ($-V\ m^{-1}$)				
	0:1	1:5	1:2	1:1	1:0
1	5.12×10^6	5.11×10^6	5.10×10^6	5.09×10^6	5.09×10^6
0.1	4.90×10^6	4.89×10^6	4.88×10^6	4.87×10^6	4.87×10^6
0.01	3.40×10^6	3.39×10^6	3.38×10^6	3.37×10^6	3.37×10^6
0.001	3.04×10^5	3.00×10^5	2.98×10^5	2.96×10^5	2.96×10^5
0.00001	13.76	13.30	12.95	12.71	12.71

3.3. Changes in the Aggregate Stability of Compound Soils

To investigate the effects of soil particle interaction forces on aggregate stability, the aggregate breaking strength of compound soils under different electrolyte concentrations is plotted in Figure 3. It can be seen that the aggregate breaking strength of compound soils followed the order of 1:0 > 1:1 > 0:1 > 1:5 > 1:2. It was observed that the breaking strength initially decreased followed by an increase in PSS under each electrolyte concentration. In other words, as the level of PSS amendment increased, the soil aggregate stability improved initially, reaching its highest point at the 1:2 ratio, before gradually weakening. The addition of Pisha sandstone improved the structural stability and erosion resistance of the compound soil. Additionally, we discovered that a low electrolyte concentration resulted in greater breaking strength. Under the condition of a high concentration (1 mol L^{-1}) of electrolyte solution, almost no small particles were released or broken.

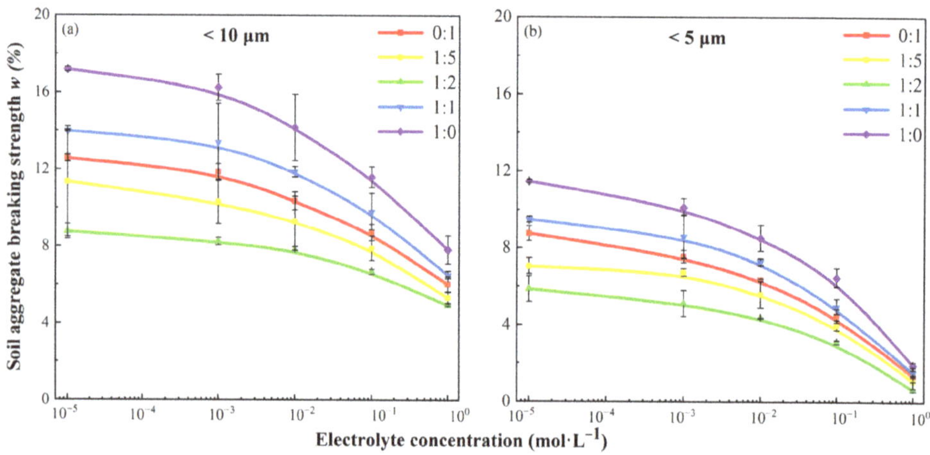

Figure 3. Changes in aggregate breaking strength under different electrolyte concentrations and compound ratios. Note: Error bars represents the standard deviation (n = 3).

3.4. P_{net} of Compound Soil Particles

The net pressure between two adjacent particle surfaces is the sum of electrostatic repulsion, van der Waals attraction, and hydration repulsion, which was calculated according to Equations (9)–(12) (Figure 4). Generally speaking, Pisha sandstone is rich in montmorillonite and clay–silt particles; the addition of Pisha sandstone enhanced the van der Waals attraction force, reduced the net repulsive force between compound soil particles, and promoted the agglomeration of aeolian sandy soil. The overall trend of the net resultant force under different addition ratios of Pisha sandstone was 1:0 < 1:1 < 1:2 < 1:5 < 0:1. The negative value indicated that the compound soil particles were net attractive, which means that the addition of Pisha sandstone enhanced the agglomeration force and structural stability of compound soil particles. For example, when the concentration was 1 mol L^{-1}, the distance of the sites corresponding to zero net pressure gradually decreased with the increase in the soft rock. At a high electrolyte concentration of 1 mol L^{-1}, when the distance between adjacent soil particles of the compound soil was greater than 2.3 nm, except for the 0:1 treatment without the addition of PSS, the net pressure between soil particles under the other four compound soil treatments with the addition of PSS was negative, showing a net attraction. Recent studies have shown that when the distance between soil particles is less than 2 nm, the hydration repulsive force plays a leading role; meanwhile, the dry soil aggregates encounter water, the distance between particles extends to 1.5~2 nm by the hydration repulsion, and the soil aggregates expand only slightly [32]. When the distance between soil particles is greater than 2 nm, van der Waals attractive force and electrostatic repulsive force play a leading role, and the agglomeration or fragmentation of soil aggregates is affected by electrostatic repulsive force, van der Waals attractive force, and hydration repulsive force [32,33]. When two adjacent soil particles of compound soil were infinitely close together, at any electrolyte concentration, the net pressure of the five compound soils was always repulsive and could reach tens of thousands of atmospheres. For instance, at the ratio of 1:2 compound soil, the net repulsive pressure at the particle distance of 0.2 nm was about 10,700 atm at the electrolyte concentration of 1 mol L^{-1}. In addition, when the electrolyte concentration was >10^{-2} mol L^{-1}, the net pressure increased substantially with decreasing electrolyte concentration at the same distance. However, when the electrolyte concentration $\leq 10^{-2}$ mol L^{-1}, the net pressure distribution curves almost overlapped. Here in, taking the compound soil with the ratio of 1:1 as an example (at a distance of 2 nm), when the electrolyte concentration decreased from 1 to 10^{-2} mol L^{-1}, the increase in net pressure was 18.12 atm. However, when the concentration decreased from 10^{-2} to 10^{-5} mol L^{-1}, the net repulsive pressure increase was only 0.42 atm. These findings showed that 10^{-2} mol L^{-1} was also the critical concentration affecting the net pressure of compound soil.

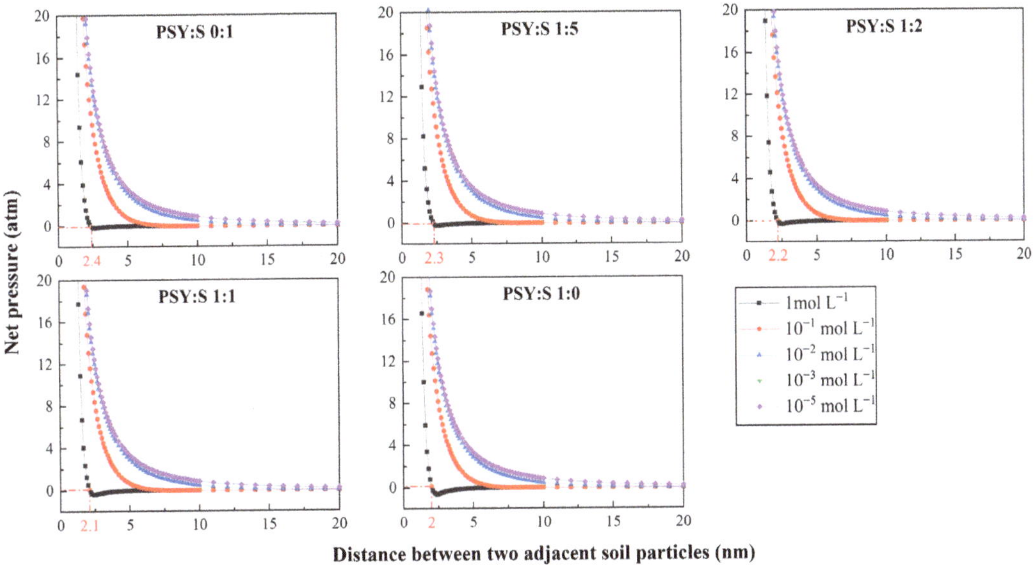

Figure 4. Distribution of net pressure (P_{net}) around compound soil particles at different electrolyte concentrations.

In order to further evaluate and analyze the influence of different compound ratios of Pisha sandstone and sandy soil on the variation of the net resultant force between compound soil particles, the distribution diagram of net resultant force at the distance of 2.0 nm and 2.4 nm between compound soil particles was drawn (Figure 5). The net resultant force between the compound soil particles under different compound ratios of Pisha sandstone and sandy soil showed a trend of 1:0 < 1:1 < 1:2 < 1:5 < 0:1. The smaller the net resultant value, the greater the net attraction between the compound soil particles. The research data showed that with the addition of Pisha sandstone, the smaller the net force between particles, the greater the net attraction. The combination of Pisha sandstone and sandy soil could promote the cementation and agglomeration of soil particles, and the net attraction between compound soil particles was enhanced. When the electrolyte concentration of the solution decreased from 1 mol L^{-1} to 10^{-2} mol L^{-1}, the net resultant force between the composite soil particles increased rapidly under different compound ratios, and the attraction between the compound soil particles weakened sharply. When the electrolyte concentration of the solution decreased from 10^{-2} mol L^{-1} to 10^{-5} mol L^{-1}, the net resultant force between the soil particles of the compound Pisha sandstone and sandy soil changed only slightly, and the change of the force between the compound soil particles tended to be gentle. The above results showed that the electrolyte concentration of 10^{-2} mol L^{-1} was the key concentration threshold for the change of the interaction between Pisha sandstone and sandy soil in the compound soil. When the electrolyte concentration of the solution was 10^0 mol L^{-1} and the particle distance of the compound soil was 2.4 nm, the net resultant force of the sandy soil without Pisha sandstone was 0.09 atm, and the net resultant force of the compound soil with Pisha sandstone added was negative under the four compound treatments (1:1, 1:2, 1:5, 0:1), and the particles were attractive. However, the aggregate stability of compound soil may be affected not only by the net resultant force between particles but also by the granulometric composition and addition ratios of the Pisha sandstone samples. In order to further clarify the influential factors of particle agglomeration of compound soil, the relationship between the net resultant force between compound soil particles and aggregate breaking strength will be further analyzed below.

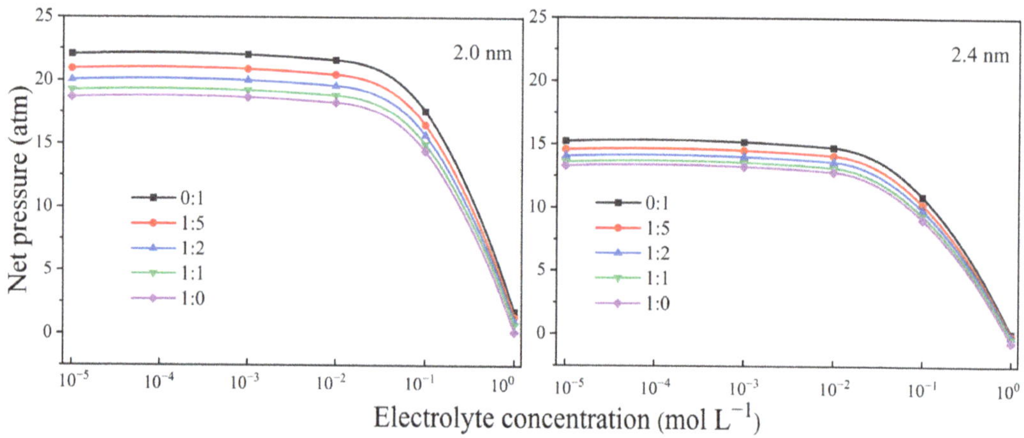

Figure 5. Net pressures at 2.0 nm and 2.4 nm distances between compound soil particles under different electrolyte concentrations and compound ratios.

3.5. Relationship between the Net Pressure (P_{net}) and the Aggregate Breaking Strength of under Different Compound Ratios

To further quantitatively analyze the combined effects of compound soil particle interaction forces and the compound ratio on aggregate stability, we established the relationship between the net pressure at a 2 nm soil particle distance and the aggregate breaking strength of the compound soils (Figure 6). The five compound soil samples with different volume ratios of Pisha sandstone and sandy soil (0:1, 1:5, 1:2, 1:1, and 1:0, *v*/*v*) were used to study the compound soil particle interaction forces and their effects on aggregate stability. Herein, the compound soils were classified into two categories based on the different Pisha sandstone mixed ratios (50% Pisha sandstone). There was a significant positive exponential relationship between the aggregate breaking strength and the soil net pressures. This indicates that the soil aggregate stability decreased exponentially with the increase in the net pressure of soil particles. Different proportions of compound soil cannot directly establish the relationship between the internal force and aggregate breaking strength. It is necessary to classify the groups of different compound proportions. When $p < 0.05$, the R^2 values of fitting curve A and fitting curve B reached more than 0.7, which indicates that both force and particle composition had important effects on aggregate stability. For different compound soil samples, when the PSS content was below 50%, the stability of the aggregates was primarily influenced by SIFs between soil particles. Conversely, when the PSS content was 50% or higher, the stability of the aggregates was affected by both the fine particle content and soil interaction forces. The addition of different ratios of Pisha sandstone samples had different effects on the composition of the compound soil particles. Therefore, our results suggest that the aggregate stability of compound soils was not only affected by compound soil particle interaction forces but also by the granulometric composition and addition ratios of the Pisha sandstone samples.

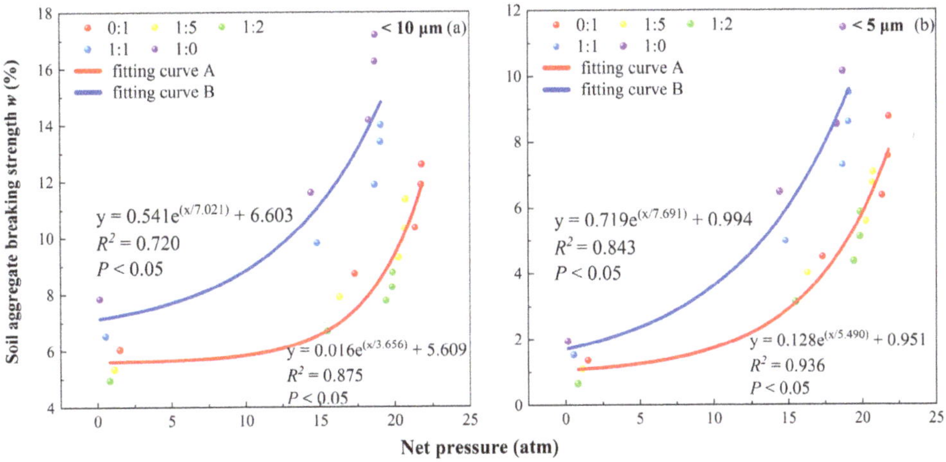

Figure 6. The relationship between the net pressure (P_{net}) and the aggregate breaking strength of compound soil under different compound ratios. Notes: Fitting curve A represents the compound soil group with PSS content < 50%. Fitting curve B represents the compound soil group with PSS content ≥ 50%.

4. Discussion

4.1. Responses of Compound Soil Properties to Soft Rock Addition

The addition of PSS can alter the particle size distribution of compound soils. As the input of PSS increased, the silt content increased while the sand content decreased (Table 1). Existing research has also mentioned a comparable trend, where the soil texture improved and the distribution of soil grades tended to be better [11]. This is mainly because sandy soil has relatively coarse particles and a high sand content, while PSS contains clay and silt with a large specific surface area. When they are combined, the fine particles fill the large pores, converting non-capillary porosity into capillary porosity, reducing soil permeability and thereby improving the soil water-holding capacity and cation adsorption capacity [12]. Our results also demonstrate a progressive increase in the carbonate content with the addition of PSS, which aligns with previous studies by Guo et al., who showed that the addition of Pisha sandstone significantly increased the calcium carbonate content of aeolian sandy soil [34]. The PSS used in our study belongs to a calcareous PSS [9]. After compounding with sandy soils, the calcium carbonate content in compound soils significantly increased, promoting the aggregate formation of sandy soils due to cementation effects [35]. After ten years of tillage management, the SOC content of compound soils increased with increasing PSS application rates (Table 1). This is consistent with prior literature [7,36], which also found significantly higher fertility in compound soils with the addition of Pisha sandstone compared to sandy soil. This result can be explained as follows: on the one hand, the addition of PSS to sandy soil increased the presence of fine particles, and Ingelmo et al. (2003) indicated that the soil organic matter (SOM) tended to concentrate around these fine particles, allowing for the formation of more organic–inorganic complexes [37]. On the other hand, the change in particle composition led to a more uniform composite soil texture, increasing the ability to retain water and fertilizer. Additionally, the ten-year farming environment also increased amount of crop residue recovery and promoted the accumulation of organic matter [6,8,12].

Pisha sandstone was rich in clay–silt particles and montmorillonite, with a large specific surface area and strong cation adsorption capacity, which led to an increase in CEC and SSA with the addition of Pisha sandstone (Table 1). These findings are consistent with prior research, which reported that the addition of PSS significantly increased the ion adsorption capacity and CEC content of aeolian sandy soil [7,38]. The increase in CEC can be attributed to the particle composition and mineral composition of the compound soil. Iturri and Buschiazzo (2014) reported that CEC is influenced by both the clay and silt content, while Hepper et al. (2006) showed that the changes in silt content positively affect CEC [39,40]. With the addition of PSS, the content of fine particles (<0.02 mm) increased in the compound soils (Table 1). PSS contains montmorillonite, which has the largest surface area and CEC among the minerals present [7,41]. In summary, the presence of fine particles and the montmorillonite content in Pisha sandstone play crucial roles in the substantial increase in CEC and SSA in aeolian sandy soil.

4.2. Effects of PSS Application on Soil Aggregate Stability and SIFs

According to the relationship between P_{net} and the aggregate breaking strength (Figures 3 and 4), the compound soils were classified into two categories based on the different mixing ratios (50% PSS). The aggregate breaking strength exhibited a significant positive exponential relationship with soil P_{net}. Therefore, our results suggest that the aggregate stability of compound soils was not only affected by SIFs but also by the granulometric composition and addition ratios of the Pisha sandstone samples. As mentioned earlier, the decrease in electrolyte concentration will lead to an increase in the net force, which in turn leads to an increase in the aggregate breaking strength and a decrease in the stability of the aggregates. The experimental results are consistent with the theoretical analysis, indicating that soil internal force is an important factor affecting the aggregate stability. Similar results are also mentioned in the literature by Hu et al. (2015) and Liu et al. (2021), who found that the electrolyte concentration of loess solution was positively correlated with the net attraction between particles [24,42]. The internal force between soil particles is closely related to the electrolyte concentration in solution, which is the key to the stability and fragmentation of the aggregate structure [32,43,44]. However, the aggregate breaking strength varies with the addition of soft rock, and the ratio of PSS to sandy soil was 1:2, which had the best effect on improving the structural stability of sandy soil aggregates. The experimental results are not consistent with the theoretical predictions, indicating that in addition to the internal force, the compound ratio and the granulometric composition of the Pisha sandstone samples are also important factors affecting the structure stability of the compound soil. Pisha sandstone is rich in montmorillonite and clay–silt particles with a large specific surface area.

Our results indicated that P_{net} tended to decrease with increasing PSS content at any given concentration, whereas the experimental data of aggregate stability showed that the 1:2 group was the most stable combination, while the 1:0 group was the least stable among all groups (Figure 3). These results may be explained by differences in the particle size distribution of the compound soils. Methodically, according to previous research [23], strong repulsive forces mainly produce single grains and micro aggregates, while the maximum particle size measured when assessing the stability of aggregates by the pipette method was 10 μm in our study, which is lower than the minimum particle size of sand. Therefore, the method is very sensitive to the sand content. As indicated in the study of Liu et al. (2021), soil with a higher sand content exhibits lower aggregate breaking strength, which could cause soil aggregates to appear more stable [42]. Similarly, in our study, compound soils with ratios of 0:1, 1:5, and 1:2 had more than 50% sandy soil content, which may lead to better aggregate stability than those with less than 50% sandy soil, such as soils with compound ratios of 1:0 and 1:1. On the other hand, according to the research of Liu et al. (2021), the pipette method is greatly affected by the difference in particle composition [42]. The smaller particles in the soil will be released after the aggregates are broken, and the stability will decrease. This is the reason why the compound soil with a high PSS content had a high breaking strength, which is not conducive to soil erosion resistance and structure improvement. From a particle composition perspective, Xu et al. (2015) found that after montmorillonite and kaolinite were compounded in different proportions, the internal pore characteristics of the compound soil structure were different, so the rate of water entering the aggregates will be different, which will further lead to different degrees of aggregate fragmentation and ultimately lead to differences in aggregate stability [15]. Therefore, these results suggest that both the SIFs and the particle composition may significantly influence the aggregate stability of compound soils.

5. Conclusions

In this study, we observed that the addition of PSS to sandy soil resulted in an increase in the $CaCO_3$ content and silt particle content, while the sand particle content decreased. Furthermore, a ten-year field study revealed that the incorporation of PSS led to an increase in SOM, CEC, and SSA. Calculations disclosed a decrease in repulsive forces and an increase in attractive force following PSS addition. The amendment of PSS caused an increase in the particle net attraction of compound soil, with an essential concentration (10^{-2} mol L^{-1}) significantly impacting SIFs in compound soils. For different compound soil samples, when the PSS content was below 50%, the stability of aggregates was primarily influenced by SIFs between soil particles. Conversely, when the PSS content was 50% or higher, the stability of aggregates was affected by both the fine particle content and soil interaction forces. The PSS and sandy soil treatment with a ratio of 1:2 had a better effect on the particle interaction force and structural stability of compound soil. In summary, the aggregate stability of compound soils is influenced by both soil interaction forces and compound soil particle composition. The preliminary findings of this paper provide a quantitative description of the interaction forces in compound soils and provide a valuable scientific basis for PSS as a soil amendment to improve sandy soil quality and erosion protection resistance. Our research results will provide important theoretical support for soil and water conservation and ecological environment construction in Mu Us Sandy Land and a new path for soil structure improvement and erosion control in similar sandy land.

Author Contributions: Conceptualization, Z.L., L.Z. and Y.Z.; methodology, Z.L., J.H. and F.H.; software, Z.L. and L.Z.; writing—original draft preparation, Z.L., F.H. and L.Z.; writing—review and editing, Z.L., R.Z. and X.L.; funding acquisition, J.H., Y.Z. and Y.S. All authors have read and agreed to the published version of the manuscript.

Funding: This work was supported by the National Natural Science Foundation of China (42307429), the Technology Innovation Center for Land Engineering and Human Settlements, Shaanxi Land Engineering Construction Group Co., Ltd. and Xi'an Jiaotong University (2024WHZ0232), the Natural Science Basic Research Program of Shaanxi (2023-JC-QN-0343 and 2023-JC-QN-0360), and the Scientific Research Item of Shaanxi Provincial Land Engineering Construction Group (DJTD-2024-1 and DJTD-2022-5).

Institutional Review Board Statement: Not applicable.

Informed Consent Statement: Not applicable.

Data Availability Statement: The data presented in this study are available on request from the corresponding author.

Acknowledgments: The authors are grateful to the Technology Innovation Center for Land Engineering and Human Settlements by Shaanxi Land Engineering Construction Group Co., Ltd. and Xi'an Jiaotong University, School of Human Settlements and Civil Engineering, Xi'an Jiaotong University, and the Institute of Land Engineering and Technology, Shanxi Provincial Land Engineering Construction Group, Xi'an, China. Special thanks go to the anonymous reviewers for their constructive comments in improving this manuscript.

Conflicts of Interest: Authors Zhe Liu, Yang Zhang, Jichang Han, Yingying Sun, Ruiqing Zhang were employed by the company Shaanxi Provincial Land Engineering Construction Group Co., Ltd. The remaining authors declare that the research was conducted in the absence of any commercial or financial relationships that could be construed as a potential conflict of interest.

References

1. Liang, P.; Yang, X.P. Landscape spatial patterns in the Maowusu (Mu Us) Sandy Land, northern China and their impact factors. *Catena* **2016**, *145*, 321–333. [CrossRef]
2. Wang, Y.C.; Wu, Y.H.; Kou, Q.; Min, D.A.; Chang, Y.Z.; Zhang, R.J. Definition of arsenic rock zone borderline and its classification. *Sci. Soil Water Conserv.* **2007**, *5*, 14–18.
3. Bazhenov, M.L.; Chauvin, A.; Audibert, M.; Levashova, N.M. Permian and Triassic paleomagnetism of the southwestern Tien Shan: Timing and mode of tectonic rotations. *Earth Planet. Sci. Lett.* **1993**, *118*, 195–212. [CrossRef]
4. Liang, Z.S.; Wu, Z.R.; Yao, W.Y.; Noori, M.; Yang, C.Q.; Xiao, P.Q.; Leng, Y.B.; Deng, L. Pisha sandstone: Causes, processes and erosion options for its control and prospects. *Int. Soil Water Conserv. Res.* **2019**, *7*, 1–8. [CrossRef]
5. Guo, J.; Shi, Y.C.; Wu, L.J. Gravity erosion and lithology in Pisha sandstone in southern Inner Mongolia. *J. Groundw. Sci. Eng.* **2015**, *3*, 45–58. [CrossRef]
6. Zhang, H.O.; Guo, Z.; Li, J.; Lu, Y.J.; Xu, Y. Improvement of aeolian sandy soil in Mu Us, China with soft montmorillonite clay stone. *Agron. J.* **2020**, *113*, 820–828. [CrossRef]
7. Guo, Z.; Hui, W.; Li, J.; Yan, J.; Zhang, H.; Wang, H. Effects of Soft Rock on Soil Properties and Bacterial Community in Mu us Sandy land, China. *PeerJ* **2022**, *10*, e13561. [CrossRef]
8. Zhang, J.; Guo, Z. Response of Soil Structure and Crop Yield to Soft Rock in Mu Us Sandy Land, China. *Sci. Rep.* **2022**, *12*, 876. [CrossRef] [PubMed]
9. Wang, N.; Xie, J.C.; Han, J.C. A Sand Control and Development Model in Sandy Land Based on Mixed Experiments of Arsenic Sandstone and Sand: A Case Study in Mu Us Sandy Land in China. *Chin. Geogr.* **2013**, *23*, 700–707. [CrossRef]
10. Guo, Z.; Han, J.C.; Xu, Y.; Lu, Y.J.; Li, J. The mineralization characteristics of organic carbon and particle composition analysis in reconstructed soil with different proportions of soft rock and sand. *PeerJ* **2019**, *7*, e7707. [CrossRef]
11. Sun, Y.; Zhang, N.; Yan, J.; Zhang, S. Effects of Soft Rock and Biochar Applications on Millet (*Setaria italica* L.) Crop Performance in Sandy Soil. *Agronomy* **2020**, *10*, 669. [CrossRef]
12. Li, Y.; Cao, Z.; Long, H.; Liu, Y.; Li, W. Dynamic analysis of ecological environment combined with land cover and NDVI changes and implications for sustainable urban–rural development: The case of Mu Us Sandy Land. *J. Clean. Prod.* **2017**, *142*, 697–715. [CrossRef]
13. Sun, Z. Effect of Soft Rock Amendment on Soil Moisture and Water Storage in Mu Us Sandy Land. *IOP Conf. Ser. Mater. Sci. Eng.* **2018**, *381*, 012044. [CrossRef]
14. Chorom, M.; Regasamy, P.; Murray, R. Clay dispersion as influenced by pH and net particle charge of sodic soils. *Soil Res.* **1994**, *32*, 1243–1252. [CrossRef]
15. Xu, C.Y.; Yu, Z.H.; Li, H. The coupling effects of electric field and clay mineralogy on clay aggregate stability. *J. Soils Sediments* **2015**, *5*, 1159–1168. [CrossRef]
16. Liu, J.F.; Wang, Z.L.; Hu, F.N.; Xu, C.Y.; Ma, R.T.; Zhao, S.W. Soil organic matter and silt contents determine soil particle surface electrochemical properties across a long-term natural restoration grassland. *Catena* **2020**, *190*, 104526. [CrossRef]
17. Hu, F.N.; Xu, C.Y.; Ma, R.T.; Tu, K.; Zhang, F.B. Biochar application driven change in soil internal forces improves aggregate stability: Based on a two-year field study. *Geoderma* **2021**, *403*, 115276. [CrossRef]
18. Tiessen, H.; Moir, J.O. Total and Organic Carbon. In *Soil Sampling and Methods of Analysis*; Carter, M.R., Ed.; CRC Press: Boca Raton, FL, USA, 1993; Volume 38, pp. 187–199.
19. Lu, K. *Analytical Methods of Soil and Agricultural Chemistry*; Scientific Research Publishing: Wuhan, China, 1999.
20. Lierop, W.V.; Mackenzie, A.F. Carbonate Determination in Organic Soils. *Can J. Soil Sci.* **1974**, *54*, 457–462. [CrossRef]
21. Li, S.; Li, H.; Xu, C.Y.; Huang, X.R.; Xie, D.T.; Ni, J.P. Particle Interaction Forces Induce Soil Particle Transport during Rainfall. *Soil Sci. Soc. Am. J.* **2013**, *77*, 1563–1571. [CrossRef]
22. Li, H.; Hou, J.; Liu, X.M.; Hou, J.; Liu, X.M.; Li, R.; Zhu, H.L.; Wu, L.S. Combined determination of specific surface area and surface charge properties of charged particles from a single experiment. *Soil. Sci. Soc. Am. J.* **2011**, *75*, 21–28. [CrossRef]
23. Bissonnais, Y.L. Aggregate stability and assessment of soil crustability and erodibility: I. Theory and methodology. *Eur. J. Soil Sci.* **1996**, *47*, 425–437. [CrossRef]
24. Yu, Z.H.; Li, H.; Liu, X.M.; Xu, C.Y.; Xiong, H.L. Influence of soil electric field on water movement in soil. *Soil Tillage Res.* **2016**, *155*, 263–270. [CrossRef]
25. Zhang, Y.K.; Tian, R.; Yang, S.S.; Guo, X.M.; Li, H. Toward an Approach for Determining the Hamaker Constant of Soft Materials Using Dynamic Light Scattering. *Colloids Surf. A Physicochem. Eng. Asp.* **2021**, *630*, 127604. [CrossRef]
26. Hamaker, H.C. The London—Van der Waals attraction between spherical particles. *Physica* **1937**, *4*, 1058–1072. [CrossRef]
27. Tuller, M.; Or, D. Water films and scaling of soil characteristic curves at low water contents. *Water Resour. Res.* **2005**, *41*, 319–335. [CrossRef]
28. Yu, Z.H.; Zhang, J.B.; Zhang, C.Z.; Xin, X.L.; Li, H. The coupling effects of soil organic matter and particle interaction forces on soil aggregate stability. *Soil Tillage Res.* **2017**, *174*, 251–260. [CrossRef]
29. Liu, Z.; Zhang, Y.; Sun, Y.; Han, J.; Hu, F.; Li, J.; Li, X. Effects of the Changes of Particle Surface Electric Field and Interaction Force on the Reclaimed Soil Aggregate Structural Stability under the Application of Different Soil Conditioners. *Agronomy* **2023**, *13*, 1866. [CrossRef]

30. Yu, Z.H.; Zheng, Y.Y.; Zhang, J.B.; Zhang, C.Z.; Ma, D.H.; Chen, L.; Cai, T.Y. Importance of soil interparticle forces and organic matter for aggregate stability in a temperate soil and a subtropical soil. *Geoderma* **2020**, *362*, 114088. [CrossRef]
31. Gong, Y.; Tian, R.; Li, H. Coupling effects of surface charges, adsorbed counterions and particle-size distribution on soil water infiltration and transport. *Eur. J. Soil Sci.* **2018**, *69*, 1008–1017. [CrossRef]
32. Hu, F.N.; Liu, J.F.; Xu, C.Y.; Du, W.; Yang, Z.H.; Liu, X.M.; Liu, G.; Zhao, S.W. Soil internal forces contribute more than raindrop impact force to rainfall splash erosion. *Geoderma* **2018**, *330*, 91–98. [CrossRef]
33. Calero, J.; Ontiveros-Ortega, A.; Aranda, V.; Plaza, I. Humic acid adsorption and its role in colloidal-scale aggregation determined with the zeta potential, surface free energy and the extended-DLVO theory. *Eur. J. Soil Sci.* **2017**, *68*, 491–503. [CrossRef]
34. Guo, Z.; Zhang, H.O.; Wang, H.Y. Soft rock increases the colloid content and crop yield in sandy soil. *Agron. J.* **2020**, *113*, 677–684. [CrossRef]
35. Pihlap, E.; Steffens, M.; Kgel-Knabner, I. Initial soil aggregate formation and stabilisation in soils developed from calcareous loess. *Geoderma* **2021**, *385*, 114854. [CrossRef]
36. Li, W.Y.; Guo, Z.; Li, J.; Han, J.C. Effects of different proportions of soft rock additions on organic carbon pool and bacterial community structure of sandy soil. *Sci. Rep.* **2021**, *11*, 4624. [CrossRef] [PubMed]
37. Ingelmo, F.; Albiach, R.; Gamon, S. Organic matter of a soil amended with composted sludges and affected by simulated processes of soil degradation. *WIT Trans. Ecol. Environ.* **2003**, *64*, 10.
38. Zhen, Q.; Zheng, J.Y.; He, H.H.; Han, F.P.; Zhang, X.C. Effects of Pisha sandstone content on solute transport in a sandy soil. *Chemosphere* **2016**, *144*, 2214–2220. [CrossRef] [PubMed]
39. Iturri, L.A.; Buschiazzo, D.E. Cation exchange capacity and mineralogy of loess soils with different amounts of volcanic ashes. *Catena* **2014**, *121*, 81–87. [CrossRef]
40. Hepper, E.N.; Buschiazzo, D.E.; Hevia, G.G.; Urioste, A.; Antón, L. Clay mineralogy, cation exchange capacity and specific surface area of loess soils with different volcanic ash contents. *Geoderma* **2006**, *135*, 216–223. [CrossRef]
41. Yukselen-Aksoy, Y.; Kaya, A. Method dependency of relationships between specific surface area and soil physicochemical properties. *Appl Clay Sci.* **2010**, *50*, 182–190. [CrossRef]
42. Liu, J.F.; Hu, F.N.; Xu, C.Y.; Wang, Z.L.; Ma, R.T.; Zhao, S.W.; Liu, G. Comparison of different methods for assessing effects of soil interparticle forces on aggregate stability. *Geoderma* **2021**, *385*, 114834. [CrossRef]
43. Feng, B.; Li, Y.L.; Li, R.; Li, H. Error analysis in calculation and interpretation of AFM tip-surface interaction forces. *Adv. Colloid Interface Sci.* **2022**, *306*, 102710. [CrossRef] [PubMed]
44. Holthusen, D.; Reeb, D.; Horn, R. Influence of potassium fertilization, water and salt stress, and their interference on rheological soil parameters in planted containers. *Soil Tillage Res.* **2012**, *125*, 72–79. [CrossRef]

sustainability

MDPI

Article

Assessing Environmental Sustainability of Phytoremediation to Remove Copper from Contaminated Soils

Juan J. Espada [1,*], Rosalía Rodríguez [1], Andrea Delgado [1], Gemma Vicente [1,2] and Luis Fernando Bautista [2,3]

1 Department of Chemical, Energy and Mechanical Technology, ESCET, Universidad Rey Juan Carlos, 28933 Móstoles, Madrid, Spain; rosalia.rodriguez@urjc.es (R.R.); a.delgadon.2016@alumnos.urjc.es (A.D.); gemma.vicente@urjc.es (G.V.)
2 Instituto de Investigación de Tecnologías para la Sostenibilidad, Universidad Rey Juan Carlos, 28933 Móstoles, Madrid, Spain; fernando.bautista@urjc.es
3 Department of Chemical and Environmental Technology, ESCET, Universidad Rey Juan Carlos, 28933 Móstoles, Madrid, Spain
* Correspondence: juanjose.espada@urjc.es; Tel.: +34-91-488-80-87

Abstract: Phytoremediation stands out as a promising technology for removing heavy metals from contaminated soils. This work focuses on studying the environmental performance of phytoremediation in removing copper from contaminated soil located in an old Spanish mine using the life cycle assessment (LCA) method. For this purpose, *Brassica juncea* (brown mustard), *Medicago sativa* (alfalfa) and their rotary cultivation were assessed along with different options for managing biomass (landfill disposal and biomass cogeneration). In addition, soil excavation and soil washing treatments were also compared to phytoremediation. *M. sativa* proved superior to *B. juncea* and their rotary cultivation, regardless of the biomass disposal option, achieving impact reductions of 30–100%. This is due to the ability of *M. sativa* to fix nitrogen, which reduces fertiliser requirements. Among the biomass management alternatives, cogeneration was superior to landfill disposal in all cases by allowing for energy recovery, thereby reducing environmental impacts by 60–100%. *M. sativa* + cogeneration is the option that presents the best environmental performance of all the studied treatments, achieving reductions up to negligible values in four of eight impact categories due to the impacts avoided by energy production. On the contrary, soil excavation is the less desirable option, followed by soil washing treatment.

Keywords: phytoremediation; phytoextraction; life cycle assessment; soil remediation; copper removal

1. Introduction

The current concern on sustainability allows for governments to take actions in order to mitigate the effects of different impacts, including greenhouse gas emissions, soil contamination, air and water pollution, depletion of natural resources, etc., on the environment and human health [1,2]. In this sense, sustainability is being progressively included in the policy priorities of many countries through the Sustainable Development Goals (SDGs) of the 2030 Agenda adopted by all United Nations Member States [3]. Concerning SDGs 3, 7 and 15, the remediation of metal-contaminated soils contributes to improve the quality of life of the population and the environment within the framework of global sustainability.

Heavy metals, like cobalt, copper or chromium, are trace elements essential for plants as micronutrients. However, an excess of these elements causes harmful effects on the growth of plant species and on human health through food chains, as plants can accumulate metals from contaminated soils [4,5]. It is well known that mining activities cause considerable adverse environmental and human effects due to harmful metal contamination [6,7]. This industry generates a large amount of waste rocks, ore dust, etc., which often contain various trace metals that cause soil contamination [8,9]. In this context, soil contamination by heavy metals is a current worldwide concern because these metals are non-biodegradable. Consequently, governments have established remediation of contaminated soil as a national priority [10–13].

In this regard, most metal-contaminated areas have been treated by two approaches: soil excavated and landfilled as hazardous waste or a soil washing procedure. However, soil is a vulnerable and almost non-renewable natural resource, so in situ and eco-friendly procedures must be addressed [11,14]. In this sense, phytoremediation is an interesting alternative due to its low price and its soil-friendly nature [15]. It is a procedure that harnesses the natural abilities of plants to absorb and eliminate pollutants such as metals, which they can tolerate and accumulate in their leaves and roots [16]. Therefore, phytoremediation emerges as an environmentally friendly method for addressing soil contamination. Particularly, the phytoextraction involves hyper-accumulator plants that can translocate heavy metals to shoots. Thus, metals are eliminated with the harvesting of these plants from the contaminated area [15].

Brasicca juncea (brown mustard) and *Medicago sativa* (alfalfa) are plants widely studied in soil phytoremediation for metal removal. *B. juncea* is a herbaceous plant belonging to the family Brassicaceae. Many studies have demonstrated that *Brassica* species (including *B. juncea*) show a rapid growth rate, high above-ground biomass production and high heavy-metal sequestration ability from contaminated media [15,17–22]. For copper, the bioaccumulation factor (BAF) of the whole mustard plant is 1.81, being 1.01 for the leaves [23,24]. In the same way, *M. sativa* is a perennial herb from the family Fabaceae or Leguminosae. It is a widely cultivated forage crop with an extensive tap-root system. It can easily uptake copper, lead and cadmium from contaminated soils, making it a plant that has recently received attention for phytoremediation studies. The BAF for copper is 3.27 for the whole alfalfa plant and 0.86 for the leaves. Additionally, the alfalfa plants can be harvested several times to obtain a higher biomass yield and show a short growth cycle [25,26].

The rotatory crop consists of a production system in which two or more species are grown successively during part or all of the life cycle. The relevant aspect of this mechanism lies in the different use of resources by the crops that make it up [27]. Following the general principles of crop rotations reported from the literature (the species with high nitrogen demand must be preceded by the cultivation of legumes; the succession of crops must be carried out with species that are not similar or analogous; in the rotation system, species with deep roots should alternate with species with more superficial roots; etc.) [28], it is concluded that the rotation design between *B. juncea* and *M. sativa* constitutes an alternative of great interest. This rotation offers a balance between the main crop, consisting of alfalfa, and the cover crop or green fertiliser, made up of brown mustard. The practice of crop rotation has been widely studied as an environmentally friendly activity that contributes to the sustainable use of agricultural land because it has the potential to address ecological problems, such as degradation of soil structure and loss of soil organic carbon [29–31]. However, it does not have to be the best procedure for soil contaminated with metals.

In the present work, the study of the continuous monoculture of *B. juncea* and *M. sativa* as well as the study of rotational cultivation of both herbs as soil phytoremediation plants are studied from an environmental view with the life cycle assessment methodology (LCA) since the environmental sustainability of soil remediation processes is a key aspect. LCA is a systematic, standardised tool to determine the environmental feasibility of soil remediation technologies as it is widely demonstrated in the literature [32–36].

The environmental impacts of *B. juncea* and *M. sativa* monocultures and rotatory-culture scenarios for soil remediation were quantified, combined with different final uses of biomass: security landfill disposal and energy recovery (this option was simulated). On the other hand, the LCA of soil washing and excavation was carried out. The former was simulated using SuperPro Designer 9.5, whereas data on the excavation process were adapted from the available resources. The LCA results of the phytoremediation-based scenarios and traditional treatments were compared. The novelty of our work is to provide LCA results on soil remediation, which are usually scarce in the literature, using an engineering approach that can be of interest to be applied to other soil remediation schemes.

2. Materials and Methods

2.1. Soil Description

The soil selected for this study is located in an old mine in Southern Spain (38°06′00″ N 3°40′03″ O), mainly dedicated to mineral extraction. As a result of this activity over time, this soil presents significant contamination by heavy metals, as reported in [37]. Among the different metals, our study is focused on copper as it can be removed by a variety of treatments and its presence in the soil can affect microbial activity as well as agriculture [38]. Therefore, the environmental assessment of treatment for its remediation is of interest. The main characteristics of the soil are taken from [37] who carried out a deep experimental analysis of the soil by collecting a total of 44 samples in an area of 2.2 km^2. Table 1 summarises the values for the soil features considered in our study, calculated as the average of the total samples.

The average Cu content shown in Table 1 exceeds the maximum level of copper in soils established by the local government regulation (595 ppm) [39]. Consequently, the amount of copper that should be removed is 77.23 ppm. Considering the particle size distribution shown in Table 1, the soil falls within clayey sand soil [40]. Finally, the average annual temperature of the soil location is 17.1 °C, the humidity is 59.3%, and the annual average precipitation value is (4410.3 m^3/hm^2) [41].

Table 1. Main characteristics of the studied soil.

Parameter	Value
pH	6.8
Density [a] (kg/m^3)	1600
OM [b] (%)	2.35
Cu content (ppm)	627.23
Sand (%)	72.5
Silt (%)	5.9
Clay (%)	21.6
Soil type [c]	Clayey sand (CS)

[a] Calculated from particle size distribution; [b] Organic matter; [c] According to Unified Soil Classification [40].

2.2. Goal and Scope of the LCA Study

The goal of this work is to analyse the use of phytoextraction with *B. juncea* and *M. sativa* to remove copper from the soil described above. For that purpose, three case studies were analysed: phytoremediation, soil washing and soil excavation. In the phytoremediation scenario, *B. juncea, M. sativa* and rotary cultivation of both species were assessed, and two end-of-life biomass scenarios were evaluated for each: security landfill disposal and energy recovery by cogeneration. In addition, copper removal using washing soil and excavation scenarios were also analysed. Table 2 shows the scenarios and case studies assessed in the present work.

Figure 1 shows a general diagram of the different case studies and the steps involved. The system boundaries encompass all inputs and outputs of the processes, whereas capital goods are excluded from the study.

The labour machinery used for cultivation operations (ploughing, harrowing, sowing and harvesting) is considered in the phytoremediation case and was calculated using the spreadsheets reported by the Spanish Government for agricultural labour [42], considering the specific characteristics of the machinery.

Table 2. Case studies considered in this work.

Case Study	Treatment			Scenario
1	Phytoremediation	*B. juncea* (BJ)	Landfill disposal Cogeneration	BJ-L BJ-C
		M. sativa (MS)	Landfill disposal Cogeneration	MS-L MS-C
		Rotatory crop (BJ + MS)	Landfill disposal Cogeneration	RC-L RC-C
2	Soil Washing			
3	Soil Excavation			

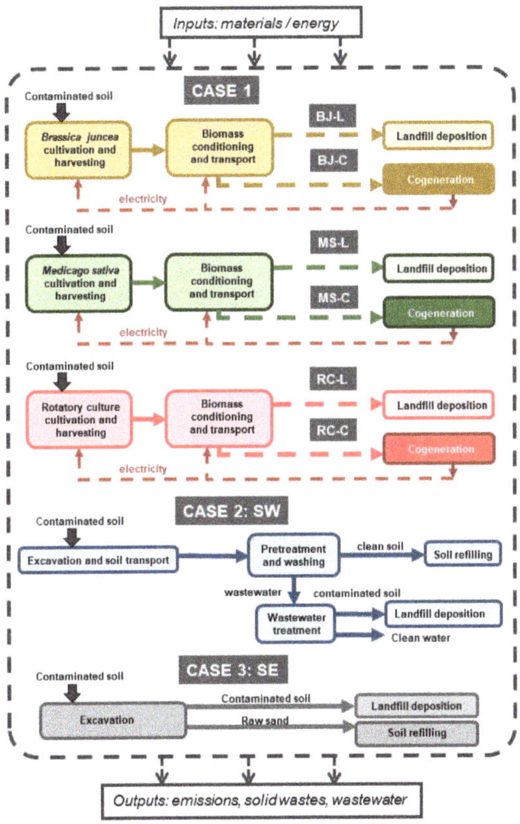

Figure 1. Battery limits for the considered case studies. Case 1: phytoremediation (BJ: *B. juncea*; MS: *M. sativa*; RC: rotary crop; L: landfill; C: cogeneration); Case 2: soil washing; and Case 3: soil excavation.

The functional unit selected for this study is the decontamination of 1 hm^2 of the considered soil (20 cm depth) up to the maximum Cu concentration allowed by government regulations. The amount of soil equivalent to the functional unit was calculated from its density (1600 kg/m^3 for this type of soil [43], see Table 1), resulting in a value of 3200 ton.

2.3. Life Cycle Inventory Analysis (LCIA)

The life cycle inventory data of all treatments were adapted from the literature or calculated by simulation using SuperPro Designer 9.5 (Intelligen Inc., Scotch Plains, NJ,

USA). Material and energy production as well as transportation vehicles were adapted from the Ecoinvent database 8.3 and Gabi Professional 2023 database (Sphera Solutions GmbH., Leinfelden-Echterdingen, Germany).

2.3.1. Case 1: Phytoremediation

The main steps involved in the phytoextraction of Cu-contaminated soils using *B. juncea*, *M. sativa* and their combination (rotary culture) are cultivation and harvesting, conditioning and transport and use of the grown biomass, as depicted in Figure 1. In all cases, energy recovery by cogeneration in a CPH system and security landfill disposal were evaluated for biomass end life. According to the flowsheet scheme reported elsewhere, the cogeneration process was simulated using SuperPro Designer 9.5 [32]. On the other hand, landfill disposal was modelled as a security deposit adapted from the LCA databases mentioned above. Table S1 (Supplementary Materials) summarises the detailed inventory of phytoremediation scenarios. The steps involved in both cases are explained below.

i. Cultivation and harvesting: The cultivation of *B. juncea* (BJ), *M. sativa* (MS) and the rotatory culture (RC) was evaluated in the different scenarios of Case 1, and the inventory data are summarised in Table S1.

-BJ: Two annual crops (including leaf cutting per culture) followed by the harvesting of the whole plant are estimated, leading to a total biomass production of 11.2 tons/hm^2·year. Considering the average value of annual rainfall in the studied area, the need for irrigation of B. juncea can be considered negligible.

-MS: This is a perennial crop belonging to the legume family and commonly used as a forage plant; its crop is considered to be annual. During the year, eight leaf cuts are considered (the first cut after 60–65 days and the others after 35–45 days each one), providing a biomass production of 19.4 tons/hm^2·year. This plant needs a total irrigation of 6000 m^3/hm^2 [44], but only 4410.3 m^3/hm^2 can be covered with the average value of annual rainfall, and consequently, an irrigation water volume of 1589.7 m^3/hm^2 should be considered.

-RC: As previously described, the rotatory crop design between BJ and MS forms an alternative of great interest. Thus, a temporal distribution of two periods of three months (from March to May and from September to November) was considered for the mustard crops. The other two three-month periods were established for alfalfa crops. In these conditions, the biomass production is 24.3 ton/(hm^2year).

As described above, the amount of copper in the contaminated soil is 672.23 mg_{Cu}/kg_{soil}, and the maximum allowed level of copper in the soil established by the local government regulation is 595 mg_{Cu}/kg_{soil} [39]. Thus, the amount of copper that should be removed is 77.23 mg_{Cu}/kg_{soil}. By considering the BAF values of *B. juncea* and *M. sativa*, the duration of the phytoremediation process using these plants was calculated, and the results are presented in Table 2. In addition, CO_2 fixation for each species was obtained from the literature [45,46] and is also shown in Table 3.

Table 3. Cultivation parameters of the different scenarios.

Parameter	*B. juncea*	*M. sativa*	Rotatory Crop
BAF_{plant}	1.81	3.27	3.27
BAF_{leaves}	1.01	0.86	0.86
Annual crops	2	1	4
Plant biomass [kg/(hm^2·year)]	7472	5249	17970
Leaf biomass [kg/(hm^2·year)]	3736	14151	6334
Total biomass [(ton/(hm^2·year)]	11.20	19.40	24.30
Accumulated Cu [kg_{Cu}/(hm^2·year)]	11.64	19.72	36.21
CO_2 fixation [kg/hm^2·year)]	2781.20	5420.00	4892.24
Years	22	13	7

In order to improve the soil conditions, particularly the porosity, a ploughing step is required. All inventory data for the labour machinery process were adapted from the Gabi Professional 2023 and the Ecoinvent 8.3 databases. As described above, mustard and alfalfa cultivations are enhanced under Mediterranean conditions. In these conditions, a seed rate of 10 kg/hm^2 and 40 kg/hm^2 is assumed for *B. juncea* and *M. sativa*, respectively, according to the literature [44,47].

Table S1 (Supplementary Materials) shows the total amount of seeds. Regarding land fertilisation requirements, the recommendation for soil fertilisation is taken from the literature [27,44,48,49]:

- *B. juncea*: compost (1507 kg/hm^2), EDTA (2806 kg/hm^2), triple superphosphate (150 kg P_2O_5/hm^2), potassium chloride (100 kg K_2O/hm^2) and urea (200 kg/hm^2).
- *M. sativa*: manure (650 kg/hm^2), borax (0.4 $kg/ton_{alfalfa}$), triple superphosphate (7.7 kg $P_2O_5/ton_{alfalfa}$), potassium chloride (7.8 kg $K_2O/ton_{alfalfa}$) and ammonium nitrate (2.7 kg $N/ton_{alfalfa}$).

The calculated values of the fertilisers for each scenario are shown in Table S1.

ii. Conditioning and transport: Biomass must be conditioned to eliminate the soil attached to the biomass. The soil amount was estimated according to the literature [50–52]. For this purpose, a vibration screening is used, and its energy requirement is calculated by simulation using SuperPro Designer 9.5 according to the flowsheet reported elsewhere [32]. The conditioned biomass is then transported to a landfill located 40 km away from the contaminated area. In the case of energy recovery, a cogeneration plant located 79 km from the contaminated area is utilised. Before transportation, the biomass is dried in open air until it reaches a moisture content of up to 20% moisture and then hayed.

iii. Biomass use: As reported by different authors, the environmental sustainability of phytoextraction-based technologies greatly depends on using cultivated biomass. In this sense, two possibilities are evaluated.

   (a) BJ-L, MS-L, RC-L: In these scenarios (see Figure 1), the growth biomass was disposed of as hazardous material in the underground landfill mentioned above; the inventory data are adapted from the Gabi Professional 2023 and the Ecoinvent 8.3 databases.

   (b) BJ-C, MS-C, RC-C: In the cogeneration scenarios (see Figure 1), the harvested biomass was valorised through a combined power and heating system (CPH), simulated using SuperPro Designer 9.5, as reported elsewhere [32]. As explained above, the biomass is transported from the contaminated area to an existing cogeneration plant in which the biomass is stoichiometrically burnt with air, obtaining steam, off-gas and ashes (mainly composed of Cu). This approach is the same as that used in the literature [32]. The molecular formula of *B. juncea* and *M. sativa* was calculated from the elementary analysis obtained from the literature [53,54]. The steam resulting from the process is thereafter expanded in a multistage turbine, yielding electricity and heat with an efficiency of 35% and 50%, respectively [32]. Finally, the ash obtained from the process is transported and disposed of in a landfill for hazardous wastes.

2.3.2. Case 2: Soil Washing

This process was simulated by using SuperPro Designer 9.5 following the scheme reported by Espada et al., 2022 [32], considering a soil treatment capacity of 10 m^3/h. Table S2 (Supplementary Materials) summarises the life cycle inventory data for this case. A scheme of the process, including all the steps involved, is depicted in Figure 1.

i. Excavation and soil transport: The removal of soil contaminated by copper requires the excavation of the soil at a depth of 0.2 m, in this case using an excavator and a skid steer. The soil surface treated was 10,000 m^2, resulting in a soil volume of 2000 m^3.

The distance from the contaminated area to the treatment plant was assumed to be the same as in the landfill scenarios.

ii. Pretreatment and washing: The soil is initially separated using a vibratory wet separator based on its particle size [32], as shown in Table 4.

Table 4. Soil fractions to be treated in Case 2.

Particle	Size (mm)	Percent (%)	Soil Volume (m^3)	Soil to Treat (kg/h)
Gravel	>4	0	0	0
Sand	1.5–4	34	679.8	5.440
	0.075–1.5	38.5	770.2	6160
Silt/Clay	<0.075	27.5	550	4400

The stream consisting of particles sized between 1.5 and 4 mm is transferred to a centrifugation step, where water removal occurs before proceeding to the subsequent washing step. On the other hand, the stream containing particles whose diameter is less than 1.5 mm passes to a hydrocyclone to achieve a more efficient separation. After that, the product stream (particle sizes between 0.075 and 1.5 mm) is conducted to a centrifugation step, while the stream whose diameters do not exceed 0.075 mm is derived to the wastewater treatment unit.

In the washing step, the streams from the aforementioned centrifugation step are mixed and treated with 1M hydrochloric acid (1M of 20% purity) using a soil:acid ratio of 1:3 (by weight) as reported in the literature [55]. The treated soil stream is centrifuged, and the resulting contaminated water stream is directed to the wastewater treatment unit. The treated soil is then used as refilling material.

iii. Wastewater treatment: The stream coming from the washing step is neutralised using sodium hydroxide. This stream and the one coming from the hydrocyclone are mixed, thickened, clarified and centrifuged to remove solid particles. The solid fraction is disposed of in a hazardous landfill (assuming the same distance as in the case of biomass phytoremediation disposal), and the clean water is recycled to the process.

2.3.3. Case 3: Excavation

The inventory data of the different steps for this case are summarised in Table S3 (Supplementary Materials). Figure 1 depicts the steps involved in this treatment.

i. Excavation: The contaminated soil is excavated using an excavator and a mini skid steer. The inventory data of this equipment were adapted from the available process from the Gabi Professional 2023 and Ecoinvent 8.3 databases and adapted to the functional unit.

ii. Landfill deposition: The excavated soil is transported to the landfill by road using a truck. The inventory data of this step were adapted from the databases mentioned above.

iii. Soil refilling: Sand from a quarry located near the contaminated site (28 km) was used to refill the excavated area. This step involves sand charge and its transport using a mini skid steer and a truck.

2.4. *Environmental Impact Assessment*

In the present work, the environmental impacts were quantified using CML 2001 and EFP methodologies, applying the mid-point approach. The impact categories evaluated are the ones that are usually considered in LCA applied to soil bioremediation [32,56]. In this sense, greenhouse gas emissions (global warming potential category, GWP), acidification and eutrophication potentials (AP and EP, respectively) as well as the toxicity-related impacts of ecotoxicity and human toxicity potentials (ETP and HTP, respectively) were quantified by using the CML 2001–August 2016 methodology. On the other hand, the water use (WU) category was calculated by using the EFP 3.1 methodology. Finally, the direct and indirect primary energy use throughout the life cycle [57] was calculated by the cumulative energy demand (CED).

3. Results and Discussion

The LCA results for all the previously described cases are summarised in Table 5. These results represent the overall values of the different impact categories assessed in this work. In this section, the LCA results for each scenario are discussed by analysing the contribution of the different steps to identify the hot spots in each case. In addition, the scenarios are compared to determine the most environmentally favourable options.

Table 5. Overall impacts for the case studies referred to the FU (1 hm^2 of decontaminated soil).

Impact Category	Case 1						Case 2	Case 3
	BJ-L	BJ-C	MS-L	MS-C	RC-L	RC-C		
CED (MJ)	$1.3 \cdot 10^7$	$7.5 \cdot 10^6$	$1.4 \cdot 10^6$	-	$7.8 \cdot 10^6$	0	$2.3 \cdot 10^7$	$1.5 \cdot 10^7$
AD (kg Sb-eq)	7.6	7.4	$6.2 \cdot 10^{-1}$	$4.0 \cdot 10^{-1}$	2.8	2.7	8.6	$5.9 \cdot 10^{-1}$
AP (kg SO_2-eq)	$2.1 \cdot 10^3$	$1.7 \cdot 10^3$	$2.8 \cdot 10^2$	-	$8.3 \cdot 10^2$	$5.3 \cdot 10^2$	$2.3 \cdot 10^3$	$2.6 \cdot 10^3$
EP (kg PO_4-eq)	$2.9 \cdot 10^3$	$2.7 \cdot 10^3$	$1.5 \cdot 10^2$	10.9	$1.0 \cdot 10^3$	$9.0 \cdot 10^2$	$8.2 \cdot 10^2$	$1.4 \cdot 10^3$
GWP (kg CO_2-eq)	$5.9 \cdot 10^5$	$3.6 \cdot 10^5$	$6.1 \cdot 10^4$	-	$2.2 \cdot 10^5$	$4.4 \cdot 10^4$	$9.3 \cdot 10^5$	$6.8 \cdot 10^5$
HTP (kg 1,4 DCB-eq) *	$4.5 \cdot 10^5$	$4.1 \cdot 10^5$	$4.4 \cdot 10^4$	-	$1.8 \cdot 10^5$	$1.5 \cdot 10^5$	$1.1 \cdot 10^5$	$2.1 \cdot 10^5$
ETP (kg 1,4 DCB-eq) *	$4.1 \cdot 10^3$	$3.8 \cdot 10^3$	$5.0 \cdot 10^2$	$1.9 \cdot 10^2$	$1.7 \cdot 10^3$	$1.5 \cdot 10^3$	$1.7 \cdot 10^3$	$2.3 \cdot 10^3$
WU (m^3)	$4.3 \cdot 10^6$	$2.3 \cdot 10^5$	$5.9 \cdot 10^3$	$9.0 \cdot 10^5$	$2.0 \cdot 10^4$	$8.8 \cdot 10^5$	$2.2 \cdot 10^7$	$5.1 \cdot 10^7$

* DCB: Dicholorobenzene.

3.1. Phytoremediation

Figure 2 depicts the contribution of each step to the studied impact categories. By analysing the phytoremediation + landfill disposal, it becomes evident that the cultivation and harvesting steps are the largest contributor to most impact categories for *B. juncea* and rotary cultivation (>80%) due to their demanding fertilisation requirements. In the former case, *B. juncea* requires not only NPK fertilisers but also additives such as EDTA in large quantities (5600 kg/hm^2), which generates remarkable impacts due to the substantial energy usage and the emission of hazardous substances during its manufacture. On the contrary, *M. sativa* cultivation contributes less than biomass disposal in most categories (by 60%, except in the AD category) because it can fix nitrogen, thus reducing N-based fertiliser consumption (urea), and this crop does not require EDTA. Regarding the phytoremediation + cogeneration steps, Figure 2 shows a similar trend to the landfill disposal scenarios, except in the case of *M. sativa* where cogeneration has less contribution compared to biomass landfill disposal. In all cases, the biomass conditioning and transport step exhibits the lowest importance in all categories, regardless of the final use of the biomass.

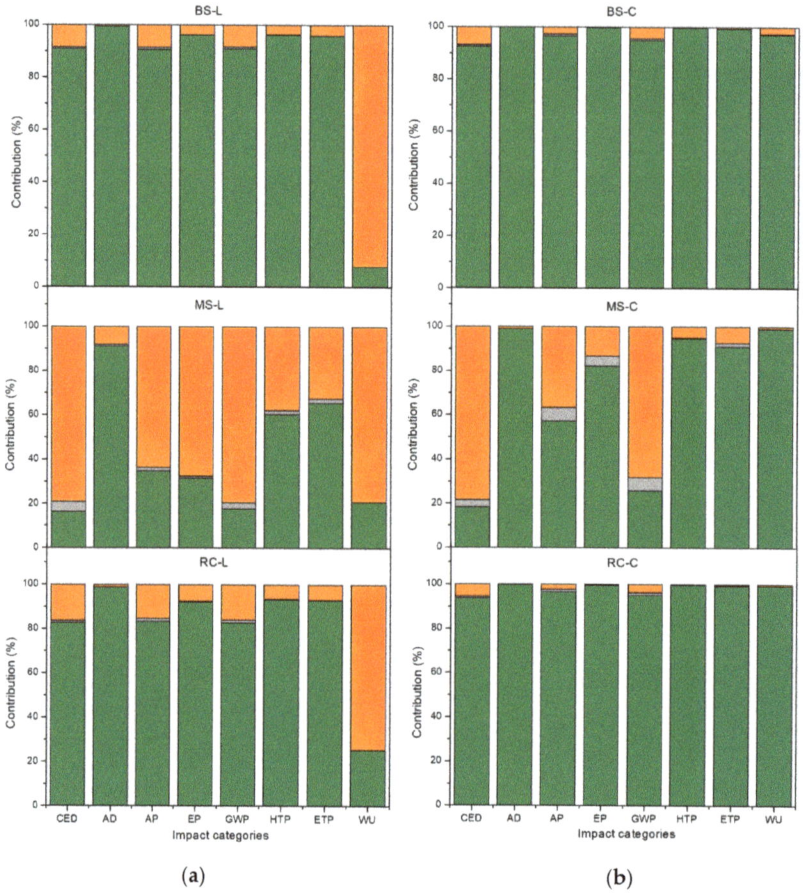

Figure 2. Contribution analysis results for phytoremediation-based scenarios combined with (**a**) landfill deposition and (**b**) cogeneration. Green bar: cultivation and harvesting; grey bar: conditioning and transport; orange bar: landfill disposal.

Figure 3 depicts the normalised comparison between all phytoremediation-based scenarios. As can be observed, *B. juncea* is the least desirable option as it presents the highest impact value for each category, regardless of the final use of the biomass. As explained above, the long treatment duration (22 years) as well as the high fertiliser requirements are the main factors affecting the environmental sustainability of this crop for the studied application. By comparing the use of *M. sativa* and rotary cultivation combined with landfill disposal, the former is environmentally better, achieving reductions by 90%, except in WU. It must be noticed that the use of *M. sativa* means double the duration of the treatment compared to rotary cultivation (13 vs. 7 years), but this fact is balanced by the lower impact produced by its cultivation, as explained above. By analysing cogeneration as the final use of the biomass, a similar trend is observed as in the case of landfill disposal, being *M. sativa* and rotary cultivation are the best options, except in the WU category. Their comparative analysis indicates that the former option is clearly superior to the latter option, achieving differences of 30–40% in most categories. Therefore, the obtained results indicate that *M. sativa* is the best option to remove copper from the studied soil regardless of the use of the biomass.

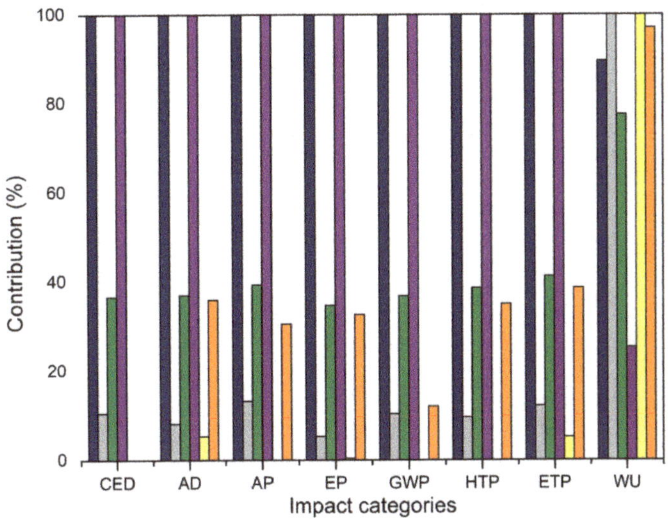

Figure 3. Normalised comparison between phytoremediation scenarios. Blue bar: BS-l; grey bar: MS-L; green bar: RC-L; purple bar: BS-C; yellow bar: MS-C; orange bar: RC-C.

Figure 4 shows the normalised comparison of *M. sativa* combined with landfill deposition or cogeneration. As can be observed, cogeneration is environmentally superior to landfill disposal in all impact categories, which agrees with the results reported by different authors. In this sense, other authors studied the environmental performance of phytoremediation-based treatments to remove Pb from contaminated soil [35]. These authors highlighted the great importance of the biomass use concerning the overall sustainability of the treatment, concluding that phytoremediation without biomass valorisation is not a feasible alternative from an environmental point of view. Furthermore, these authors stated the superiority of using energy recovery as biomass valorisation compared to landfill disposal due to the energy recovery because of the avoided impacts related to heat production. The same conclusions were made by other authors [58], who assessed different scenarios for industrial phytoremediation of a soil contaminated with heavy metals using *C. sativa*. These authors concluded that biomass energy recovery is the most feasible option when using phytoremediation-based treatments. These results were also observed elsewhere [32] for Pb removal from soils using *F. arundinacea*, achieving reductions of 20–80% in all the selected categories.

By considering the impact of reductions achieved by cogeneration over landfill disposal, it can be observed that they ranged between 40 and 100%, which indicates that electricity and heat production is clearly beneficial from an environmental point of view, not only related to energy use (CED, GWP or AP) but also to toxicity-related ones (HTP). The scarcity of impact category results in the literature makes comparing the results shown in this work difficult. Suer et al. (2011) reported impact category results showing the remarkable reductions obtained when combining phytoremediation + energy recovery using *Salix viminalis* to remove BTEX from a soil [59]. Other authors calculated CED for different scenarios, including energy recovery by incineration and anaerobic digestion, ranging from 20 to 35 GJ/hm^2 [58]. These values are lower than those obtained in our work when using landfill disposal but higher than those obtained by the best option using *M. sativa* (Table 2, scenario MS-C). Regarding carbon footprint (GWP) results for arsenic (As), lead (Pb) and Thallium (Tl) removal from a soil using *L. albus*, *B. juncea* and *H. annuus*, some authors reported a value of ~$3 \cdot 10^5$ kg CO_2-eq./hm^2 for biomass landfill disposal, similar to the values for our scenarios including this biomass end use [33]. Concerning scenarios including energy recovery, these authors reported a value of ~$25 \cdot 10^5$ kg CO_2-eq./hm^2, similar to those obtained in our work for the scenarios including cogeneration (Table 4, BJ-C and RC-C) but higher than our best scenario using energy recovery (Table 4, MS-C). Other authors studied the same categories as those analysed in this work, showing that cogeneration was superior to landfill disposal for all impact categories [32]. These comparisons can be used qualitatively to have an idea of the sustainability of the phytoremediation treatments assessed in this work. However, they must be considered with caution, since the conditions of the study (soil characteristics, removed contaminant, etc.) are particular in each case. Despite this and considering these limitations, the sustainability of our best option *M. sativa* + cogeneration (MS-C) is a promising alternative to remove copper from the studied soil.

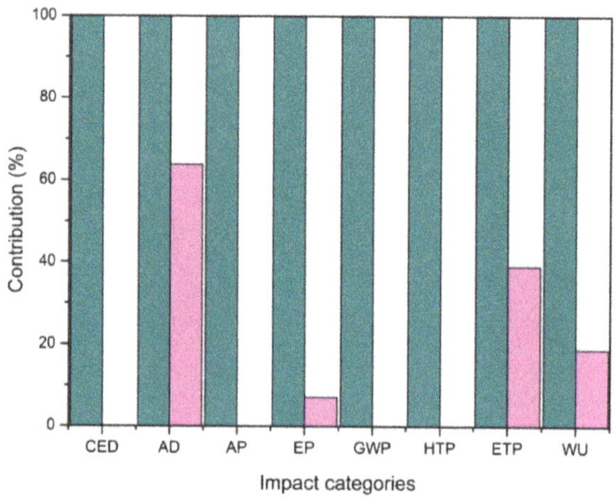

Figure 4. Comparative analysis between the phytoremediation treatments using *M. sativa*. Green bar: MS-L; pink bar: MS-C.

3.2. Soil Washing

Figure 5 shows the relative contribution to the impact categories of the different steps involved in the washing treatment (Case 2). As can be observed, soil landfill disposal is the largest contributor in most categories, except for CED, AD and GWP. This fact is due to the large amount of soil to be disposed of in the security landfill as hazardous material (by 40% of the treated soil). The pretreatment and washing steps are the second contributor in most categories, being the first one for AD (~58%). This contribution is mostly due to using a large amount of hydrochloric acid to solubilise the copper from the contaminated soil. Finally, wastewater treatment contributes 39% to the AD category due to the impacts of using a large amount of sodium hydroxide, whose manufacture and use produces high impacts concerning energy/materials consumption and emissions of hazardous substances. Other authors assessed the carbon footprint of soil washing treatment to remove Pb from contaminated soil, reporting that the most critical contributing steps to the treatment were the transport of soil and wastewater treatment [45]. In addition, the use of HCl in the washing process contributes less than in our case. This is because Pb removal requires a smaller amount of acid than in our case. However, the results reported by these authors are not directly comparable to those obtained in this work since these authors include a previous pretreatment of the soil, which is not included in our work. Nevertheless, the overall carbon footprint value obtained by these authors (~$9 \cdot 10^5$ $kgCO_2$-eq./hm^2) is similar to that obtained in our case (Table 4, Case 2: $9.3 \cdot 10^5$ kg CO_2-eq./hm^2), despite the different characteristics of the treatment, the soil and the removed metal. Finally, our results agree with the ones reported elsewhere [32] in the sense that the soil to be disposed of and the use of chemicals (HCl and NaOH) contribute largely to most impact categories.

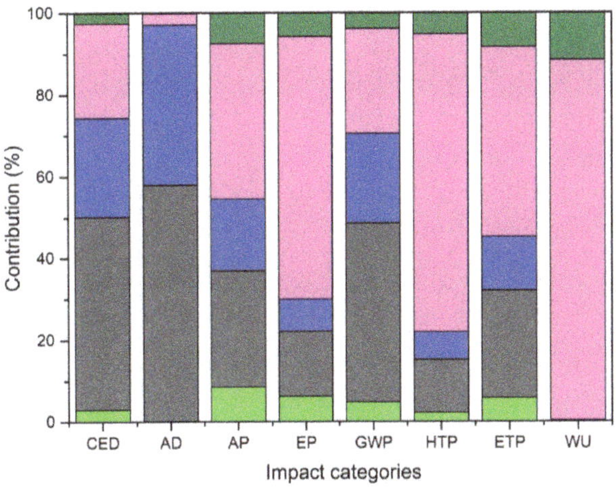

Figure 5. Contribution analysis results for washing treatment (Case 2). Dark green: refilling; pink bar: landfill; blue bar: wastewater; grey bar: pretreatment + washing; green: excavation + transport.

3.3. Soil Excavation

The analysis of the relative contributions to the impact categories associated with the excavation treatment (Case 3) is represented in Figure 6. As inferred, landfill disposal is the most significant contributor (80–95%) to all the impact categories. In this sense, the large amount of contaminated soil (3200 tons) implies a higher consumption of materials and energy for manufacturing containers to store the soil in the secure deposit, resulting in significant contributions of this step across all impact categories. This result agrees well with the work reported elsewhere [59], which analysed phytoremediation (using *Salix viminalis*) + biomass disposal (in a sanitary landfill and in a security deposit), obtaining that the landfill/deposit was the most important contributor (>80%) evaluated for GWP, AP, EP and CED, regardless of the type of landfill. This result is of interest as similar findings are reported by other authors for the treatment of Pb-contaminated soils through excavation [32].

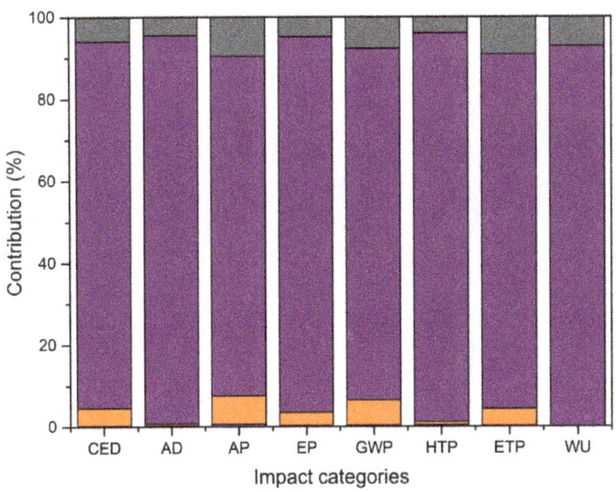

Figure 6. Contribution analysis results for excavation treatment (Case 3). Grey bar: refilling; purple bar: landfill; orange bar: transport; blue bar: excavation.

3.4. Comparative Analysis

Finally, the best phytoremediation-based scenarios (MS-L and MS-C) were compared to the rest of the treatments studied in this work: soil washing (Case 2) and soil excavation (Case 3). Figure 7 depicts the normalised comparison between these treatments from which it can be inferred that the phytoremediation-based scenarios using *M. sativa* (MS-C and MS-L) are clearly superior to the rest from an environmental point of view, achieving reductions between 80 and 100% in the studied categories. Comparing the MS-L and MS-C scenarios, the latter leads to negligible impact values in some categories because energy recovery avoids impacts related to heat and electricity production from biomass. In this sense, the dramatic reduction in energy consumption (CED) and carbon footprint (GWP) must be highlighted. These findings are consistent with other studies examining metal removal from soils. In this sense, other authors [59] reported the comparison of soil excavation and phytoremediation + energy recovery (biofuel), obtaining that the latter was clearly superior for several impact categories (energy consumption, GWP, AP and EP), obtaining reductions of 99% in all cases. Concerning the carbon footprint (GWP) of phytoremediation, soil washing and soil excavation, it is reported in the literature that the former was the most suitable option, achieving reductions >99% and >80% with respect to soil excavation and soil washing treatment [45]. Finally, similar results for most of the impact categories calculated in this work are reported elsewhere [32], revealing phytoremediation (using *F. arundinacea* + biomass cogeneration) as the most suitable option to remove Pb from soil. In this case, it must be noticed that the achieved reductions were lower for most categories (by 75–90%), mainly due to the longer duration of the process (31 vs. 13 years), which implies more demanding requirements that cannot be fulfilled entirely by the energy recovered from biomass.

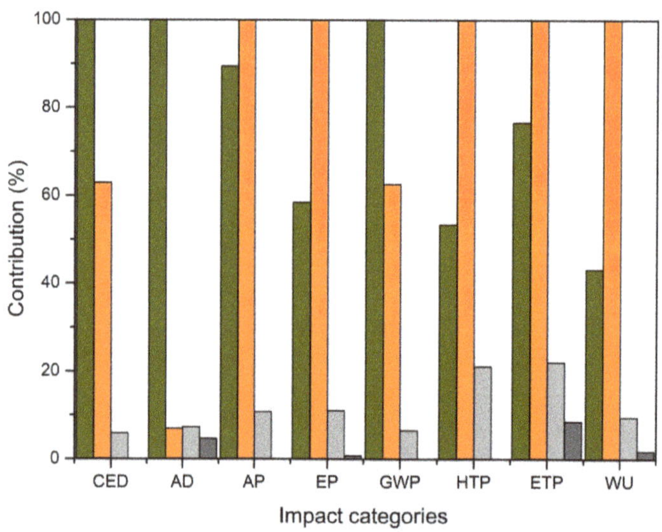

Figure 7. Comparative analysis between the best phytoremediation treatments and the other soil remediation processes (Cases 2 and 3). Dark grey bar: MS-C; light grey bar: MS-L; olive bar: Case 2; orange bar: Case 3.

As inferred from the above discussion, our results agree well with those reported in the literature, despite the differences regarding soil type, plant species and removed metal. These results, although qualitative, indicate the suitability of phytoremediation using *M. sativa* combined with biomass cogeneration to remove copper. In this sense, the results of our work can be a starting point for considering this alternative for further research in the field.

4. Conclusions

Research on environmental sustainability concerning metal removal from soils is ongoing. Nevertheless, systematic studies on this issue are scarce and often applied to a limited number of environmental impacts. In this work, the life cycle assessment (LCA) was applied to environmentally compare phytoremediation-based treatments with other alternatives to remove copper from contaminated soil. For this purpose, different species (*B. juncea*, *M. sativa* and their rotatory crop) were assessed, and they were combined with landfill disposal or cogeneration for biomass use. Experimental and simulation data were used to build the life cycle inventory from which the LCA results were calculated for several impact categories.

The main findings reported in this work are the following:

- *M. sativa* is superior to *B. juncea* and their rotary cultivation as the phytoremediation species from an environmental perspective due to its lower nitrogen fertiliser requirements. This fact balances the longer duration of phytoextraction using *M. sativa* for rotary cultivation, reducing the environmental impact by 30–100% in the analysed categories.
- The contribution of biomass use to the overall impact is more significant in the case of using landfill disposal compared to cogeneration, primarily due to the large amount of biomass to be disposed of as well as the possibility of recovering energy from cogeneration, which reduces environmental impact values.
- The combination of *M. sativa* + cogeneration is clearly superior to the other phytoremediation-based treatments analysed in this work due to the plant's lower fertiliser requirements, its ability to remove copper and the possibility of recovering energy to fulfil most of the process requirements.
- The soil washing treatment involves significant contributions from soil landfill disposal and the use of chemicals (HCl and NaOH required in the process). On the other hand, soil excavation treatment presents soil disposal as the primary contributor to all the studied impacts.
- The combination of *M. sativa* + cogeneration is the option that presents the best environmental performance, achieving reductions up to negligible values in four of eight impact categories (CED, AP, GWP and HTP) due to the impacts avoided by energy production. On the contrary, soil excavation is the least desirable option, followed by the soil washing treatment to remediate the soil studied in this work.

The main limitation of this work is the lack of actual field data on applications of *M. sativa* for copper removal. However, the results reported in this work can be a first step to advancing the research on soil phytoremediation. In this sense, inventory data and LCA results can be used as a benchmark for further studies in this field since they provide the analysis of a wide number of impacts, which is usually scarce in the literature. The main engineering application of our work is to provide LCA results that can help to identify environmental hotspots and to extrapolate our approach to other phytoremediation applications.

Supplementary Materials: The following supporting information can be downloaded at: https://www.mdpi.com/article/10.3390/su16062441/s1, Table S1: Inventory data of Case 1 referred to the FU (1 hm^2 of treated soil). Table S2: Inventory data of Case 2 referred to the FU (1 hm^2 of decontaminated soil). Table S3: Inventory data of Case 3 referred to the FU (1 hm^2 of decontaminated soil).

Author Contributions: Conceptualisation, methodology and writing—original draft preparation: J.J.E. and R.R. Writing—review and editing: L.F.B. and G.V. Methodology and investigation: A.D. All authors have read and agreed to the published version of the manuscript.

Funding: The software used in this work to carry out LCA was acquired through the project ALGA2BIOPROJET (PID2020-114943RB-I00) financed by Ministerio de Ciencia e Innovación (Spanish Government).

Data Availability Statement: Data is contained within the article and Supplementary Materials.

Conflicts of Interest: The authors declare no conflicts of interest.

References

1. Abad-Segura, E.; González-Zamar, M.D. Sustainable Economic Development in Higher Education Institutions: A Global Analysis within the SDGs Framework. *J. Clean. Prod.* **2021**, *294*, 126133. [CrossRef]
2. Buerkle, A.; O'Dell, A.; Matharu, H.; Buerkle, L.; Ferreira, P. Recommendations to Align Higher Education Teaching with the UN Sustainability Goals—A Scoping Survey. *Int. J. Educ. Res. Open* **2023**, *5*, 100280. [CrossRef]
3. Leal Filho, W.; Shiel, C.; Paço, A.; Mifsud, M.; Veiga Ávila, L.; Londero Brandli, L.; Molthan-Hill, P.; Pace, P.; Azeiteiro, U.M.; Ruiz Vargas, V.; et al. Sustainable Development Goals and Sustainability Teaching at Universities: Falling behind or Getting Ahead of the Pack? *J. Clean. Prod.* **2019**, *232*, 285–294. [CrossRef]
4. Rathika, R.; Srinivasan, P.; Alkahtani, J.; Al-Humaid, L.A.; Alwahibi, M.S.; Mythili, R.; Selvankumar, T. Influence of Biochar and EDTA on Enhanced Phytoremediation of Lead Contaminated Soil by *Brassica juncea*. *Chemosphere* **2021**, *271*, 129513. [CrossRef] [PubMed]
5. Covre, W.P.; Pereira, W.V.d.S.; Gonçalves, D.A.M.; Teixeira, O.M.M.; do Amarante, C.B.; Fernandes, A.R. Phytoremediation Potential of Khaya Ivorensis and Cedrela Fissilis in Copper Contaminated Soil. *J. Environ. Manag.* **2020**, *268*, 110733. [CrossRef] [PubMed]
6. Sdiri, W.; AlSalem, H.S.; Al-Goul, S.T.; Binkadem, M.S.; Ben Mansour, H. Assessing the Effects of Treated Wastewater Irrigation on Soil Physico-Chemical Properties. *Sustainability* **2023**, *15*, 5793. [CrossRef]
7. Rodríguez de Castro, A.; Ben Abdelwahed, A.; Bertin, H. Enhancing Pollutant Removal from Contaminated Soils Using Yield Stress Fluids as Selective Blocking Agents. *J. Contam. Hydrol.* **2023**, *255*, 104142. [CrossRef]
8. Ghazaryan, K.; Movsesyan, H.; Ghazaryan, N.; Watts, B.A. Copper Phytoremediation Potential of Wild Plant Species Growing in the Mine Polluted Areas of Armenia. *Environ. Pollut.* **2019**, *249*, 491–501. [CrossRef]
9. Ma, Y.; Yan, G.; Scheuermann, A.; Li, L.; Galindo-Torres, S.A.; Bringemeier, D. Discrete Microbubbles Flow in Transparent Porous Media. In *Unsaturated Soils: Research & Applications*; CRC Press: Boca Raton, FL, USA, 2020; pp. 1219–1223.
10. Khalid, S.; Shahid, M.; Niazi, N.K.; Murtaza, B.; Bibi, I.; Dumat, C. A Comparison of Technologies for Remediation of Heavy Metal Contaminated Soils. *J. Geochemical Explor.* **2017**, *182*, 247–268. [CrossRef]
11. Gluhar, S.; Kaurin, A.; Lestan, D. Soil Washing with Biodegradable Chelating Agents and EDTA: Technological Feasibility, Remediation Efficiency and Environmental Sustainability. *Chemosphere* **2020**, *257*, 127226. [CrossRef]
12. Alaboudi, K.A.; Ahmed, B.; Brodie, G. Phytoremediation of Pb and Cd Contaminated Soils by Using Sunflower (*Helianthus annuus*) Plant. *Ann. Agric. Sci.* **2018**, *63*, 123–127. [CrossRef]
13. Wang, Z.; Zhang, J.; Wen, T.; Liu, X.; Wang, Y.; Yang, H.; Sun, J.; Feng, J.; Dong, S.; Sun, J. Highly Effective Remediation of Pb(II) and Hg(II) Contaminated Wastewater and Soil by Flower-like Magnetic MoS_2 Nanohybrid. *Sci. Total Environ.* **2020**, *699*, 134341. [CrossRef]
14. Al-Baldawi, I.A.; Yasin, S.R.; Jasim, S.S.; Abdullah, S.R.S.; Almansoory, A.F.; Ismail, N.I. Removal of Copper by Azolla Filiculoides and Lemna Minor: Phytoremediation Potential, Adsorption Kinetics and Isotherms. *Heliyon* **2022**, *8*, e11456. [CrossRef] [PubMed]
15. Bilal Shakoor, M.; Ali, S.; Hameed, A.; Farid, M.; Hussain, S.; Yasmeen, T.; Najeeb, U.; Aslam Bharwana, S.; Hasan Abbasi, G. Citric Acid Improves Lead (Pb) Phytoextraction in *Brassica napus* L. by Mitigating Pb-Induced Morphological and Biochemical Damages. *Ecotoxicol. Environ. Saf.* **2014**, *109*, 38–47. [CrossRef] [PubMed]
16. Alikasturi, A.S.; Kamil, M.Z.A.M.; Shakri, N.A.A.M.; Serit, M.E.; Rahim, N.S.A.; Shaharuddin, S.; Anuar, M.R.; Radzi, A.R.M. Phytoremediation of Copper in Mineral, Distilled and Surface Water Using Limnocharis Flava Plant. *Mater. Today Proc.* **2019**, *19*, 1489–1496. [CrossRef]
17. Belouchrani, A.S.; Mameri, N.; Abdi, N.; Grib, H.; Lounici, H.; Drouiche, N. Phytoremediation of Soil Contaminated with Zn Using Canola (*Brassica napus* L.). *Ecol. Eng.* **2016**, *95*, 43–49. [CrossRef]
18. Dikšaitytė, A.; Kniuipytė, I.; Žaltauskaitė, J. Drought-Free Future Climate Conditions Enhance Cadmium Phytoremediation Capacity by Brassica Napus through Improved Physiological Status. *J. Hazard. Mater.* **2023**, *452*, 131181. [CrossRef] [PubMed]
19. Włóka, D.; Placek, A.; Smol, M.; Rorat, A.; Hutchison, D.; Kacprzak, M. The Efficiency and Economic Aspects of Phytoremediation Technology Using *Phalaris arundinacea* L. and *Brassica napus* L. Combined with Compost and Nano SiO_2 Fertilization for the Removal of PAH's from Soil. *J. Environ. Manag.* **2019**, *234*, 311–319. [CrossRef] [PubMed]
20. Guan, C.; Fu, W.; Zhang, X.; Li, Z.; Zhu, Y.; Chen, F.; Ji, J.; Wang, G.; Gao, X. Enhanced Phytoremediation Efficiency of PHE-Contaminated Soil by Rape (*Brassica napus* L.) Assisted with PHE-Degradable PGPR through Modulating Rhizobacterial Communities. *Ind. Crops Prod.* **2023**, *202*, 117057. [CrossRef]
21. Chi, G.; Fang, Y.; Zhu, B.; Guo, N.; Chen, X. Intercropping with *Brassica juncea* L. Enhances Maize Yield and Promotes Phytoremediation of Cadmium-Contaminated Soil by Changing Rhizosphere Properties. *J. Hazard. Mater.* **2024**, *461*, 132727. [CrossRef]
22. Bakshi, M.; Kumar, A. Co-Application of TiO_2 Nanoparticles and Hyperaccumulator *Brassica juncea* L. for Effective Cd Removal from Soil: Assessing the Feasibility of Using Nano-Phytoremediation. *J. Environ. Manag.* **2023**, *341*, 118005. [CrossRef]

23. Zunaidi, A.A.; Lim, L.H.; Metali, F. Transfer of Heavy Metals from Soils to Curly Mustard (*Brassica juncea* (L.) Czern.) Grown in an Agricultural Farm in Brunei Darussalam. *Heliyon* **2021**, *7*, e07945. [CrossRef]
24. Chigbo, C.; Batty, L.; Bartlett, R. Interactions of Copper and Pyrene on Phytoremediation Potential of *Brassica juncea* in Copper-Pyrene Co-Contaminated Soil. *Chemosphere* **2013**, *90*, 2542–2548. [CrossRef]
25. Chen, L.; Beiyuan, J.; Hu, W.; Zhang, Z.; Duan, C.; Cui, Q.; Zhu, X.; He, H.; Huang, X.; Fang, L. Phytoremediation of Potentially Toxic Elements (PTEs) Contaminated Soils Using Alfalfa (*Medicago sativa* L.): A Comprehensive Review. *Chemosphere* **2022**, *293*, 133577. [CrossRef]
26. Li, Y.; Dong, Q.; Wu, D.; Yin, Y.; Du, W.; Guo, H. A 24-Epibrassinolide Treatment and Intercropping Willow with Alfalfa Increase the Efficiency of the Phytoremediation of Cadmium-Contaminated Soil. *Sci. Total Environ.* **2023**, *854*, 158471. [CrossRef]
27. Pereyra, T.W.; Pagliaricci, H.R.; Ohanian, A.E.; Bonvillani, M.J. Producción de Biomasa Aérea y Uso Equivalente de La Tierra En Intercultivos de Alfalfa (*Medicago sativa* L.). *Pastos Forrajes* **2013**, *36*, 177–183.
28. Flores-Calvete, G.; Pereira-Crespo, S.; Valladares, J. Rotaciones Forrajeras Ecológicas. *Vaca Pint.* **2019**, *14*, 134–142.
29. Zhang, Y.; Li, M.; Su, A.; Lv, X.; Qiu, Y.; Xu, Y. Co-Planting Improves the Phytoremediation Efficiency of Combined Phenanthrene and Copper Co-Contaminated Soils. *J. Clean. Prod.* **2023**, *382*, 135380. [CrossRef]
30. Tsai, H.W.; Lee, Y.C. Effects of Land Use Change and Crop Rotation Practices on Farmland Ecosystem Service Valuation. *Ecol. Indic.* **2023**, *155*, 110998. [CrossRef]
31. Liu, J.; Chen, K.; Zhang, R.; Hou, T.; Yang, P.; Zhang, K.; Yang, F.; Pan, J. Effects of Nitric Oxide on the Phytoremediation of Swine Wastewater by Pistia Stratiotes under Copper Stress. *J. Clean. Prod.* **2023**, *424*, 138680. [CrossRef]
32. Espada, J.J.; Rodríguez, R.; Gari, V.; Salcedo-Abraira, P.; Bautista, L.F. Coupling Phytoremediation of Pb-Contaminated Soil and Biomass Energy Production: A Comparative Life Cycle Assessment. *Sci. Total Environ.* **2022**, *840*, 156675. [CrossRef]
33. Vocciante, M.; Caretta, A.; Bua, L.; Bagatin, R.; Franchi, E.; Petruzzelli, G.; Ferro, S. Enhancements in Phytoremediation Technology: Environmental Assessment Including Different Options of Biomass Disposal and Comparison with a Consolidated Approach. *J. Environ. Manag.* **2019**, *237*, 560–568. [CrossRef]
34. Harris, T.M.; Hottle, T.A.; Soratana, K.; Klane, J.; Landis, A.E. Life Cycle Assessment of Sunflower Cultivation on Abandoned Mine Land for Biodiesel Production. *J. Clean. Prod.* **2016**, *112*, 182–195. [CrossRef]
35. Vigil, M.; Marey-Pérez, M.F.; Martinez Huerta, G.; Álvarez Cabal, V. Is Phytoremediation without Biomass Valorization Sustainable?—Comparative LCA of Landfilling vs. Anaerobic Co-Digestion. *Sci. Total Environ.* **2015**, *505*, 844–850. [CrossRef]
36. Witters, N.; Mendelsohn, R.; Van Passel, S.; Van Slycken, S.; Weyens, N.; Schreurs, E.; Meers, E.; Tack, F.; Vanheusden, B.; Vangronsveld, J. Phytoremediation, a Sustainable Remediation Technology? II: Economic Assessment of CO_2 Abatement through the Use of Phytoremediation Crops for Renewable Energy Production. *Biomass Bioenergy* **2012**, *39*, 470–477. [CrossRef]
37. Cortada, U.; Hidalgo, M.C.; Martínez, J.; Rey, J. Impact in Soils Caused by Metal(Loid)s in Lead Metallurgy. The Case of La Cruz Smelter (Southern Spain). *J. Geochem. Explor.* **2018**, *190*, 302–313. [CrossRef]
38. Lee, H.; Sam, K.; Coulon, F.; De Gisi, S.; Notarnicola, M.; Labianca, C. Recent Developments and Prospects of Sustainable Remediation Treatments for Major Contaminants in Soil: A Review. *Sci. Total Environ.* **2024**, *912*, 168769. [CrossRef]
39. Decreto 18/2015, Por el que se Aprueba el Reglamento que Regula el Régimen Aplicable a los Suelos Contaminados. *Bol. Of. Junta Andal.* **2015**, *38*, 42–61.
40. *D 2487-83*; Classification of Soils for Engineering Purposes: Annual Book of ASTM Standards. American Society for Testing and Materials: West Conshohocken, PA, USA, 1985.
41. Agencia Estatal de Meteorología. Available online: https://www.aemet.es/es/eltiempo/prediccion/municipios/linares-id23055 (accessed on 5 March 2023).
42. Ministerio de Agricultura Pesca y Alimentación—Spanish Government. Consulta de Hojas de Cálculo de Costes de Maquinaria. Available online: http://www.mapama.gob.es/es/ministerio/servicios/informacion/plataforma-de-conocimiento-para-el-medio-rural-y-pesquero/observatorio-de-tecnologias-probadas/maquinaria-agricola/hojas-calculo-maqui.aspx (accessed on 5 March 2023).
43. Antúnez, B.A.; Felmer, S.; Vidal, S.; Marcelo Morales, J.R.; Fuentes, F.; Francisco y Coz, E. Propiedades Físico-Hídricas Del Suelo En El Cultivo Del Maíz Grano. *Boletín INIA—Inst. Investig. Agropecu.* **2015**, *312*, 31–50.
44. Córdoba, E.M.; Sastre, F.; Sánchez-Noriega, A.; González-Verdejo, C.I. *Guías de Cultivo Serie Leguminosas: La Alfalfa*; Junta de Andalucia: Malaga, Spain, 2015; pp. 1–18.
45. Vocciante, M.; de Folly D'Auris, A.; Franchi, E.; Petruzzelli, G.; Ferro, S. CO_2 Footprint Analysis of Consolidated and Innovative Technologies in Remediation Activities. *J. Clean. Prod.* **2021**, *297*, 126723. [CrossRef]
46. Alfalfaspain Ventajas de La Alfalfa. Available online: https://www.alfalfaspain.es/ventajas-de-la-alfalfa (accessed on 5 March 2023).
47. García Ruiz, J.R.; García López, J. Estudios Agronómicos Sobre Camelina Sativa y otras Especies Oleaginosas en Andalucía Occidental. Junta de Andalucía. In *Informe técnico Servifapa*; Junta de Andalucia: Malaga, Spain, 2012.
48. Zhang, Y.; Xie, D.; Ni, J.; Zeng, X. Optimizing Phosphate Fertilizer Application to Reduce Nutrient Loss in a Mustard (*Brassica juncea* var. *tumida*)-Maize (*Zea mays* L.) Rotation System in Three Gorges Reservoir Area. *Soil Tillage Res.* **2019**, *190*, 78–85. [CrossRef]
49. Mir, M.R.; Mobin, M.; Khan, N.A.; Bhat, M.A.; Lone, N.A.; Bhat, K.A.; Razvi, S.M.; Wani, S.A.; Wani, N.; Akhter, S.; et al. Effect of Fertilizers on Yield Characteristics of Mustard (*Brassica juncea* L.). *J. Phytol.* **2010**, *2010*, 20–24.

50. Fox, R.L.; Lipps, R.C. Subirrigation and Plant Nutrition I. Alfalfa Root Distribution and Soil Properties. *Soil Sci. Soc. Am. J.* **1955**, *19*, 468–473. [CrossRef]
51. Jennings, J.A.; Nelson, C.J. Zone of Autotoxic Influence around Established Alfalfa Plants. *Agron. J.* **2002**, *94*, 1104–1111. [CrossRef]
52. Andrade Fumachi, J.K. *Abonamiento Con Dosis de Gallinaza y Su Influencia En Las Características Agronómicas y Rendimiento de Brassica juncea L. "Mostaza de Hoja"*; UNAP: Iquitos, Peru, 2022.
53. Chen, J.; Bian, X.; Rapp, G.; Lang, J.; Montoya, A.; Trethowan, R.; Bouyssiere, B.; Portha, J.F.; Jaubert, J.N.; Pratt, P.; et al. From Ethyl Biodiesel to Biolubricants: Options for an Indian Mustard Integrated Biorefinery toward a Green and Circular Economy. *Ind. Crops Prod.* **2019**, *137*, 597–614. [CrossRef]
54. Izydorczyk, G.; Skrzypczak, D.; Kocek, D.; Mironiuk, M.; Witek-Krowiak, A.; Moustakas, K.; Chojnacka, K. Valorization of Bio-Based Post-Extraction Residues of Goldenrod and Alfalfa as Energy Pellets. *Energy* **2020**, *194*, 116898. [CrossRef]
55. Park, B.; Son, Y. Ultrasonic and Mechanical Soil Washing Processes for the Removal of Heavy Metals from Soils. *Ultrason. Sonochem.* **2017**, *35*, 640–645. [CrossRef]
56. Visentin, C.; da Silva Trentin, A.W.; Braun, A.B.; Thomé, A. Application of Life Cycle Assessment as a Tool for Evaluating the Sustainability of Contaminated Sites Remediation: A Systematic and Bibliographic Analysis. *Sci. Total Environ.* **2019**, *672*, 893–905. [CrossRef]
57. Huijbregts, M.A.J.; Hellweg, S.; Frischknecht, R.; Hendriks, H.W.M.; Hungerbühler, K.; Hendriks, J. Cumulative Energy Demand as Predictor for the Environmental Burden of Commodity Production. *Environ. Sci. Technol.* **2010**, *44*, 2189–2196. [CrossRef]
58. Todde, G.; Carboni, G.; Marras, S.; Caria, M.; Sirca, C. Industrial Hemp (*Cannabis sativa* L.) for Phytoremediation: Energy and Environmental Life Cycle Assessment of Using Contaminated Biomass as an Energy Resource. *Sustain. Energy Technol. Assessments* **2022**, *52*, 102081. [CrossRef]
59. Suer, P.; Andersson-Sköld, Y. Biofuel or Excavation?—Life Cycle Assessment (LCA) of Soil Remediation Options. *Biomass Bioenergy* **2011**, *35*, 969–981. [CrossRef]

MDPI AG
Grosspeteranlage 5
4052 Basel
Switzerland
Tel.: +41 61 683 77 34

Sustainability Editorial Office
E-mail: sustainability@mdpi.com
www.mdpi.com/journal/sustainability

www.ingramcontent.com/pod-product-compliance
Lightning Source LLC
Chambersburg PA
CBHW040747250926
51850CB00007BB/1431
9783725853991